ALLE ZEIT WACH
1842

Helmut Lydtin Peter Trenkwalder

CALCIUM ANTAGONISTS

A Critical Review

With a Foreword by
A. Fleckenstein

With 38 Figures

Springer-Verlag Berlin Heidelberg New York
London Paris Tokyo Hong Kong

Helmut Lydtin, M.D.
Professor of Medicine, University of Munich
Medical Director, Starnberg Regional Hospital
Head of Department of Internal Medicine
Osswaldstrasse 1, D-8130 Starnberg, FRG

Peter Trenkwalder, M. D.
Senior Medical Officer
Department of Internal Medicine, Starnberg Regional Hospital
Osswaldstrasse 1, D-8130 Starnberg, FRG

With the assistance of

Claudia Trenkwalder, M. D.
Department of Neurology, Klinikum Großhadern,
University of Munich,
Marchioninistrasse 15, D-8000 München 70
(Sect. 4.7 Calcium Antagonists in Neurological Disease)

Extended and revised version of the German edition:
Calcium-Antagonisten

ISBN-13: 978-3-642-74889-9 e-ISBN-13: 978-3-642-74887-5
DOI: 10.1007/978-3-642-74887-5

Library of Congress Cataloging-in-Publication Data. Lydtin, Helmut. [Calcium-Antagonisten. English] Calcium antagonists: a critical review / Helmut Lydtin, Peter Trenkwalder; with a foreword by A. Fleckenstein. p. cm. Translation of: Calcium-Antagonisten. Includes bibliographical references. ISBN-13: 978-3-642-74889-9 (U.S.:alk.paper) 1. Calcium – Antagonists. I. Trenkwalder, Peter, 1955– . II. Title. [DNLM: 1. Calcium Channel Blockers – pharmacokinetics. 2. Calcium Channel Blockers – pharmacology. 3. Calcium Channel Blockers – therapeutic use. 4. Calcium – physiology. QV 276 L983c] RM666.C243L9313 1990 615'.7 – dc20 DNLM/DLC for Library of Congress 89-21686 CIP

Softcover reprint of the hardcover 1st edition 1990

2125/3130-543210 – Printed on acid-free paper

Foreword

Our research at the Freiburg Physiological Institute over the past 25 years has demonstrated that specific calcium antagonists interfere with all physiological and pathophysiological reactions involving the heart and vessels in which an elevated inflow of calcium ions across the potential-dependent slow membrane channels into the cells is a decisive factor. From the outset, our therapeutic interest has centred on the possibility of inhibiting excessively high levels of transmembrane calcium influx and thereby reducing the pathogenic consequences of intracellular calcium overload. For example, all excessive calcium influx evokes contractile hyperactivity, which manifests itself in the myocardial fibres as hyperkinetic reactions and an uneconomic rise in oxygen demand and, analogously, in vascular smooth muscle cells as increased tonus and spasm. The therapeutic use of calcium antagonists to inhibit excessive mechanical and metabolic activity in the myocardium and to achieve prophylaxis and lysis in smooth muscle spasm was therefore a logical step.

The highest potencies of calcium antagonists were found in the vascular smooth musculature (coronary, cerebral, mesenteric and renal arteries and other systemic resistance vessels). Consequently, calcium antagonists were recognized as the agents of choice for numerous indications for vascular therapy.

It is well established that calcium ions are also essential for the generation of impulses in cardiac pacemaker cells and for the conduction of electrical charges. Calcium antagonists therefore possess, at least at higher dosage levels, in addition to their inhibitory action on the mechanical development of myocardial tone, direct negative chronotropic and negative dromotropic effects. Moreover, ectopic pacemakers reveal in general an even stronger calcium dependence (i. e. depressibility by calcium antagonists) than the regular nomotopic pacemakers of the sinus node and the atrioventricular (AV) node. This selectivity of excitatory inhibition of pathological impulse activity bestows on certain antiarrhythmic and antifibrillatory potencies on calcium antagonists. Thus as long ago as 1972 calcium antagonists, in particular verapamil, were designated as group IV in the widely accepted classification scheme for antiarrhythmic agents.

The therapeutically exploitable actions of calcium antagonists are by no means restricted to providing protection against calcium-dependent functional irregularities. Every intracellular calcium overload of a relatively high degree or protracted duration has cytotoxic sequelae. There can now be no doubt that, beside hypoxia, calcium overload of myocardial fibres is one of the most important factors in the aetiology of disseminated or confluent cardiac necroses. Calcium antagonists provide a spectacular level of protection against such calcium-induced structural lesions. This was first shown experimentally in 1968, and today their successful addition to cardioplegic solutions also prevents such histopathological structural damage in the clinical situation.

A last, and perhaps decisive, step in the therapeutic application of calcium antagonists has so far been carried out only in animal experiments, namely the prevention of severe morphological (calcium-overload-induced) structural damage to the arteries and arterioles, which, depending on age and in particular on the influence of risk factors (e.g. diabetes, nicotine, hypertension) develops progressively. In these experimental studies long-term treatment with suitable calcium antagonists has brought to light dramatic anticalcinotic and antiatherogenic effects. We were thus able, for example, to maintain the full integrity of the vascular walls of three types of hypertensive rats – Okamoto rats (SHR); sodium-loaded (salt-sensitive) Dahl rats, and rats with Goldblatt's hypertension – for a period of many months by using calcium antagonists, while the untreated control animals died prematurely of their vascular lesions.

A fundamental task of basic physiological research is the elucidation of such pathophysiological connections, to provide a stimulus to clinical medicine. Undoubtedly, the long-term treatment of hypertensive patients with calcium antagonists will in future provide an opportunity to go beyond the measurement of blood pressure as a means of therapeutic control, and to apply more demanding techniques to directly determine the structural status of the arterial and arteriolar walls. This book reveals the avalanche of new data and insights in the field of calcium antagonism that has already emerged from the joint efforts of experimental and clinical research. Such cooperation will be even more imperative in the future.

Professor Lydtin was one of the first clinicians who clearly recognized, and put to practical application, the therapeutic potential of calcium antagonists. This lends special weight to the critical analysis presented here.

Freiburg Professor A. FLECKENSTEIN, M.D.

Preface

Today calcium antagonists are employed worldwide in the management of hypertension, cardiac arrhythmias, certain forms of cardiovascular disease and vascular spasm. Their application in numerous other fields, including neurology, oncology, gynaecology and urology, is also discussed. Parallel to this development, the concept of calcium antagonism has been a stimulus in nearly all areas of physiology, pathophysiology and pharmacology. In 1964 Fleckenstein first reported that prenylamine and verapamil inhibited the coupling of excitation and contraction in the heart, i.e. excitation-contraction coupling, in the same manner as a decrease in extracellular calcium. This demonstration of an excitation-contraction decoupling allowed development of the concept of calcium antagonism. At that time one could hardly have anticipated the initially slow and then rapid development of "calcium-antagonistic substances" during the next two decades, nor their growing therapeutic importance not only in the cardiovascular field. Fleckenstein's energy, persistence and clearsightedness contributed greatly to this development. In 1967 he and his co-workers experimentally identified the first specific calcium antagonists verapamil and gallopamil as representatives of a new class of drugs significantly different from β-receptor blockers. Two years later nifedipine joined the group of specific calcium antagonists. Subsequently numerous further 1,4-dihydropyridines with calcium-antagonistic action were synthesized. Diltiazem, a substance with a distinctly different chemical structure, was developed in 1975, and today ranks worldwide, along with verapamil and nifedipine, among the most widely prescribed calcium antagonists.

All calcium antagonists interfere with the uptake of calcium into the myocardial cell. Here, as well as in smooth muscle cells, they inhibit excitation-contraction coupling in concentrations at which other concomitant effects (e.g. inhibition of rapid sodium influx) are negligible. Today, this definition is internationally accepted.

The volume of current literature concerning these drugs is almost overwhelming: monographs, reviews and symposia reports compete with countless individual papers to produce such a flood of information that it has become increasingly difficult for the practising physician to make rational therapeutic decisions. Furthermore, the advertising campaigns of some pharmaceutical companies often exacerbate this confusion by drawing rash and prejudiced clinical extrapolations from limited experimental research data.

Table 1. Historical review: significant discoveries and steps in the development of calcium antagonists

1882	Dependence of myocardial contraction on calcium ions (S. Ringer)
1913	Calcium deprivation impairs contractile force in the ventricular myocardium more than the bioelectrical activity (Mines)
1960	Pharmacological action of prenylamine (Lindner)
1962	Pharmacological action of verapamil (Haas and Härtfelder)
1963	Application of verapamil in coronary heart disease (Tschirdewahn and Klepzig)
1964	Verapamil and prenylamine inhibit, in the same manner as a calcium deficit, excitation-contraction coupling in the myocardium (Fleckenstein)
1965	Use of verapamil as antiarrhythmic agent in atrial flutter (Bender)
1967	Introduction of the terms "calcium antagonist" and "calcium antagonistic" (Fleckenstein)
1966–1969	Differentiation of the calcium antagonists verapamil, gallopamil and prenylamine from the β-receptor blockers. Definition of calcium antagonism and calcium antagonists as a new pharmacological class (Fleckenstein)
1967	Synthesis of gallopamil (D 600) (Dengel)
1968–1969	Cardioprotective action of calcium antagonists in animal experiments (Fleckenstein)
1969	Treatment of arterial hypertension with verapamil (Bender and Brittinger)
1969	Pharmacological testing of nifedipine (Bossert, Vater, Kroneberg)
1969–1970	Identification of nifedipine as a specific calcium antagonist (Fleckenstein)
1968–1972	Action of calcium antagonists on smooth muscle (Grün and Fleckenstein, Haeusler)
1970–1972	Mechanism and site of action of calcium antagonists: Significance of the slow calcium channel (Fleckenstein, published in *Calcium and the heart* 1970–1971)
1971	Pharmacological action of diltiazem (CRD 401) described (Sato et al.)
1974	Treatment of paroxysmal supraventricular tachycardias with verapamil (Krikler and Spurrell)
1975	Identification of diltiazem as a calcium antagonist (Nakajima et al.)
1976	Use of calcium antagonists for hypertrophic cardiomyopathy (Kaltenbach)
1980	Editorial in *American Journal of Cardiology* (Braunwald)

Subsequent worldwide use of calcium antagonists in a wide range of indications (Chap. 4) and synthesis of numerous successor substances to the "classical" calcium antagonists verapamil, nifedipine and diltiazem (Sect. 2.5).

This situation stimulated us to write a book, first in German and then in English, with the goal of providing our fellow physicians in various disciplines with a concise, clear, critical and objective guide to the proven physiological and pharmacological principles concerning calcium antagonists, and to their use in day-to-day clinical practice. Evaluation of the vast expanding literature in this field presented perhaps the greatest challenge, and revision and updating of the English version was necessary up to the time of printing.

The preliminary chapters of the book discuss the physiological role of calcium at the cellular and subcellular level as an introduction to the specific pharmacology of the calcium antagonists. The following sections deal with the action of these drugs on the individual organ systems, and then with their integrated effects on the whole body, to provide a rational basis for their use in practice. Therapy with calcium antagonists is reviewed for each indication from the standpoint of nosological classification, therapeutic goals and underlying pathophysiology. To supplement, and perhaps also in contrast to some textbooks of pharmacology and clinical medicine, step-by-step guidelines are given for the use of calcium antagonists in every indication, together with dosages, therapeutic alternatives, important drug interactions and side effects. This guide was based on a critical analysis of the international literature available up to the summer of 1989.

One of us was involved in some early physiological and pharmacological clinical studies on calcium antagonists at the end of the 1960s. We have worked together for over 7 years in the same clinic and write foremost from the standpoint of experience with our own patients, for it is our conviction that, beyond renewed scholastic dogmatism, medicine can remain an open empirical system in the service of the patient only through open discussion, self-criticism and the willingness to question even "statistically proven realities".

We thank our colleagues, both within and outside the hospital, for their trust, for their suggestions and for their criticisms. Only many years of working closely together has made it possible for us to write a book on the application of calcium antagonists in practice.

We thank F. Bender M. D., Professor of Medicine at the University of Münster, for kindly providing us with original ECG recordings (see Sect. 4.3), and A. Wischnik M. D., Munich and Mannheim, for literature on tocolysis.

Our special thanks go to Marianne Hildinger and Ulrike Zunner for their unflagging labours on the German manuscript, and to Alan Huet Owen and Dianne Owen-Lydtin for the translation and the preparation of the English version. Finally, we need to say a word of thanks to Springer-Verlag, to Dr. Thiekötter, Mr. Oehm and Mrs. Gründler for their constant encouragement and help in preparing this book.

Starnberg, Summer 1989 H. Lydtin, P. Trenkwalder

Contents

1 Physiological Significance of Calcium: Calcium Antagonism

1.1 Physiological Significance of Calcium

Calcium ions (Ca^{2+}) fulfil both in the animal and human organism a wide range of functions. They are involved in the maintenance of hemostasis, a broad array of enzymatic reactions, the metabolism of bone, electrical activation of various excitable cells, and the coupling of electrical activation to cellular secretion and muscular contraction (Antman et al. 1980). Katz (1985c) distinguishes between structural functions of calcium (e.g. in bone) and its capacity to regulate cellular functions. Berridge (1985) even speaks, by analogy with cAMP, of calcium as a "second messenger" (e.g. in excitation–contraction coupling or insulin secretion). An up-to-date review of the "calcium messenger system" is provided by Rasmussen (1986).

In the cardiovascular system calcium participates in cardiac excitation and impulse conduction, as well as in contraction of the myocardium and smooth muscle. Indirectly, calcium regulates cardiac energy consumption and vasomotor tone (Braunwald 1982).

By way of introduction, essential features of electrophysiology and excitation–contraction coupling in the heart and smooth muscle will be presented. More specialized information may be found in contemporary textbooks of physiology and pathophysiology (e.g. Ganong, *Review of medical physiology;* Guyton, *Medical physiology;* Sodeman, *Pathologic physiology;* Price and Wilson, *Pathophysiology*) and in several surveys on this topic (Braunwald 1982, 1984; Morad and Tung 1982; Bayer 1984; Opie 1984). In this context discussion tends still to be extremely controversial, and much is explained only in terms of hypothetical models.

1.1.1 Action Potential in Cardiac Pacemaker Cells

Action potential in so-called cardiac pacemaker cells (excitation of impulse in the sinoatrial node, conduction of impulse through the AV node) differs in several respects from the action potential of the myocardium in the atrium and ventricle (Sect. 1.1.2.1). As will be seen from Fig. 1, three phases of action potential in pacemaker cells can be distinguished: the so-called phase 4 corresponds to spontaneous diastolic depolarization (cf. the phase of stable resting potential in the myocardium); this is specific to cardiac pacemaker cells, and it is observed in the myocardium only in ischaemia, hypoxia or exposure to catecholamines (causing arrhythmias) (Antman et al. 1980; Braunwald 1982). The speed of spontaneous depolarization, that is, of the angle of ascent, determines the time required to reach the threshold value (about -60 mV), and therewith the frequency of the

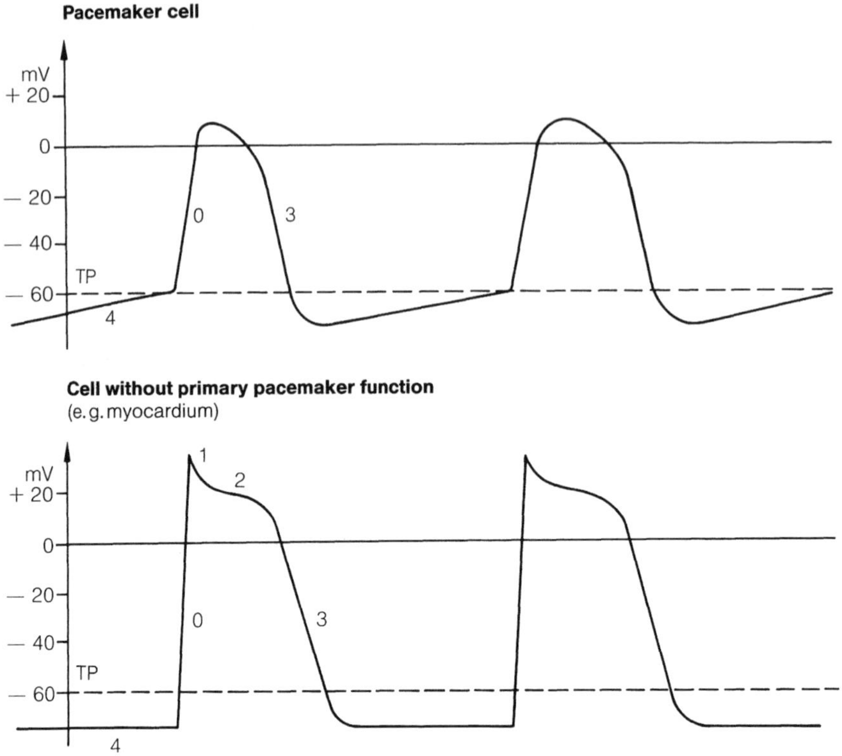

Fig. 1. Action potential in cardiac pacemaker cells and in cells with no primary pacemaker function (e.g. myocardium). *Numerals* indicate the phases of action potential (for explanation see text). *TP* signifies threshold potential

pacemaker cell. The subsequent rapid depolarization phase (phase 0) takes a much slower course than that in skeletal muscle or the myocardium and is determined by the so-called slow inward current, which depends primarily on an influx of calcium ions. In physiological conditions, sodium ions are involved here to less than 1%. In phase 3 repolarization occurs through an outward flow of potassium.

1.1.2 Excitation–Contraction Coupling in the Myocardium

The term excitation–contraction coupling is intended to mean all steps between electrostimulation of a membrane and resultant muscular contraction. It is now undisputed that calcium plays a decisive role in this process: the free intracellular calcium concentration controls via regulatory proteins (troponin–tropomyosin–calmodulin) the interaction of the contractile proteins (actin and myosin). Calcium is also directly involved in the generation of action potential, that is, in the specific electrical stimulation of the myocardial membrane. For that reason, in the following discussion, myocardial action potential, the cellular and subcellular movements of calcium, together with their influence on intracellular calcium concentration, as well as the molecular mechanism of muscular contraction, are dealt with as separate conceptual entities.

1.1.2.1 Action Potential in the Myocardial Cell

The action potential shown in Fig. 1 is found both in the myocardium (atrium and ventricle) and in the conduction system of the bundle of His and of the Purkinje fibres. There are some notable differences to the action potential of pacemaker cells. A stable diastolic resting potential of about −90 mV (phase 4) is followed by a fast depolarization phase (phase 0), brought about by a fast sodium inflow. The latter can be experimentally inhibited by tetrodotoxin (venom of the Japanese balloon fish, *Fugu*) or by an increased extracellular concentration of potassium chloride, and by such antiarrhythmic agents as lidocaine and quinidine (class I, according to Vaughan-Williams and Singh). A brief phase of early repolarization (phase 1), brought about by a transient potassium outflow, is followed by the typical "plateau" of the myocardial action potential. During this plateau phase the principal event is a slow inward calcium current, which can be inhibited by the presence of manganese, cobalt or nickel ions or by organic calcium antagonists. Additional participation of sodium and potassium currents in plateau formation is under discussion (Carafoli 1984). In the fast repolarization phase (phase 3) there is an outward potassium current, as in pacemaker cells.

1.1.2.2 Anatomy of the Myocardial Cell

Myocardial cells are normally 40–100 µm long and 10–20 µm wide. They consist of cell membrane, cell nucleus, cell organelles (mitochondria, sarcoplasmic reticulum) and striated myofibrils (Fig. 2). The cell membrane (also known as the sarcolemma) curls inward in regular intervals into the cell interior. These invaginations form the transverse tubular system (TTS), which plays a role in the transmission of excitation. The cell membrane is externally joined to a network of glycoproteins and glycolipids, the so-called glycocalyx, which functions as a superficial calcium depot. Inside the cell the sarcoplasmic reticulum (SR) forms a complex network which, although not directly communicating with the extracellular space, is located in the immediate vicinity of the TTS of the cell membrane (lateral cisterns). The SR releases calcium during systole and takes it up again in diastole. Mitochondria, too, can store calcium, but their function first and foremost is to provide energy, by synthesizing energy-rich phosphate (ATP). The decisive subunit of the myofibrils, which is responsible for the myocardial contraction process, is known as the sarcomere. Individual sarcomeres are kept separate from one another by Z-lines (intercalated discs). A sarcomere consists of contractile proteins, thin actin filaments and thick myosin filaments, which by forming transverse bridges displace the filaments and thereby perform the act of contraction per se. Additionally, regulatory proteins (troponin, tropomyosin) are superposed on the actin filaments and control muscle tone.

1.1.2.3 Cellular and Subcellular Calcium Movements

The concentration of free intracellular calcium in the cytoplasm of the myocardial cell is the result of several calcium movements across the cell membrane and within the cell.

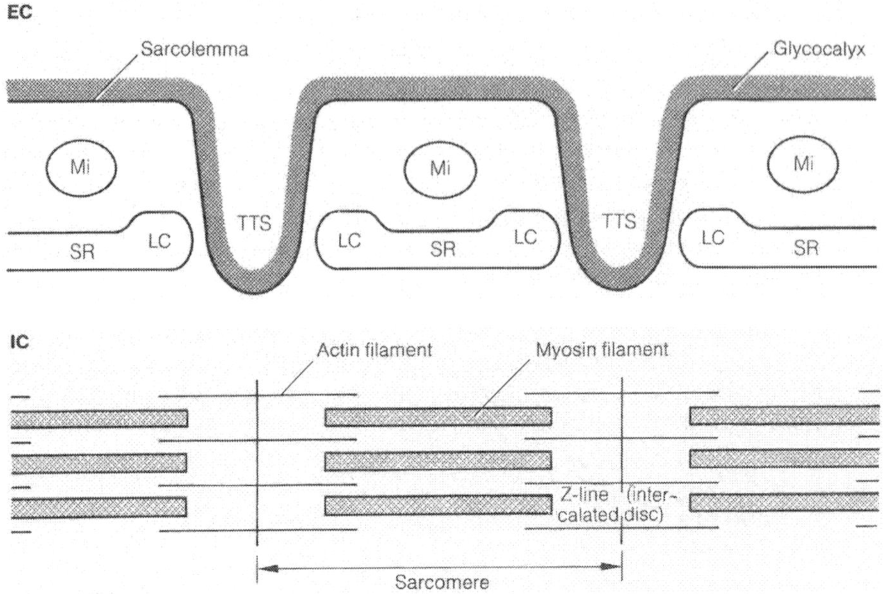

Fig. 2. Anatomy of the myocardial cell. *EC* extracellular, *IC* intracellular, *Mi* mitochondria, *TTS* transverse tubular system, *SR* sarcoplasmic reticulum, *LC* lateral cisterns. Further explanation in text

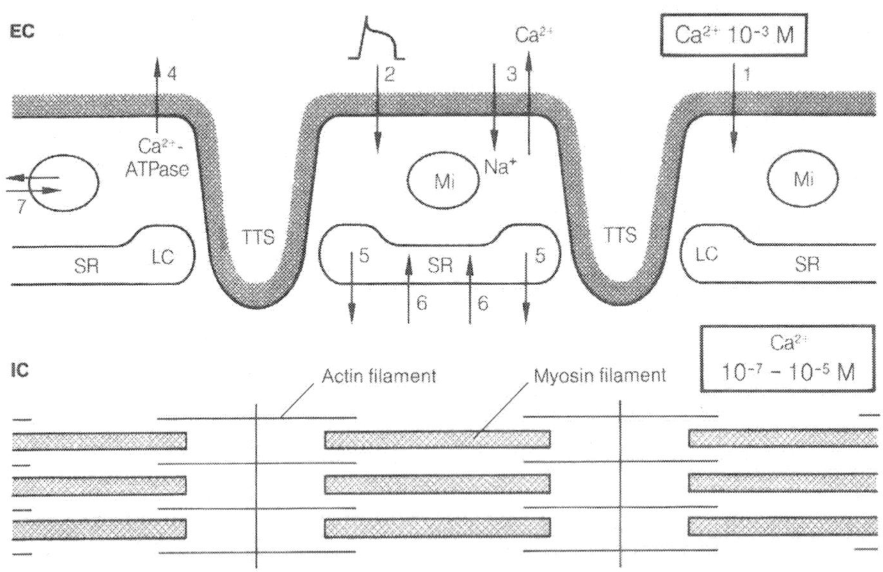

Fig. 3. Cellular and subcellular calcium movements in the myocardial cell. *EC* extracellular, *IC* intracellular, *Mi* mitochondria, *SR* sarcoplasmic reticulum, *TTS* transverse tubular system. *Numerals 1* to *7* refer to calcium movements described in the text

Figure 3 summarizes the seven types of experimentally more or less well documented calcium movements (more detailed accounts in Bayer 1984; Opie et al. 1984):

1. **Calcium "leakage current"**, a passive calcium inward current through the extracellular–intracellular concentration gradient (extracellular calcium concentration 10^{-3} *M*, intracellular 10^{-5} to 10^{-7} *M*).

2. **Slow inward calcium current** during the formation of the myocardial action potential (Sect. 1.1.2.1) as a result of the fast (depolarizing) inward sodium current. This slow inward calcium current can be modified by β-sympathomimetics and β-receptor blockers and also by calcium antagonists.

3. Electrically neutral (i.e. two sodium ions against one calcium ion) and electrogenic (i.e. three sodium ions against one calcium ion) **sodium–calcium exchange,** which lowers the calcium concentration, especially in systole (Carafoli 1984). The driving force behind this sodium–calcium exchange is sodium inflow due to the extracellular–intracellular concentration gradient. This gradient is maintained by the function of Na^+, K^+-ATPase. If sodium–potassium exchange is inhibited, e.g. through digitalis glycosides, the intracellular sodium concentration rises, the transmembranal sodium gradient declines, and the outward flow of calcium decreases. Increasing the intracellular calcium concentration potentiates contractility, providing an explanation for the positive inotropic action of digitalis glycosides (Katz 1985a).

4. **Calcium-ATPase,** an energy-dependent calcium pump, which pumps calcium out of the cell, especially in diastole (Carafoli 1984).

Calcium movements 1–4 take place across the cell membrane (sarcolemma). The intracellular calcium movements between cell organelles and cytoplasm are described below:

5. **Calcium release from the sarcoplasmic reticulum** through calcium-induced calcium release (Fabiato and Fabiato 1978) or depolarization-induced calcium release: both mechanisms are probably complementary (Fabiato and Fabiato 1977). In the one case the increase in intracellular calcium concentration (through the action of the slow inward calcium current) releases calcium from the sarcoplasmic reticulum; in the other case depolarization switches from the transverse tubular system of the cell membrane to the sarcoplasmic reticulum and results in calcium release.

6. **Calcium uptake into the sarcoplasmic reticulum** by energy-dependent calcium-ATPase: the latter is controlled by a special regulatory protein (phospholamban), thus making possible regulation of (late) systolic calcium uptake, and thereby also influencing myocardial relaxation. Since the calcium uptake additionally influences the resultant calcium release from the sarcoplasmic reticulum, contraction can also be indirectly modified (Katz 1984).

7. Calcium uptake and calcium release by the **mitochondria** do not directly participate in the contraction and relaxation process, although long-term they play a role in the regulation of the intracellular calcium content (Carafoli and Crompton 1978).

The sum of calcium movements leads during systole to an increase in intracellular calcium concentration, from 10^{-7} *M* (mean value at end of diastole) to 10^{-5} or 10^{-4} *M* (by comparison, extracellular calcium concentration remains constant at 10^{-3} *M*). Although the increase in intracellular calcium concentration is initiated by the slow inward calcium current, quantitatively only 2% of this increase is due to calcium moved across the cell membrane. By far the greatest proportion is released from the sarcoplasmic reticulum (about 80%–90%). Analogously, at the end of systole the sarcoplasmic reticulum is also the chief site of calcium uptake (88% according to Opie 1984).

1.1.2.4 Molecular Mechanism of Muscle Contraction

As described in Sect. 1.1.2.2, the contraction mechanism per se requires formation of a transverse bridge between actin and myosin filaments. By means of the regulator proteins troponin (troponin complex) and tropomyosin, the state of contraction of the myocardial cell can be controlled via intracellular calcium concentration. During the systolic rise in calcium concentration above 10^{-7} *M*, calcium binds to the troponin complex and thereby annuls the inhibitory action of tropomyosin on the interaction between actin and myosin. The resultant actomyosin complex is, on one hand, a contractile substrate and, on the other, an enzyme (actomyosin–ATPase), which supplies energy to the contraction process (Bayer 1984; Burkart and Heierli 1984; Gevers 1984a, b).

Through resumed uptake of calcium into the sarcoplasmic reticulum at the end of systole, intracellular calcium concentration decreases, thereby permitting dissociation of the calcium–troponin bond. Consequently, the transverse bridges dissolve, and the muscle can relax. The basic mechanisms of excitation–contraction coupling can be represented in a simplified scheme (Fig. 4).

1.1.3 Excitation–Contraction Coupling in Smooth Muscle

The concept of excitation–contraction coupling, which has been developed from study of the myocardium, has been applied also to smooth muscle. The fundamental differences of the models entertained at the present time are briefly given below (Fig. 5):

1. In cardiac muscle, calcium inflow occurs principally dependent on depolarization of the cell membrane through fast sodium inflow: accordingly there are

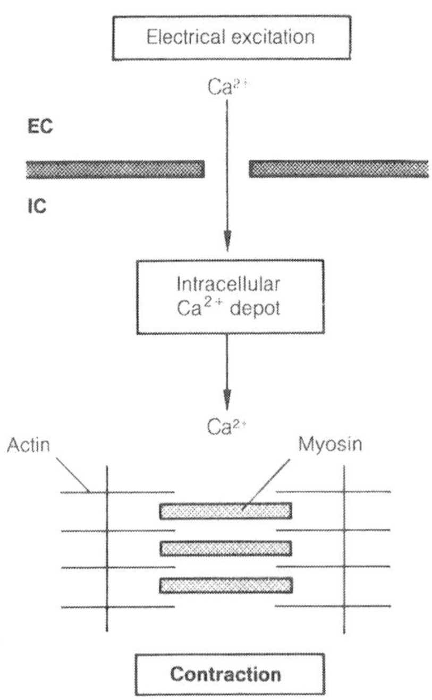

Fig. 4. Synopsis of excitation–contraction coupling. (After Burkart and Heierli 1984)

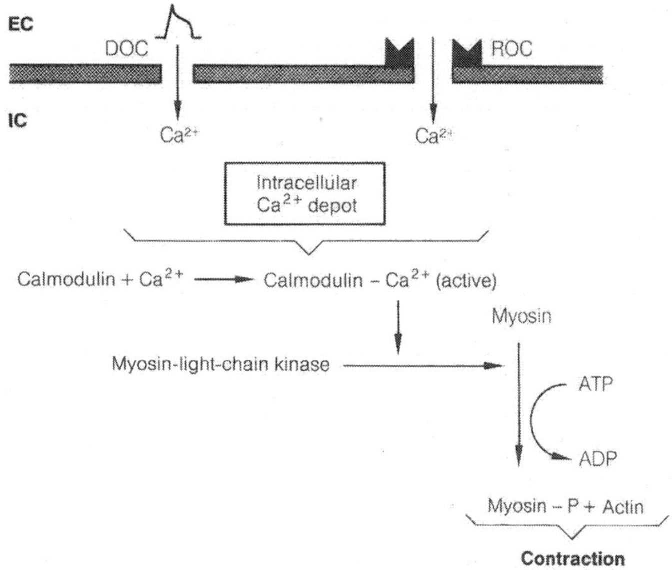

Fig. 5. Excitation–contraction coupling in the smooth muscle cell. (After Krämer and Kübler 1983) *DOC* depolarization-operated calcium channel, *ROC* receptor-operated calcium channel, *Myosin-P* myosin phosphate, *EC* extracellular, *IC* intracellular

so-called depolarization-operated calcium channels (DOC or VOC voltage-operated channel). In smooth muscle, calcium influx is thought to be additionally controlled via receptor-operated channels (ROC receptor-operated channel). These channels are activated by binding noradrenaline to specific α_1 and α_2 receptors, or by appropriately specific receptor binding of serotonin, acetylcholine, histamine or angiotensin.

2. In smooth muscle there are short-term (phasic) and longer-term (tonic) contractions: in vascular smooth muscle of the resistance vessels spontaneous, myogenic contractions contribute to the resting vascular tone (Folkow 1949). Since resistance vessels offer the greatest resistance to flow, their myogenic activity and vasomotor tone determines peripheral resistance (Folkow and Neil 1971; Opie 1984; Murphy and Gertshoffer 1984; Rosendorff 1984; Schrör 1985). Myogenic activity and hence vasomotor tone are dependent on a continuous inflow of calcium from the extracellular space; intracellular calcium depots play only a subordinate role in vascular smooth muscle (Casteels and Droogmanns 1983). It is self-evident that the intracellular calcium concentration is subject to steady-state control. By contrast, in other organ systems with smooth muscle (e.g. bronchial system, gastrointestinal tract), release from intracellular calcium depots plays a variable role in the free intracellular calcium concentration (Katz et al. 1985b).

3. As in cardiac muscle, the two contractile proteins actin and myosin are responsible for muscular contraction; in smooth muscle the increase in the intracellular calcium concentration activates the intracellular regulator protein, calmodulin. In turn, calmodulin promotes the transfer of a phosphate group to

myosin (so-called myosin phosphorylation through myosin-light-chain-kinase), thereby enabling an interaction between actin and myosin, and contraction takes place (Fig. 5). Myosin-light-chain-kinase is also regulated through a cAMP-dependent enzyme, whereby the relaxing action of β-sympathomimetics on smooth muscle is explained (phosphorylation hypothesis after Adelstein 1983). Complementarily to the mechanism of muscular contraction described here, further models are also under discussion (details in Gevers 1984a, b).

4. β-Sympathomimetics exercise a contraction-intensifying (positive inotropic) effect on the heart; in smooth muscle they have an inhibitory effect, which explains the vasodilating action of these substances. This phenomenon can be explained by an organ-specific distribution of β_1 and β_2 receptors with varying influence on intracellular calcium concentration in cardiac muscle and the vascular smooth muscle cell.

1.2 Calcium Antagonism

Theoretically, calcium antagonism is the result of one of two basic mechanisms (Cohen et al. 1984):

1. Inhibition of the intracellular action of calcium: this is the mode of action of the so-called calmodulin antagonists (e.g. trifluoperazine).

2. Inhibition of calcium availability: this can occur through inhibition of the slow inward calcium current in specific calcium channels, which is today referred to as calcium antagonism in the narrower sense (Fleckenstein 1971). Theoretically, a reduced calcium availability would also be possible through an increase in the outward calcium current and/or through increased uptake into intracellular depots. Both mechanisms today enjoy less favour in experimental research. Moreover, at the present time, the theorizing is of no therapeutic significance.

Although Fleckenstein's concept of "calcium antagonists" has meanwhile become adopted worldwide, various synonyms continue to be used, for instance, "calcium channel blockers" (Katz 1979), "slow channel blockers", "calcium entry blockers (or inhibitors or antagonists)", "calcium modulators".

To avoid confusion, the term "calcium antagonists" is used exclusively in this book, although the term "calcium channel blockers" probably describes the action of these substances more accurately (Katz et al. 1985b).

1.2.1 Definition of Calcium Antagonists

Before the definitive principle of action of calcium antagonists was demonstrated, Fleckenstein defined a specific calcium antagonist as "a substance which so predominantly exerts its inhibitory effects upon the calcium-dependent functions in myocardium and smooth muscle that, in this dosage range, all other pharmacodynamic properties are more or less negligible" (Fleckenstein 1982, 1984).

In this context calcium antagonists imitate the effect of calcium deficiency or deprivation (Fig. 6). Accordingly, the effects of calcium antagonists can be nullified by increasing the extracellular concentration (with a few exceptions).

Today the definition of a calcium antagonist connotes that the substance specifically and predominantly inhibits the slow inward current of calcium ions.

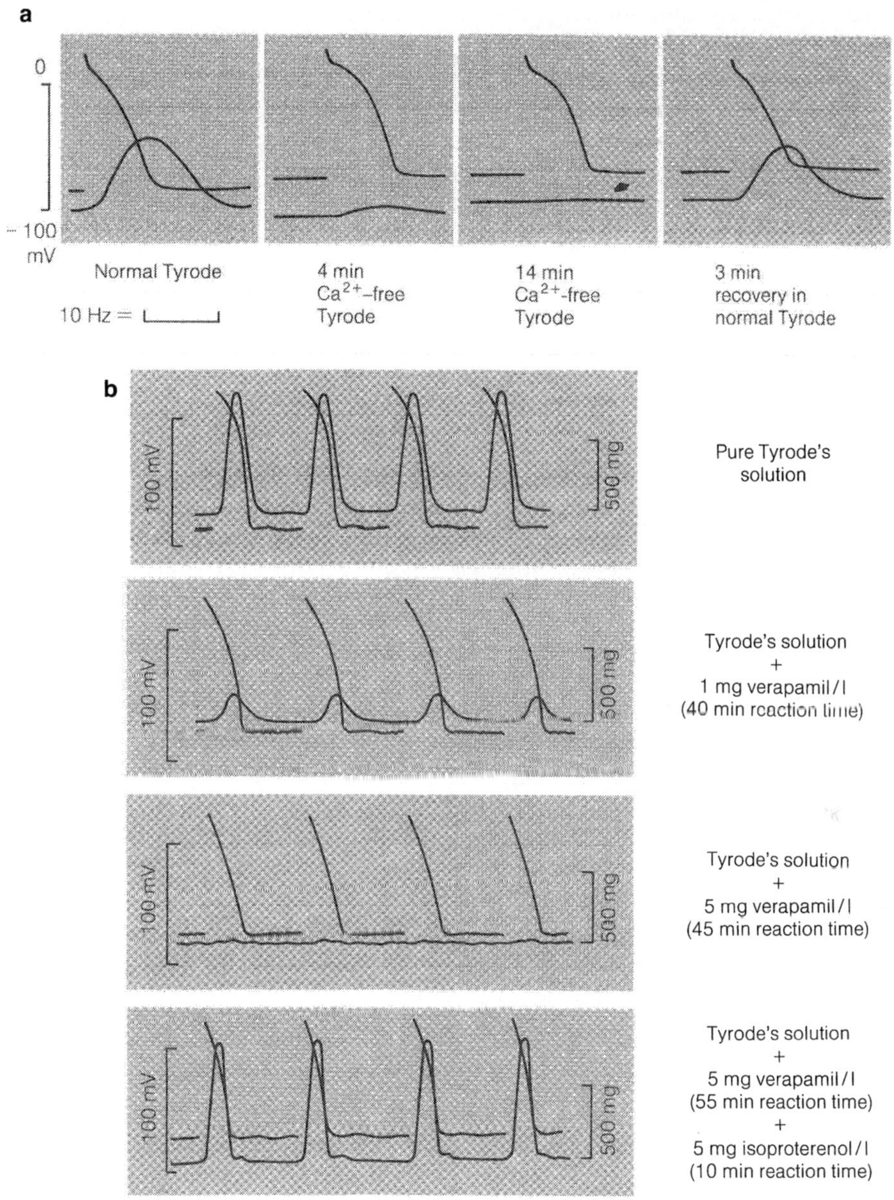

Fig. 6. Excitation–contraction decoupling via **a** calcium withdrawal, **b** calcium antagonists (verapamil): cancellation of contractility of an isolated rabbit papillary muscle without significant influence on action potential. Calcium supply or isoproterenol restores contractility. (After Fleckenstein 1980)

Other effects, e.g. inhibition of the fast inward sodium current, should play no role.

1.2.2 Basic Effects on the Heart

1. Calcium antagonists lower myocardial contractility without any readily recognizable influence on the action potential ("excitation–contraction decoupling"; negative inotropic effect).

2. Calcium antagonists lower the activity of nomotopic (sinoatrial node, AV node) and possibly that of ectopic pacemakers; i.e. they possess a negative chronotropic, negative dromotropic and antiarrhythmic action (predominantly in the region of the atrium and the AV node). (Details in Fleckenstein 1980, 1982, 1984; Antmann et al. 1980; Opie 1984, and in Sect. 4.3.)

Fleckenstein additionally discusses the proposition that calcium antagonists can prevent catecholamine-induced myocardial necroses by inhibition of intracellular calcium overload, so-called "cardioprotection" (Fleckenstein 1982; see also Sect. 3.1.4).

1.2.3 Basic Effects on Smooth Muscle

The basic effects described apply primarily also to vascular smooth muscle.

1. Calcium antagonists inhibit both phasic (acetylcholine-, serotonin-, histamine-, noradrenaline- or digitalis-glycoside-induced) and tonic (potassium-depolarization-induced) contractility (Fleckenstein-Grün 1983). In general, inhibition of depolarization-operated calcium channels prevails (DOC, VOC, Sect. 1.1.3), but receptor-operated channels (ROC) are also affected (Rosendorff 1984; Katz 1985b).

2. Calcium antagonists partially lower spontaneous bioelectrical activity (as in cardiac pacemaker cells, calcium is needed in smooth muscle for automatic formation of action potentials; details in Fleckenstein 1982).

1.2.4 Mechanism and Site of Action

The essential mechanism of action of calcium antagonists was elucidated by Fleckenstein between 1969 and 1972 (Fleckenstein 1971): calcium antagonists inhibit the slow, transmembranal inward calcium current through the so-called slow calcium channels. Biochemically, calcium channels are macroproteins in the cell membrane, which selectively regulate the passage of calcium ions through the membrane. The degree of opening of these channels is, on one hand, dependent on membrane potential, and on the other hand, on the presence of specific agonists or antagonists. It is also proposed that, in addition to inhibition of calcium inflow, calcium antagonists reduce calcium binding to the cell membrane and/or a displacement of calcium from superficial depots in the cell membrane (glycocalyx) (Fleckenstein 1982; Nayler et al. 1984).

The essential site of action of calcium antagonists thus appears to be the cell membrane (sarcolemma complex): in physiological dosages intracellular actions play no role. Within and at the cell membrane there are different binding sites ("receptors") for the various calcium antagonists. Thus, calcium antagonists of the nifedipine type (dihydropyridines) are bound to the external surface of the cell membrane, whereas those of the verapamil type are bound to the inside of the cell membrane; a third binding site for diltiazem is under discussion (Krebs 1984; Dompert and Traber 1984; Snyder and Reynolds 1985; Schrör 1985). Calcium antagonists of the verapamil type (verapamil, gallopamil) and diltiazem inhibit calcium channels in their open state; that is, they act upon the depolarized cell membrane. Thus, their action intensifies with an increasing heart rate, because more calcium channels in the activated ("open") state can be blocked per unit time. In contrast, the substances of the dihydropyridine group act on the closed calcium channel. Their action is independent of heart rate and of membrane potential (depolarization).

Individual calcium antagonists have, in addition to their calcium antagonistic action, certain further properties, although these manifest themselves predominantly at dosages above the physiological or therapeutic level:

1. Bepridil, gallopamil, verapamil, perhexiline, lidoflazine, and perhaps also diltiazem inhibit the fast sodium channel (Opie 1984; Krebs 1984).
2. Verapamil blocks α_1 receptors (Motulsky et al. 1983).
3. Verapamil blocks cardiac α_2 receptors; postsynaptic α_2 receptor blocking with inhibition of receptor-operated channels appears to be even in smooth muscle a principle of action of all calcium antagonists (Opie 1984).
4. Bepridil acts intracellularly on the sarcoplasmic reticulum; calmodulin antagonism is a subject of discussion in the context of felodipine, flunarizine, bepridil and nimodipine. (For more details, see Opie 1984; Sperelakis 1984; Lugnier et al. 1984.)
5. Verapamil, gallopamil, diltiazem and bepridil are thought to have a nonspecific sympatholytic action (so-called noncompetitive sympathetic antagonism) (Singh et al. 1984).

In Summary. Calcium antagonists in physiological dosage act principally at the cell membrane and here inhibit the slow inward calcium current into the cell.

1.2.5 Classification of Calcium Antagonists

The classification presented is guided, on the one hand, by consideration of chemical structure, but, on the other hand, according to predominant action (modified according to recent studies by Nayler and Horowitz 1983; Schrör 1985; Singh et al. 1985).

1. Verapamil Group

Chemically papaverine derivatives; so-called central calcium antagonists with predominant cardiac action on the myocardium, and the pacemaker and conduction system.

Representatives: verapamil
gallopamil (D 600)
tiapamil
(for more recent derivatives see Chap. 6)

2. Diltiazem
Chemically a benzothiazepine; a calcium antagonist acting on vascular smooth muscle with simultaneous action in the heart on the myocardium, and the pacemaker and conduction system.

3. Dihydropyridine Group (Nifedipine Group)
Chemically dihydropyridines; so-called peripheral calcium antagonists, with predominant action on vascular smooth muscle.
Representatives: nifedipine
nitrendipine
nimodipine
nisoldipine
niludipine
nicardipine
felodipine
isradipine
(for more recent derivatives see Chap. 6)

4. Other Calcium Antagonists
a) Piperazine group:
lidoflazine
cinnarizine [1]
flunarizine [1]
b) prenylamine
c) fendiline
d) perhexiline
e) etafenone
f) cyclandelate
g) bepridil

A scheme of classification put forward by the WHO in 1987 (Table 2) distinguishes between selective and nonselective blocking of slow calcium channels. Moreover, flunarizine-like substances (type IV) and prenylamine-like substances (type V) form groups of their own. It remains to be seen to what extent this reclassification of calcium antagonists will be adopted and will justify itself worldwide.

Alternative classification schemes "according to clinical efficacy" (Hopf et al. 1982; Schneider et al. 1984) appear less appropriate, since they only inadequately address the different chemical structures and the various sites of action.

---

[1] These substances have lately been juxtaposed to the other calcium antagonists as "calcium overload blockers" (selective blockers of transmembranal calcium flooding; van Zwieten 1985). They are thought to inhibit pathological cellular calcium overload in ischaemic conditions (calcium influx diffusely across the cell membrane), but in physiological conditions manifest only a weak action upon the specific calcium influx via the so-called slow calcium channels. This is thought to explain the assumedly selective action in the brain and the virtually nonexistent action on the heart and the peripheral vascular smooth muscle. An additional blockade of histamine-1 receptors (antihistaminic effect) is postulated for cinnarizine and flunarizine (Palmer et al. 1981; Tinkelmann 1985).

Table 2. WHO classification of calcium antagonists (after Vanhoutte 1987)

Class	Type	Designation	Substances
A	I	Verapamil type	Verapamil, gallopamil, tiapamil
	II	Nifedipine type	Nifedipine, nitrendipine, nicardipine, nimodipine, nisoldipine, felodipine, isradipine
	III	Diltiazem type	Diltiazem
B	IV	Flunarizine type	Flunarizine, cinnarizine
	V	Prenylamine type	Prenylamine, fendiline
	VI	Others	Perhexiline, caroverine, etc.

Types I–III: selective blocking of slow calcium channels (class A)
Types IV–VI: no selective blocking of slow calcium channels (class B)

References

Adelstein RS (1983) Regulation of contractile proteins by phosphorylation J Clin Invest 72: 1863–1866

Antman EM, Stone PH, Muller JE, Braunwald E (1980) Calcium channel blocking agents in the treatment of cardiovascular diseases Part I: Basic and clinical electrophysiologic effects Ann Intern Med 93: 875–885

Bayer R (1984) Grundprozesse der elektromechanischen Kopplung im Myokard In: Riecker G (ed) Herzinsuffizienz Springer Berlin pp 87–122

Berridge MJ (1985) Calcium: a universal second messenger Triangel 24: 79–90

Braunwald E (1982) Mechanism of action of calcium channel blocking agents N Engl J Med 307: 1618–1627

Braunwald E, Sonnenblick EH, Ross J (1984) Contraction of the normal heart In: Braunwald E (ed) Heart disease – a textbook of cardiovascular medicine Saunders Philadelphia pp 409–446

Burkart F, Heierli B (1984) Hämodynamik, Koronardurchblutung und Sauerstoffbedarf des normalen und insuffizienten Herzens In: Riecker G (ed) Herzinsuffizienz Springer Berlin pp 123–175

Carafoli E, Crompton M (1978) The regulation of intracellular calcium by mitochondria Ann NY Acad Sci 307: 269–284

Carafoli E (1984) How calcium crosses plasma membranes including the sarcolemma In: Opie LH (ed) Calcium antagonists and cardiovascular disease Raven Press New York pp 29–42

Casteels R, Droogmanns G (1983) The role of calcium in contraction of vascular smooth muscle In: Stone PH, Antman EM (eds) Calcium channel blocking agents in the treatment of cardiovascular disorders Futura Press Mount Kisco pp 59–70

Cohen CJ, Janis RA, Taylor DG, Scrabine A (1984) Where do calcium antagonists act? In: Opie LH (ed) Calcium antagonists and cardiovascular disease Raven Press New York pp 151–164

Dompert WU, Traber I (1984) Binding sites for dihydropyridine calcium antagonists In: Opie LH (ed) Calcium antagonists and cardiovascular disease Raven Press New York pp 175–180

Fabiato A, Fabiato F (1977) Calcium release from the sarcoplasmic reticulum Circ Res 40: 119–129

Fabiato A, Fabiato F (1978) Calcium-induced release of calcium from the sarcoplasmic reticulum of skinned cells from adult human, dog, cat, rabbit, rat and frog hearts and from fetal and newborn rat ventricles Ann NY Acad Sci 307: 491–522

Fleckenstein A (1971) Specific inhibitors and promotors of calcium action in the excitation – contraction coupling of heart muscle and the role in prevention or production of myocardial lesions In: Harris P, Opie LH (eds) Calcium and the heart Academic Press London pp 135–188

Fleckenstein A (1980) Steuerung der myokardialen Kontraktilität, ATP-Spaltung, Atmungsintensität und Schrittmacherfunktion durch Calcium-Ionen – Wirkungsmechanismus der Calcium-Antagonisten In: Fleckenstein A, Roskamm H (eds) Calcium-Antagonismus Springer Berlin pp 1–27
Fleckenstein A (1982) Calcium-Antagonismus – ein neues Prinzip in der kardiovasculären Therapie In: Roskamm H, Reindell H (eds) Herzkrankheiten Springer Berlin pp 777–799
Fleckenstein A (1984) Calcium antagonism: history and prospects for a multifaceted pharmacodynamic principle In: Opie LH (ed) Calcium antagonists and cardiovascular disease Raven Press New York pp 9–15
Folkow B, Neil E (1971) Circulation Oxford University Press, London
Gevers W (1984a) The mechanisms of myocardial contraction in: Opie LH (ed) The heart Grune & Stratton London pp 98–107
Gevers W (1984b) Calcium and the contractile mechanism in heart and smooth muscle In: Opie LH (ed) Calcium antagonists and cardiovascular disease Raven Press New York pp 67–74
Hopf R, Becker HJ, Kober G, Dowinsky S, Kaltenbach M (1982) Therapie der Angina pectoris mit Calcium-Antagonisten Herz 7: 221–234
Katz AM, Reuter H (1979) Cellular calcium and cell death Am J Cardiol 44: 188–190 (Editorial)
Katz AM (1984) Calcium fluxes across the sarcoplasmatic reticulum In: Opie LH (ed) Calcium antagonists and cardiovascular disease Raven Press New York pp 53–66
Katz AM (1985a) Effects of digitalis on cell biochemistry: sodium pump inhibition J Am Coll Cardiol 5: 16 A
Katz AM, Hager WD, Messineo FC, Pappano AJ (1985b) Cellular actions and pharmacology of the calcium channel blocking drugs Am J Med 79 (Suppl 4 A): 2–10
Katz AM (1985c) Editorial Triangel 24: 75–78
Krebs R (1984) Calcium antagonists: new vistas in theoretical basis and clinical use In: Opie LH (ed) Calcium antagonists and cardiovascular disease Raven Press New York pp 347–358
Lugnier C, Follenius A, Gerard D, Stocket JC (1984) Bepridil and flunarizine as calmodulin-inhibitors Eur J Pharmacol 98: 157–158
Morad HJ, Tung L (1982) Ionic events responsible for the cardiac resting and action potential Am J Cardiol 49: 584–594
Motulsky HJ, Snavely MD, Hughes RJ, Insel PA (1983) Interaction of verapamil and other calcium channel blockers with alpha 1– and alpha 2 adrenergic receptors Circ Res 52: 226–231
Murphy RA, Gertshoffer WT (1984) Cell calcium and contractile system regulation in arterial smooth muscle In: Opie LH (ed) Calcium antagonists and cardiovascular disease Raven Press New York pp 75–84
Nayler WG, Horowitz DR (1983) Calcium antagonists: a new class of drugs Pharmacol Ther 20: 203–262
Nayler WG, Dillon JS, Dayly MJ (1984) Cellular sites of action of calcium antagonists and β-adrenoreceptor blockers In: Opie LH (ed) Calcium antagonists and cardiovascular disease Raven Press New York pp 181–192
Opie LH (1984) The heart Grune & Stratton London pp 40–87 und 230–260
Opie LH, Nayler W, Gevers W (1984) Calcium fluxes In Opie LH (ed) The heart Grune & Stratton London pp 88–97
Palmer JH, Worth PHL, Extan-Smith AN (1981) Flunarizine: a once-daily therapy for urinary incontinence Lancet 2: 279–281
Rasmussen H (1986) The calcium messenger system N Engl J Med 314: 1094–1101 und 1164–1170
Rosendorff C (1984) Calcium channel blockers and hypertension In: Opie LH (ed) Calcium antagonists and cardiovascular disease Raven Press New York pp 323–332
Schneider W, Bussmann WD, Kaltenbach M (1984) Aspekte zur Therapie von Herz-Kreislauf-Erkrankungen mit Calcium-Antagonisten und Betarezeptorenblockern Dtsch Ärztebl 81: 1349–1355
Schrör K (1985) Physiologische Grundlagen, Pharmakologie und Prinzipien der klinischen Anwendung In: Hossmann V, Grötz J, Schrör K (eds) Kalzium Antagonisten und zerebrale Erkrankungen Springer Berlin pp 1–39

Singh BN, Josephson MA, Nademanee K (1984) Elektrophysiologische und hämodynamische Wirkungen von Calcium-Antagonisten: klinischer Überblick In: Althaus U, Burckhardt D Vogt E (eds) Calcium Antagonismus – Internat. Symposium on Calcium-Antagonism Universimed Frankfurt/Main pp 57–85

Singh BN, Baky S, Nademanee K (1985) Second generation calcium antagonists: search for greater selectivity and versatility Am J Cardiol 55: 214B–221B

Snyder SH, Reynolds IJ (1985) Calcium antagonist drugs: receptor interactions that clarify therapeutic effects N Engl J Med 313: 995–1002

Sperelakis N (1984) Properties of calcium-dependent slow action potentials: Their possible role in arrhythmias In: Opie LH (ed) Calcium antagonists and cardiovascular disease Raven Press New York pp 277–291

Tinkelman DG (1985) Calcium channel blocking agents in the prophylaxis of asthma Am J Med 78 (Suppl 2 B): 35–38

Vanhoutte PM (1987) The expert committee of the world health organization on classification of calcium antagonists: the viewpoint of the raporteur Am J Cardiol 59: 3A–8A

Van Zwieten PA (1985) Calcium-Antagonisten – Terminologie, Klassifizierung und Vergleich Arzneimittelforschung 35 (I): 298–301

2 Chemical Structure and Pharmacokinetics of Calcium Antagonists

2.1 Chemical Structure

2.1.1 Verapamil Group

The substances of this group are chemically derivatives of papaverine. As the first of these substances, verapamil was synthesized in 1962 (Haas and Härtfelder 1962). The basic chemical structures of verapamil and subsequent derivatives are shown in Fig. 7. In contrast to the 3,4-dimethoxy substitution of verapamil, the successor compound gallopamil (D 600) shows a 3,4,5-trimethoxy substitution on the benzene ring. Both substances are optically active; according to recent investigations (Schrör 1985; Hugenholtz 1985), *S*(−)enantiomers of verapamil and gallopamil have a more powerful effect than the *R*(+)derivatives.

Another successor substance of verapamil, tiapamil (Ro-11-1781), by contrast, shows no stereoselectivity, which would indicate that a chiral (asymmetric) C atom is not an essential structural element of the verapamil group. The substances of the verapamil group and their substituents are listed in Table 3, and the structural formulae of verapamil, gallopamil and tiapamil are shown in Fig. 9a.

Verapamil and gallopamil are not photosensitive and are soluble in water and ethanol. A racemate of both stereoisomers is commercially available, but biological activity rests predominantly within the domain of the *S*(−)enantiomer. Tiapamil is currently undergoing clinical trial and is not yet commercially available.

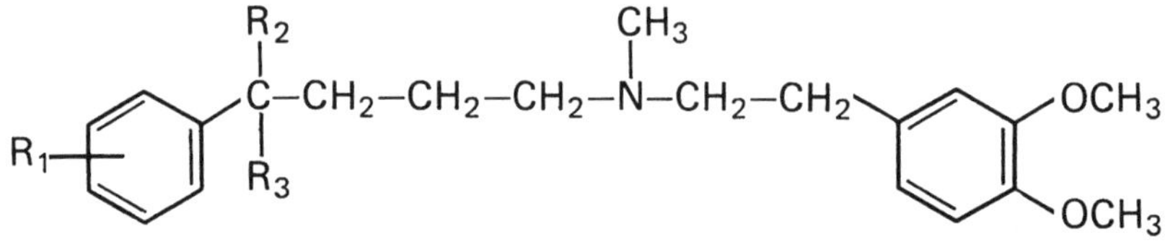

Fig. 7. Basic chemical structure of verapamil and its analogues (for substituents R_1–R_3 see Table 3)

Table 3. Substances of the verapamil group and their chemical substituents (basic chemical structure in Fig. 7)

Substance	R_1	R_2	R_3
Verapamil	3,4-$(OCH_3)_2$	$CH(CH_3)_2$	CN
Gallopamil	3,4,5-$(OCH_3)_3$	$CH(CH_3)_2$	CN
Tiapamil	3,4-$(OCH_3)_2$	$-SO_2-CH_2-CH_2-CH_2-SO_2-$	

2.1.2 Diltiazem

Diltiazem is a benzothiazepine derivative and is structurally unrelated to any other calcium antagonist. The structural formula is shown in Fig. 9b. The alkylaminoalkyl substitution at the nitrogen atom is of decisive significance for its biological activity. Since the structure of diltiazem possesses two chiral (asymmetric) C atoms, two diastereomer pairs (four enantiomers) are possible. The dextrorotatory (+)-*cis* enantiomer is commercially available. The substance was synthesized at the beginning of the 1970s in Japan (Sato et al. 1971). It is not photosensitive, and while readily soluble in water, only slightly so in alcohol.

2.1.3 Dihydropyridine Group

The basic chemical structure of the 1,4-dihydropyridines is shown in Fig. 8. Nifedipine, synthesized by Vater et al. at the end of the 1960s, can be taken as a reference substance (Bossert and Vater 1971; Vater et al. 1972). Hydrogenation of the pyridine rings is indispensable to the biological activity of the dihydropyridines; nonhydrogenated pyridines are ineffective. Biological activity in the vascular system (vasodilation) and in the heart (negative inotropic action) is crucially determined by the substituent in position 4 (R_3), where an aromatic or heteroaromatic ring is required (Meyer 1984a; Meyer et al. 1985).

With nifedipine as a prototype (structural formula Fig. 9a), several substances were synthesized by different substitutions on the pyridine ring. These differ in effective strength, duration of action, and also partly in organ selectivity. In comparison with nifedipine with its symmetric substitution in the 3 and 5 positions (R_2 and R_4), a reinforced vasodilating or antihypertensive activity is obtained with different substituents at these positions: this principle has been applied, for instance, in nitrendipine, nimodipine, nisoldipine, nicardipine and felodipine.

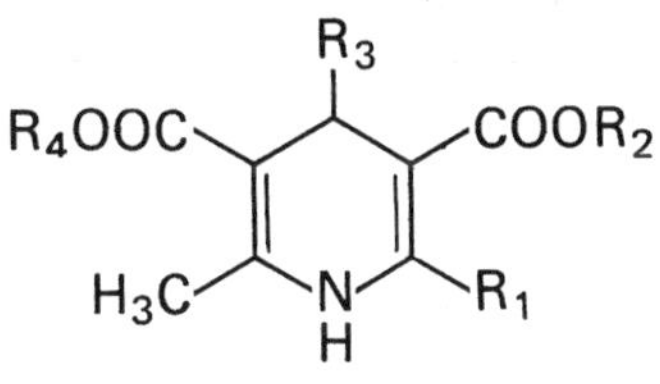

Fig. 8. Basic chemical structure of 1,4-dihydropyridines (for substituents R_1–R_4 see Table 4)

Table 4. Substances of the dihydropyridine group and their chemical substituents (basic chemical structure in Fig. 8). (After Meyer 1984a; Schrör 1985)

Substance	R_1	R_2	R_3	R_4
Nifedipine	CH_3	CH_3	$2\text{-}NO_2-C_6H_4$	CH_3
Nicardipine	CH_3	$CH_2-CH_2-N\begin{smallmatrix}CH_2-C_6H_5\\ CH_3\end{smallmatrix}$	$3\text{-}NO_2-C_6H_4$	CH_3
Nitrendipine	CH_3	C_2H_5	$3\text{-}NO_2-C_6H_4$	CH_3
Nimodipine	CH_3	$CH_2-CH_2-OCH_3$	$3\text{-}NO_2-C_6H_4$	$CH-(CH_3)_2$
Nisoldipine	CH_3	CH_3	$2\text{-}NO_2-C_6H_4$	$CH_2-CH-(CH_3)_2$
Felodipine	CH_3	C_2H_5	$2,3\text{-}Cl_2-C_6H_3$	CH_3
Niludipine	CH_3	$C_3H_7OCH_2CH_2$	$3\text{-}NO_2-C_6H_4$	$C_3H_7OCH_2CH_2$
Isradipine	CH_3	C_2H_5	$2,3\text{-}N_2O-C_6H_3$	C_2H_5

a) Verapamil group

Verapamil

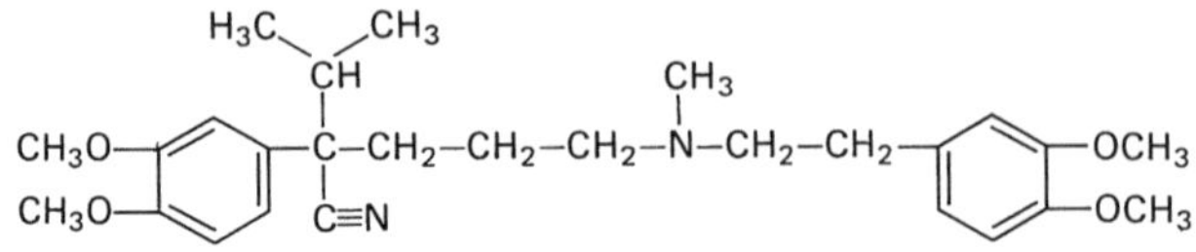

Gallopamil (D 600)

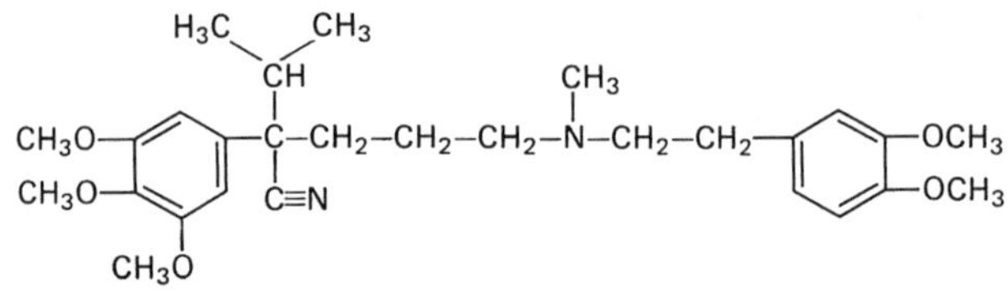

Tiapamil (Ro–11–1781)

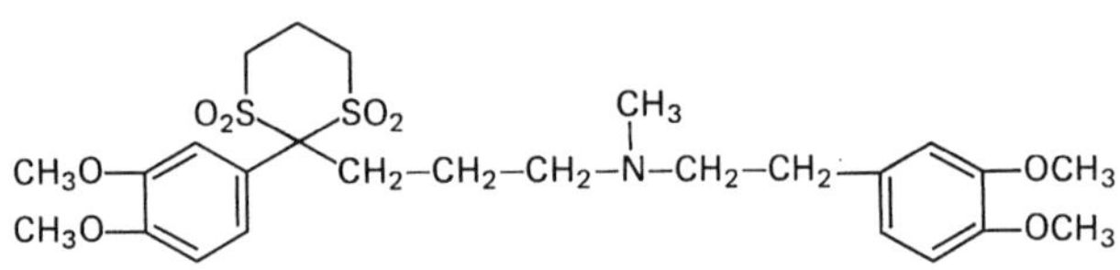

b) Diltiazem

Diltiazem

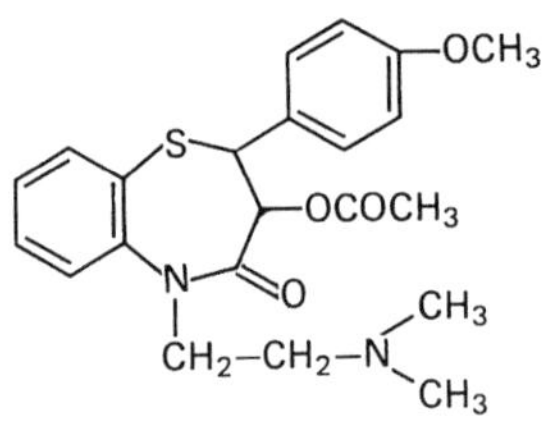

c) Dihydropyridine group

Nifedipine

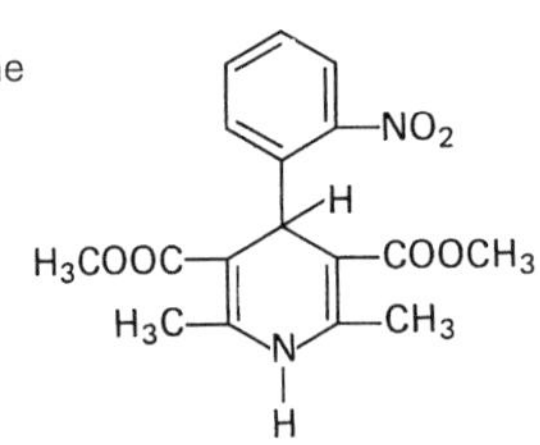

Nitrendipine

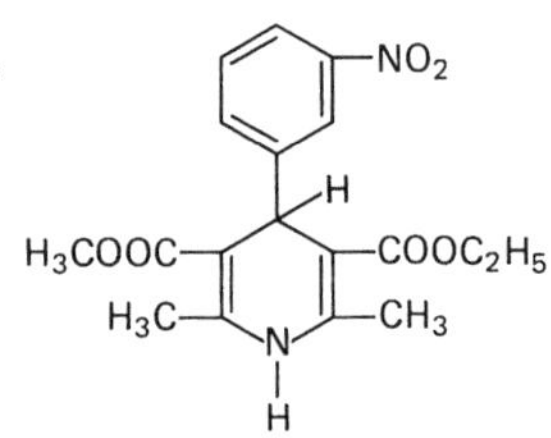

Nimodipine

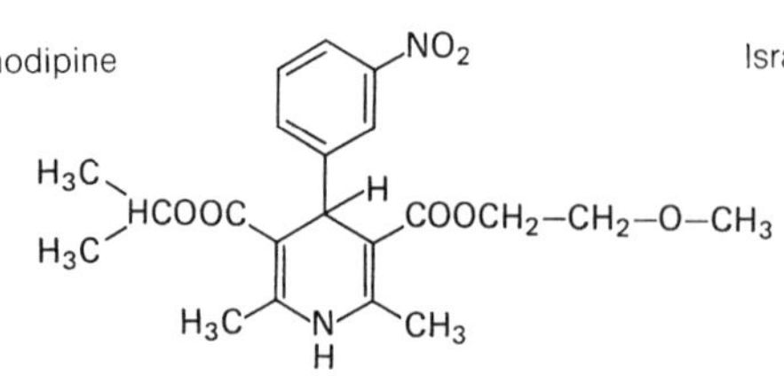

Isradipine

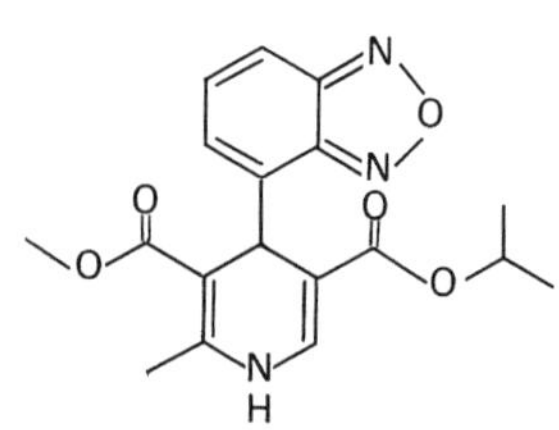

Felodipine

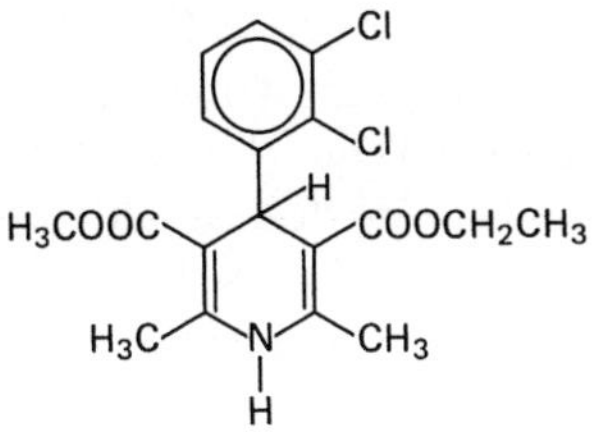

Nisoldipine

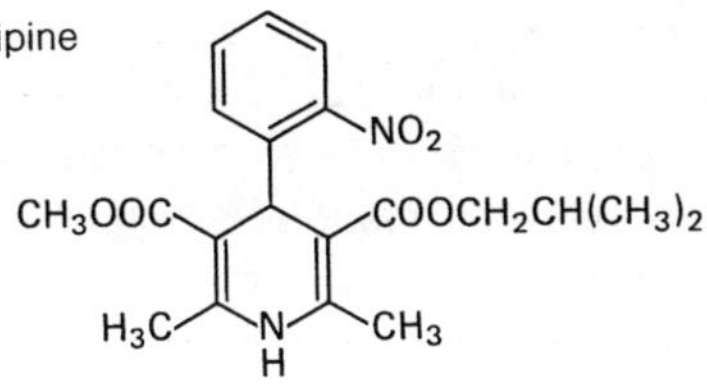

d) Other calcium antagonists

Lidoflazine

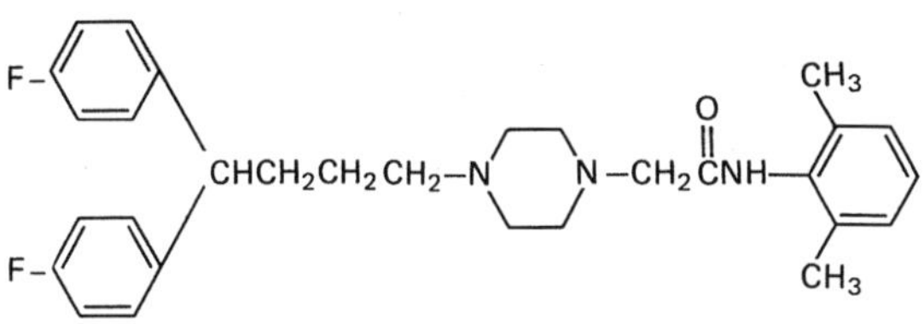

Flunarizine

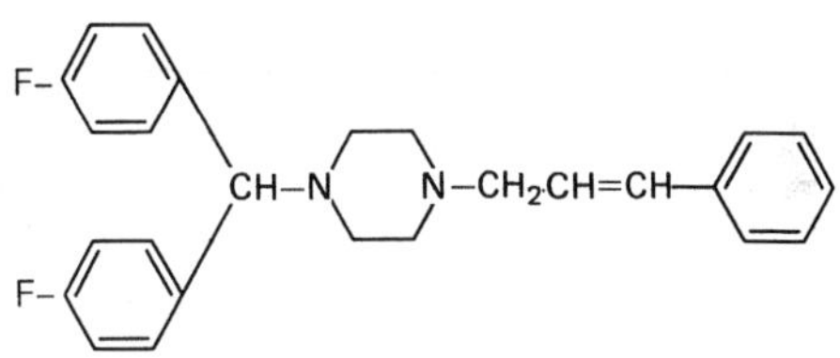

Cinnarizine

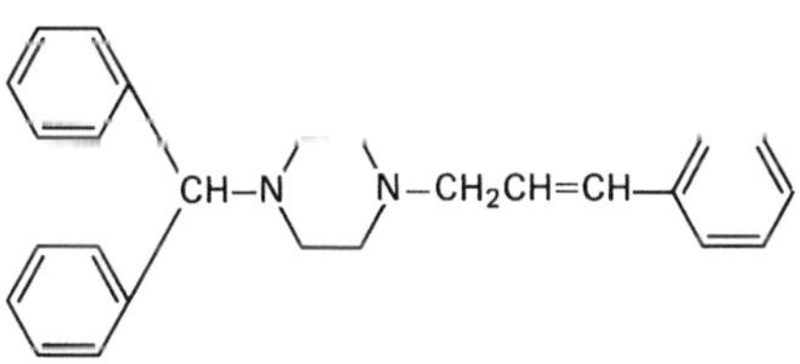

Cyclandelate

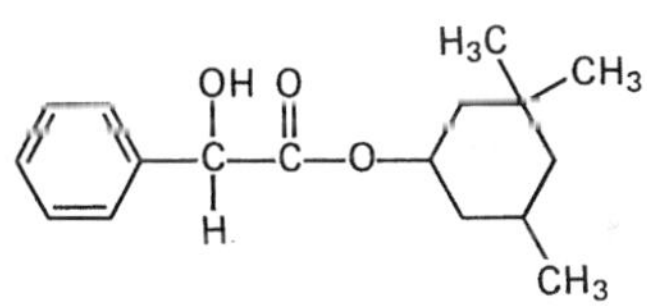

Fendiline

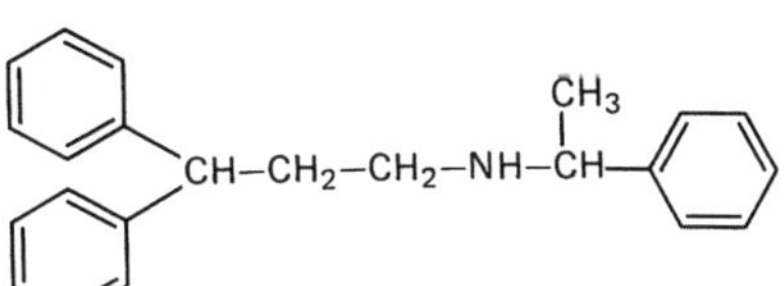

Prenylamine

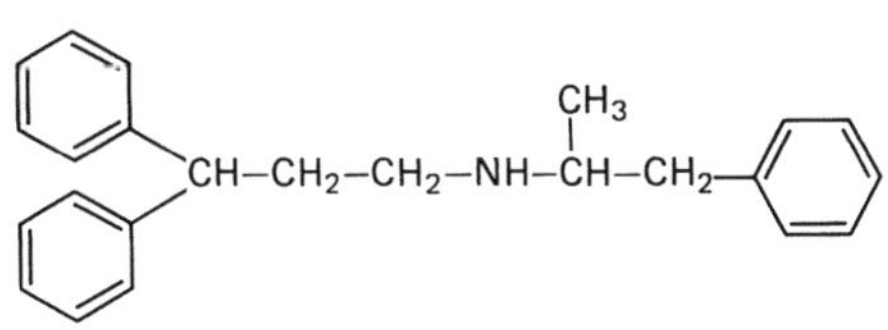

Perhexiline

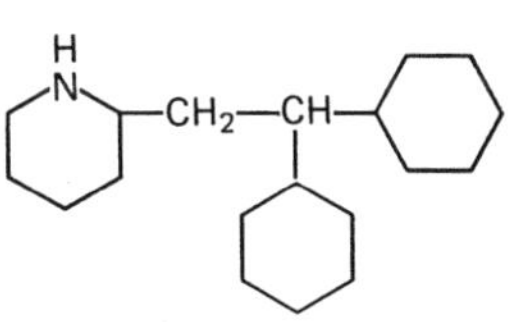

Bepridil

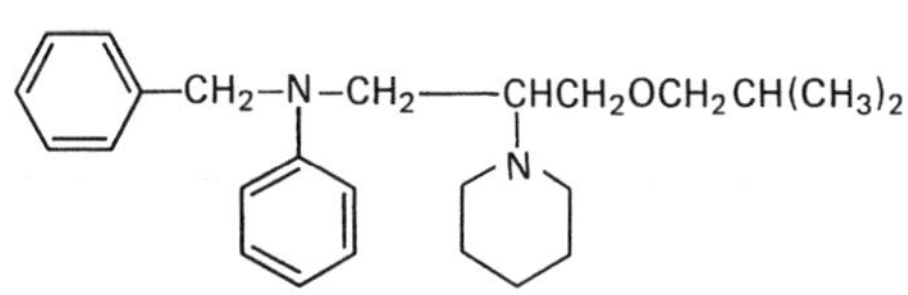

Fig. 9a–d. Structural formulae of the calcium antagonists

Asymmetrical substitution also influences the duration of action (longer elimination half-life of nitrendipine, nisoldipine and felodipine than of nifedipine) and organ selectivity. Table 4 summarizes the substances of the dihydropyridine group and their corresponding chemical substituents. Apart from nifedipine, nimodipine, nitrendipine, nicardipine and felodipine are now also commercially available (for structural formulae see Fig. 9c). Nimodipine and nitrendipine were synthesized by Meyer in the 1970s and are optically active racemates of two optically active enantiomers (chirality at C atom 4; Meyer et al. 1984b; through asymmetrical substituents in positions 3 and 5). Felodipine (synthesized in 1977), nisoldipine (structural formulae in Fig. 9c) and nicardipine have also been introduced in various countries.

Nifedipine and the other dihydropyridines are, with the exception of felodipine, photosensitive; both UV light and ordinary daylight give rise to inactive catabolites. The substances must therefore be protected against light, and parenteral forms of application (nifedipine, nimodipine) must be shielded from all light. All dihydropyridines are odourless and tasteless. They are virtually insoluble in water, but soluble in alcohol, chloroform and acetone.

2.1.4 Other Calcium Antagonists

The piperazine derivatives lidoflazine, cinnarizine and flunarizine are chemically diphenylalkylamines, and their structural formulae are shown in Fig. 9d. Flunarizine, being lipophilic, readily penetrates into the cerebrospinal fluid.

Prenylamine, already synthesized at the beginning of the 1960s, contains the active agent *N*-(3,3-diphenylpropyl)-*N*(α-methylphenethyl)-amine. While the substance exhibits optical activity because of an asymmetric carbon atom, both stereoisomers have the same efficacy. Prenylamine is readily soluble in ethanol, but only slightly soluble in water (Eichstädt 1984).

Fendiline (so-called demethylprenylamine) was created by splitting off a methyl group; it has similar chemical properties. The structural formulae of both substances are given in Fig. 9d.

Perhexiline, like the dihydropyridines and diltiazem, chemically belongs to the *N*-heterocycles; the structural formula is given in Fig. 9d. Cyclandelate is a mandelic acid derivative – chemically 3,3,5-trimethylcyclohexylmandelic acid ester. Bepridil (for structural formula see Fig. 9d), a calcium antagonist with additional effect on the fast sodium channel and with an extremely long elimination half-life (around 42 h), is still being clinically tested.

2.2 Pharmacokinetics

By means of pharmacokinetic data, drugs can be characterized with respect to their distribution within the organism, their metabolism and their elimination. Knowledge of pharmacokinetic parameters is an essential prerequisite for selection of the appropriate form of application (i.e. oral, sublingual, buccal, intravenous), for determination of the onset of action and the dosage interval, and for adjustment of the dose in restricted hepatic or renal function. All pharmacokinetic data must obviously be evaluated in terms of inter- and intraindividual variations, and in the light of such critical factors as whether they are derived from healthy or sick subjects, from young or old patients, or from single- or multiple-dose administration. Moreover, since patients frequently take several medicaments concomitantly, the appropriate drug interactions must also be taken into account (Sect. 5.8). In theory, changes in the absorption quota (changed stomach emptying and/or duration of passage), in bioavailability (changed hepatic enzyme induction or inhibition, or changed hepatic blood flow), in plasma protein binding (competitive protein binding with a rise of free active agent), and in renal elimination of metabolites (affecting filtration, secretion and reabsorption) are possible. There have been to date very few systematic investigations of drug interactions in calcium antagonist therapy, but established and clinically relevant interactions are discussed in Sect. 5.8.

Commercially available calcium antagonists exhibit similarity in a wide range of pharmacokinetic parameters despite the heterogeneity exhibited by their chemical structure (described already). All substances are almost completely (on average more than 90%) absorbed from the gastrointestinal tract (measured as radioactivity). Further common features of all calcium antagonists are a high degree of protein binding (80%–99%), hepatic degradation with formation of active and inactive metabolites, and very low renal elimination of the unchanged basic substance (on average less than 1%), except for verapamil and diltiazem.

Individual elimination half-life ($t_{1/2}$) values make it necessary in most cases (when sustained-release medication is not used) to spread the daily dose over several appropriately smaller separate doses – with the exception of nitrendipine, nisoldipine, felodipine and flunarizine (Table 5).

There are considerable differences in the bioavailability of the calcium antagonists (here defined as the ratio of the areas under the plasma concentration curves after oral and intravenous application). Given the virtually total absorption of the substances, the lower bioavailability of oral doses is due to the so-called first-pass effect, i.e. the phenomenon that the liver, immediately after their absorption from the gastrointestinal tract, binds and metabolizes substances to varying degrees during the first "pass". The first-pass effect of the gastrointestinal mucosa and liver amounts, for individual calcium antagonists, to between 30% and 95%, corresponding to a bioavailability of 5%–70% (Table 5). Even for an individual substance, the percentage of the first-pass effect or the bioavailability have no fixed values; in addition to intra- and interindividual variations, special consideration must be given to age, hepatic function, concomitant medication, and the duration of dosage (single- or multiple-dose administration).

The essential pharmacokinetic data for the calcium antagonists discussed in this book are summarized in Table 5. Further discussion is confined to the properties of individual substances relevant to practical therapy. Special questions on pharmacokinetics are dealt with in recent reviews (Schomerus et al. 1976; Stone et al. 1980; Freedman 1984; Schrör 1985; Hugenholtz 1985).

2.2.1 Verapamil Group

Given orally, **verapamil** is absorbed to more than 90%. Given intravenously, an effect is discernible within 2 min. With single-dose administration, bioavailability is around 20%, and the elimination half-life is between 3 and 7 h. The latter is markedly age-dependent and can extend to 7–8 h in elderly patients (Abernethy et al. 1986). With long-term oral therapy the first-pass

Table 5. Pharmacokinetic data relating to calcium antagonists. (After Opie 1980; Piepho 1981; Holmes et al. 1984; Eichstädt 1984; Freedman 1984; Follath 1984; Hugenholtz 1985; Schrör 1985; Klein 1985; Goodman and Gilman 1985; Edgar 1988; Saltiel et al. 1988)

Substance	Absorption	Bioavailability	Onset of action	Peak effect	$t_{1/2}$	Protein binding	Active metabolites
Verapamil	>90%	20% (50%–60% [a]; 35%–40% [c])	2 min i.v. 2 h oral	10–15 min 1 h oral 3–5 h oral (SR)	3–7 h (14 h[a]; 10 h[c])	90%	Norverapamil (20%)[f]
Gallopamil	90%	15%–25%		1–2 h	3–4 h	90%	
Tiapamil	95%	20%–25%			2.5 h	78%	
Diltiazem	>90%	40%–50%	1 min i.v. 15 min oral	5–10 min i.v. 3–4 h oral	3 h i.v. 6–9 h oral	80%	Desmethyldiltiazem (~20%)[f] Desacetyldiltiazem (~40%)[f]
Nifedipine	>90%	45%–70%	1 min i.v. 3 min sublingual[g] 20 min oral	1–2 h oral 3–4 h oral (SR) ½–1 h sublingual[g]	2 h i.v. 4 h oral	90%	
Nitrendipine	90%	10%–30% (44% [a])		1½ h oral	6–15 h (17 h[a]; 24 h[b])	98%	
Nimodipine	95%	3%–10%		3 min i.v. ½–1½ h oral	2–3 h	98%	
Nisoldipine	>90%	3%–10%			11–12 h	99%	M9 (10%–25%)[f]
Felodipine	>95%	15%		3–5 h oral (SR)	10 h i.v. 17–25 h oral	99%	
Lidoflazine	>90%	<5% [e]		2–4 h[d]	16–24 h		
Cinnarizine	90%	40%–60%		2–4 h[d]	3–8 h	80%	
Flunarizine	90%	85% (?)[e]		2–4 h[d]	4–19 days (!)	90%	
Prenylamine	90%	20%–30% [e]		1–2 h	Phase I 7 h; overall 3 days	>95%	
Fendiline	>90%	20%		2–4 h	18–22 h	90%	
Perhexiline	>90%	?		12–24 h	3–12 days (!)	>90%	
Bepridil					42 h		

[a] Restricted hepatic function; [b] restricted renal function; [c] multiple dosage/long-term therapy; [d] peak plasma level; [e] data provided by pharmaceutical companies; [f] percentage of effective strength of basic substance; [g] so-called sublingual administration (biting a capsule and swallowing its content; van Harten et al. 1987b); SR = sustained release form

effect is reduced (with an increase in bioavailability to 35%–40%), and elimination half-life thereby increases to about 10 h. This permits long-term therapy with 2–3 doses a day. In patients with restricted hepatic function (e.g. in liver cirrhosis), elimination half-life is increased to about 14 h, and bioavailability can rise to 50%–60%. The rise in plasma concentration is also due to decreases in clearance and in the volume of distribution. This phenomenon is seen with most calcium antagonists (Stieren et al. 1983; Eichler et al. 1985). Restricted hepatic function lowers the first-pass effect; it thereby increases bioavailability and extends the elimination half-life owing to retarded metabolism of the substance under consideration. Careful dose adjustment (reduction) and/or extension of intervals between doses is therefore necessary in restricted hepatic function; this applies also in markedly restricted renal function (changed distribution volume and whole-body clearance), and in the treatment of elderly patients. Verapamil is broken down in the liver into a range of metabolites, and of these only norverapamil is biologically active, accounting for about 20% of the effect of the basic substance (Opie 1980; Eichelbaum and Somogyi 1981; Freedman 1984; Schrör 1985). A sustained-release form of verapamil is available, with a bioavailability thought to be slightly less than that of the normal oral form (relative bioavailability approximately 90%; Follath 1984). The aim of this preparation is to attain stable serum concentrations of verapamil for 12 h. Recently a sustained-release (SR) form with a spread of effect over 24 h has become commercially available (dose 240 mg). In general, however, reduced bioavailability is to be expected with SR preparations, since with a slower supply of substance to the liver, a greater proportion of it is metabolized. Higher dosage is therefore usually necessary.

Gallopamil and **tiapamil** behave pharmacologically in a manner similar to verapamil; bioavailability is between 15% and 25% and the elimination half-life is about 3 h.

2.2.2 Diltiazem

Diltiazem is absorbed to more than 90% when orally administered, and the bioavailability of the substance is around 45%. When tablets (60 or 90 mg) are taken, the onset of action occurs after 15 min, and the effect reaches its maximum level after 3–4 h. The elimination half-life is 6–7 h, in elderly patients sometimes longer. In long-term oral therapy, although the elimination half-life remains unchanged, the possibility of cumulation of diltiazem under such conditions is a subject of current discussion. Diltiazem is largely metabolized in the liver. The first metabolites to appear are desmethyldiltiazem and desacetyldiltiazem. Although both still exhibit a reduced action (20% or 40% of that of the mother substance diltiazem), their contribution to the total effect can, owing to their significantly lower plasma concentrations in comparison with diltiazem, be safely ignored. Metabolites of diltiazem are excreted in equal amounts via the kidneys and the biliary system. In cases of restricted hepatic function, the maintenance dose of diltiazem should be reduced (Piepho et al. 1981; Zells and Kinney 1982; Kölle et al. 1983; Just et al. 1985).

2.2.3 Dihydropyridine Group

After oral and sublingual administration, **nifedipine** is also almost totally (>90%) absorbed. The observable onset of action is 1 min with intravenous administration, after "sublingual" administration (biting a capsule and swallowing its content) 3 min, and after oral ingestion (capsules, tablets) 20 min (Lydtin and Lohmöller 1972; Stone et al. 1980). Of late, as an alternative to sublingual administration of capsules, nifedipine has become available in liquid form (as drops); this presentation is photosensitive, and the drops must be ingested as soon as they are taken from the container (Stotzingen and von Pösl 1986). Maximum plasma concentration, and therewith also peak effect, is observed 2 h after oral ingestion. Here there are considerable interindividual variations; after oral ingestion, peak plasma levels can be attained in a period of 1–6 h. The bioavailability of nifedipine is cited in the literature as between 45% and 70%. Of available compounds, it is therefore the calcium antagonist with the highest bioavailability. The elimination half-life following intravenous administration is approximately 2 h, whereas following oral administration it is about 4 h. The drug must therefore either be infused, or in the case of oral administration, given several times a day. If commercially available sustained-release preparations are given, it seems possible to spread the total daily dose over two separate doses (Corvol et al. 1983). At present, a "continuous-release" form is being tested, which is reported to provide adequate levels of action over 24 h (dosage once daily: Vetrovec et al. 1989). Since the bioavailability of sustained-release preparations is usually markedly lower than that of non-sustained-release forms, the dose must be appropriately increased. Nifedipine is broken down in the liver into several ineffective metabolites, and is excreted mainly through the kidneys. No cumulation has been observed so far in long-term therapy (Horster 1975; Eichelbaum 1983; Freedman 1984; Schrör 1985).

Nimodipine and **nitrendipine** behave similarly to nifedipine, but their bioavailability is significantly lower: 3%–10% for nimodipine and 10%–30% for nitrendipine. The elimination half-life for nitrendipine is 6–15 h, which permits the spreading of the total daily dose over two separate doses. In hepatic diseases nitrendipine shows raised bioavailability (around 45%) and an extension of the elimination half-life to around 17 h. In restricted renal function the elimination half-life for nitrendipine is increased on average to about 24 h. Consequently, restricted hepatic and/or renal function calls for dosage reduction of nitrendipine (Raensch et al. 1984, 1985).

Felodipine has a bioavailability of 15% (Edgar et al. 1985); the elimination half-life is 17–25 h, and it increases with age. Treatment with two separate doses a day is possible (non-sustained-release form). A sustained-release form with the same bioavailability is held to ensure effective plasma levels over 24 h (one daily dose, maximum effect after 3–5 h; Hedner et al. 1987; Saltiel et al. 1988).

Nisoldipine has a bioavailability of 3%–10%, which rises to around 15% in the presence of liver cirrhosis. The elimination half-life is about 11–12 h (in liver cirrhosis about 15–20 h). The substance is broken down in the liver, where inter alia a weakly effective metabolite M9 (around 10%–25% of the effect of the mother substance) is formed. The total daily dose is usually spread over two separate doses (Ahr et al. 1987; van Harten et al. 1987a; Scherling et al. 1987).

2.2.4 Other Calcium Antagonists

Only a few pharmacokinetic studies concerning **lidoflazine** are available. The substance is absorbed to more than 90%. Bioavailability is probably under 5%, and the elimination half-life is reported to be 16–24 h. The question of drug cumulation in long-term therapy is not yet clarified (Schrör 1985).

Cinnarizine is absorbed to more than 90%. Bioavailability (40%–60%) is markedly higher than that of lidoflazine. Peak plasma level of the substance is reached after 2–4 h. The elimination half-life is 3–8 h (Morrison et al. 1979). Protein binding is 80%.

Pharmacokinetic studies in humans are also inadequate for **flunarizine.** The bioavailability of the substance is reported to reach 85% (communication from the Janssen company). Since under continuous therapy plasma levels of the substance vary interindividually by a factor of 10 or more, a variable first-pass effect in the liver is postulated (Heykants et al. 1979). Consequently, a variable bioavailability is held to explain these considerable variations in plasma levels from one individual to another. In long-term therapy with 10 mg/day, there is a steady rise in plasma concentration during the first 5–6 weeks; only thereafter are uniform concentrations (steady-state) achieved. This phenomenon is probably related to the extremely long elimination half-life of flunarizine of approximately 18 days (Holmes et al. 1984), with fluctuations according to Holmes (1986, personal communication) of 4–19 days.

Prenylamine is absorbed to about 90% (Opie 1980), and the bioavailability is 20%–30% (communication from the Albert Roussel company; Eichstädt 1984). Thus the substance behaves similarly to verapamil. Elimination exhibits a fast initial phase with a half-life of about 7 h, and a slower second phase. The overall elimination half-life is around 70 h. With long-term medication there is a cumulation of the substance in the first few days, but after 14 days a stable plateau is reached.

Fendiline has a bioavailability of about 20%, and maximum effect is achieved after 2–4 h. The elimination half-life is 18–22 h (Klein 1985).

Perhexiline is almost totally absorbed after oral ingestion. Data on the bioavailability of the substance are not available. The elimination half-life exhibits considerable variations from one individual to another and is estimated to be 3–12 days (Opie 1980), although in a few isolated reports extreme values of 30 days are claimed.

Bepridil, a substance which at present is still undergoing clinical testing, distinguishes itself pharmacokinetically by an elimination half-life of almost 2 days (about 42 h; Hill et al. 1985).

The bioavailability values given in Table 5 are mean values in each case. The range of scatter (i.e. average value plus or minus simple standard deviation) within a population is substantial, however (diltiazem 44% ± 10%, verapamil 19% ± 12%, nifedipine 45% ± 28%; according to Goodman and Gilman 1985); in the case of low bioavailability (verapamil), the percentage range of variation is higher. The effect of a single dose in a patient can only be unreliably predicted; pharmacokinetic data can only be considered a rough guide. The definitive dosage must always be titrated in terms of the drug's clinical and pharmacodynamic effects (Sect. 2.2.5).

2.2.5 Dosage Guidelines for Calcium Antagonists: Individualized Dosage

As can be seen from Table 5, the pharmacokinetic data of all calcium antagonists exhibit considerable interindividual differences. Not only absorption and metabolism of the substances used, but also body weight, i.e. distribution volume, time of intake and concomitant therapy with other pharmaceuticals vary from patient to patient. For this reason, dosage guidelines given in this book (and in other publications or in data sheets issued by pharmaceutical companies) can serve only as a rough orientation for the doctor responsible for treatment. The definitive dosage must always be individually assessed and must accommodate itself in symptomatic therapy (e.g. angina pectoris, Raynaud's syndrome, migraine) to the clinical and pharmacodynamic action of the substance, i.e. to the therapeutic efficacy.

On the other hand, in the treatment of asymptomatic patients (e.g. secondary prevention following acute myocardial infarction), dosage must strictly adhere to the precise data stated in the literature. As with every prophylactic therapy, several independent critical studies on large groups of patients must be available before general therapeutic recommendations may be given. For calcium antagonists, there are for any indication as yet inadequate data to ensure safe prophylactic treatment of asymptomatic patients (see for instance Sect. 3.4; 4.1.4; 4.4.3).

In Summary. Calcium-antagonistic therapy is at present still predominantly symptom orientated. The individual dosage must therefore be guided by the clinical outcome, i.e. by the degree of therapeutic success.

2.2.6 Generics: Sustained-Release Preparations

Because of the limited patent protection conceded to new pharmaceuticals, there are now already generic agents, some available as monosubstances and some as combinations, for various calcium antagonists. In many cases there are no precise pharmacokinetic data or clinical efficacy studies on these so-called analogue preparations. Moreover, the galenical properties of the analogue preparations are not always comparable with the original preparation (tablets, coated tablets, capsules with fluid content, sustained-release preparations etc.). The action of the analogue preparations can be ascertained from therapeutic success in treating symptomatic patients, and dosage adjustment may be necessary when an analogue is substituted in the regimen.

In the treatment of asymptomatic patients by reference to an intervention study (e.g. primary or secondary prophylaxis of acute myocardial infarction), the physician should adhere precisely to the data given and the substances used in the literature.

In Summary. The therapeutic success of any drug therapy should be reviewed at regular intervals: when there is change of preparation employed, especially when generic substances are substituted, close attention must be given to the possibility of a change in clinical or pharmacodynamic action.

As stated in Sect. 2.2, sustained-release preparations mostly exhibit a limited bioavailability and must therefore usually be given in higher dosages. On the other hand, through use of sustained-release preparations, peaking of plasma levels is prevented, to some extent, thereby possibly reducing the incidence of side effects (Chap. 5). Generally, the haemodynamic response depends not only on the plasma concentration but also on the change in concentration over time (dCp/dt). The incidence of tachycardia following dihydropyridines (e.g. felodipine) is reduced when the input rate is lowered (Edgar 1988). Similar observations have been made concerning the relationship between diltiazem and changes in the PQ interval. In dosage guidelines of any kind, especially in the case of dosages taken from clinical studies, careful attention must be given to whether the relevant data refer to normal preparations or sustained-release preparations. Here, too, the prescribing physician should always check and, where necessary, appropriately adjust the selected dosage according to the degree of therapeutic success.

References

Abernethy DR, Schwartz JB, Todd EL, Luchi R, Snow E (1986) Verapamil pharmacodynamics and disposition in young and elderly hypertensive patients Ann Intern Med 105: 329–336

Ahr, G, Wingender W, Kuhlmann J (1987) Pharmacokinetics of nisoldipine In: Hugenholtz PG, Meyer J (eds) Nisoldipine 1987 Springer Berlin Heidelberg New York Tokyo pp 59–66

Arznei-Telegramm (1985) Retardierung von Nifedipin (Adalat retard) ein Fortschritt? Arznei-Telegramm 1: 3–4

Bossert F, Vater W (1971) Dihydropyridine, eine neue Gruppe stark wirksamer Koronartherapeutika Naturwissenschaften 58: 578

Bossert F, (1975) The chemistry of nifedipine In: Lochner W, Braasch W, Kroneberg G (eds) 2nd International Adalat Symposium Springer Berlin pp 20–26

Braunwald E (1984) Heart disease, a textbook of cardiovascular medicine Saunders Philadelphia

Calcium Antagonisten: Klinische Pharmakokinetik (1985) Arzneimitteltherapie 3: 30

Corvol P, Marre M, Banzet O, Colin JN, Thibonnier M, Singlers E, Alexandre JM (1983) Acute antihypertensive effects and pharmacokinetic parameters of a new galenic form of nifedipine In: Kaltenbach M, Neufeld HN, (eds) 5th Adalat Symposium Excerpta Medica Amsterdam pp 158–162

Edgar B (1988) Clinical pharmacokinetics of felodipine Ph. D. thesis, University of Göteborg

Edgar B, Hoffmann KJ, Lundborg P, Regardh CG, Rönn O, Weidolf L (1985) Pharmacokinetic and pharmacodynamic effects of felodipine in healthy subjects Drugs 29 (Suppl 2): 9–15

Eichelbaum M, Somogyi A (1981) Verapamil distribution in health and the diseased state In: Zanchetti A, Krikler DM (eds) Calcium antagonism in cardiovascular therapy. Experience with verapamil Excerpta Medica Amsterdam pp 64–71

Eichelbaum M (1983) Pharmakokinetik von Diltiazem, Nifedipin und Verapamil In: Magometschnigg D (ed) Kalziumantagonisten in der Therapie der arteriellen und pulmonalen Hypertonie Uhlen Wien pp 21–27

Eichler HG, Mabin TA, Commerford PJ, LLoyd EA, Beck W, Opie LH (1985) Tiapamil, a new calcium antagonist: hemodynamic effects in patients with acute myocardial infarction Circulation 71: 779–786

Eichstädt H (1984) Calciumantagonisten In: Roskamm H (ed) Handbuch der inneren Medizin IX/3 Koronarkrankheiten Springer Berlin pp 1069–1113

Fleckenstein A (1983) History of calcium antagonists Circ Res 52 (Suppl I): 3–16

Follath F (1984) Konzentrationswirkungsbeziehung, Bioverfügbarkeit und Wirkdauer von Verapamil In: Althaus U, Burckhardt D, Vogt E (eds) Calcium-Antagonismus – Internat. Symposium on Calcium-Antagonism Universimed Frankfurt/Main pp 50–56

Freedmann SB (1984) Pharmakokinetik von Calcium-Antagonisten In: Althaus U, Burckhardt D, Vogt E (eds) CalciumAntagonismus – International Symposium on Calcium Antagonism Universimed Frankfurt/Main pp 30–49

Goodman L, Gilman A (1985) The pharmacological basis of therapeutics Macmillan New York

Haas H, Härtfelder G (1962) Alpha-Isopropyl-alpha N-Methyl-N homoveratryl-alphaamino' propyl 3,4,-dimethoxyphenylacetonitril, eine Substanz mit coronargefäßerweiternden Eigenschaften Arzneimittelforschung 12: 549–558

Hedner T, Elmfeldt D, Dahlöf C, Sjögren F (1987) Comparison of antihypertensive effect and pharmacokinetics of conventional and extended release felodipine tablets in patients with arterial hypertension Drugs 34 (Suppl 3): 125–131

Heykants J, De Creé J, Hörig C (1979) Steady-state plasma levels of flunarizine in chronically treated patients Arzneimittelforschung 29 (II): 1168–1171

Hill JA, O'Brien JT, Alpert JS, Gore JM, Zusman RM, Christensen D, Boucher CA, Vetrovec G, Borer JS, Friedman C, Mack R, Conti R, Pepine CJ (1985) Effect of bepridil in patients with chronic stable angina: results of a multicenter trial Circulation 71: 98–103

Holmes B, Brogden RN, Heel RC, Speight TM, Avery GS (1984) Flunarizine: a review of its pharmacodynamic and pharmacokinetic properties and therapeutic use Drugs 27: 6–44

Horster FA (1975) Pharmacokinetics of nifedipine C14 in man In: Hashimoto K, Kimira F, Kobayaska T (eds) 1st International Nifedipine-Adalat Symposium Univ of Tokyo Press Tokyo pp 67 70

Horster FA (1975) Pharmacokinetics of nifedipine – C14 in man In: Lochner W, Braasch W, Kroneberg G (eds) 2nd International Adalat Symposium Springer Berlin pp 124–127

Hugenholtz PG (1985) Calcium antagonists In: Abshagen U (ed) Clinical pharmacology of antianginal drugs Springer Berlin pp 459–538

Just H, Wollschläger H, Drexel H, Kasper W (1985) Introduction: Recent developments in the clinical use of calcium antagonists, with particular reference to diltiazem In: Just H, Schröder JS (eds) Advances in clinical application of calcium antagonistic drugs International Diltiazem Workshop Excerpta Medica Amsterdam pp 1–10

Kaltenbach M, Hopf R (1983) Gallopamil Springer Berlin

Klein G (1985) Pharmakokinetik von Calcium-Antagonisten unter besonderer Berücksichtigung von Fendilin In: Lüthy E (ed) Calmodulin und Calciumantagonisten Fischer Stuttgart pp 3–24

Kölle EU, Ochs HR, Vollmer KO (1983) Pharmacokinetic model of diltiazem Arzneimittelforschung 33: 972–977

Lydtin H, Lohmöller R (1972) Hämodynamische Wirkungen von Bay a 1040 vor und nach Atropingabe Verh Dt Gesellsch Inn Medizin 78: 1554–1559

Meyer H (1984a) Structure – activity relationships in calcium antagonists In: Opie LH (ed) Calcium antagonists and cardiovascular disease Raven Press New York pp 165–173

Meyer H, Bossert F, Wehinger E, Schelling D (1984b) Chemistry of Nitrendipine and its metabolism In: Scriabine A, Vanov S, Deck K (eds) Nitrendipine Urban & Schwarzenberg Baltimore pp 1–9

Meyer H, Wehinger E, Bossert F, Böshagen H, Franckowiak G, Goldmanns S, Seidel W, Stoltefuss J (1985) Chemistry of dihydropyridines In: Fleckenstein A,von Bremen C, Groß R, Hoffmeister F (eds) Cardiovascular effects of dihydropyridine-type Ca-antagonists Springer Berlin pp 90–103

Morrison PJ,Bradbrook HD, Rogers HJ (1979) Plasma cinnarizine levels resulting from oral administration as capsule or tablet formulation investigated by gas-liquid chromotography Br J Clin Pharmacol 7: 349–352

Opie LH (1980) Calcium Antagonists In: Opie LH (ed) Drugs and the heart Lancet London pp 27–38

Piepho RW, Bloedow DC, Lacz JP, Simons MA, Runser DJ, Dimmit DC, Browne RK (1981) Pharmakokinetik von Diltiazem In: Bender F, Greeff K (eds) Calcium Antagonisten zur Behandlung der Angina pectoris, Hypertonie und Arrhythmie 1. Dilzem Symposium Excerpta Medica Amsterdam pp 59–67

Raensch KD, Sommer J (1984) Pharmacokinetic and metabolism of nitrendipine In: Scriabine A, Vanov S, Deck K (eds) Nitrendipine Urban & Schwarzenberg Baltimore pp 409–422

Raensch KD, Sommer J Graefe KH (1985) Pharmacokinetic and metabolism of nimodipine In: Betz E, Deck K, Hoffmeister F (eds) Nimodipine Schattauer Stuttgart pp 147–159

Saltiel E, Gray Ellrodt A, Monk JP, Langley MS (1988) Felodipine – a review of its pharmacodynamic and pharmacokinetic properties, and therapeutic use in hypertension Drugs 36: 387–428

Sato M, Nagao T, Yamaguchi I, Nakajima H, Kiyomoto A (1971) Pharmacological studies on a new 1,5–benzothiazepine derivate (CRD-401) I. Cardiovascular actions Arzneimittelforschung 21: 1338–1343

Scherling D, Ahr G, Karl W (1987) Biotransformation of nisoldipine in man In: Hugenholtz PG, Meyer J (eds) Nisoldipine 1987 Springer Berlin Heidelberg New York pp 85–88

Schomerus M, Spiegelhalder B, Stieren B, Eichelbaum M (1976) Physiologic disposition of verapamil in man Cardiovasc Res 10: 605–612

Schrör K (1985) Kalzium Antagonisten – Physiologische Grundlagen, allgemeine und spezielle Pharmakologie In: Hossmann V, Grötz J, Schrör K (eds) Kalziumantagonisten und zerebrale Erkrankungen Springer Berlin pp 1–39

Singh NB, Bakey S, Nadmanee K (1985) Second generation calciumantagonists: search for greater selectivity and versatility Am J Cardiol 55: 214 B-221 B

Stieren B, Bühler V, Hege HG, Hollmann M, Neun H, Schlepper M Weymann J (1983) Pharmakokinetik und Metabolismus von Gallopamil In: Kaltenbach M, Hopf R (eds) Gallopamil Springer Berlin pp 90–96

Stone PH, Antman EM, Muller JE, Braunwald E (1980) Calcium channel blocking agents in the treatment of cardiovascular disorders: part II hemodynamic effects and clinical application Ann Intern Med 93: 886–904

Stotzinen W, von Pösl HJ (1986) Nifedipin in Tropfenform ist gut wirksam Praxis-Kurier 41: 21–23

van Harten J, van Brummelen P, Wilson JHP, Breimer DD (1987a) Pharmacokinetics and haemodynamic effects of nisoldipine in patients with liver cirrhosis In: Hugenholtz PG, Meyer J (eds) Nisoldipine 1987 Springer Berlin Heidelberg New York pp 76–79

van Harten J, Burggraaf K, Danhof M, van Brummelen P, Breimer DD (1987b) Negligible sublingual absorption of nifedipine Lancet 2: 1363–1365

Vater W, Kroneberg G, Hoffmeister F, Kaller H, Meng K, Oberdorf A, Puls W, Schloßmann K, Stoepel K (1972) Zur Pharmakologie von 4-(2-Nitrophenyl)-2,6-dimethyl 1,4-dihydropyridin-3.5-dicarbonsäuredimethylester (Nifedipine, Bay a 1040) Arzneimittelforschung 22: 1–14

Vetrovec GW (1989) Once-daily therapy for angina pectoris with nifedipine gastrointestinal therapeutic system Am J Med 86 (Suppl 1 A): 28–32

Zelis RF, Kinney EL (1982) The pharmacokinetics of diltiazem in healthy man Am J Cardiol 49: 529–532

3 Pharmacological Effects of Calcium Antagonists

3.1 Cardiovascular System

3.1.1 Preliminary Remarks

According to Gauer and Henry (review by Gauer and Henry 1963; Gauer 1972), the cardiovascular system consists of a high-pressure (arterial) and a low-pressure compartment (Fig. 10). The high-pressure compartment embraces the left ventricle in systole and the arterial vascular system as far as the capillary bed. The postcapillary venules, veins, right side of the heart, pulmonary circulation and left atrium (and in diastole the left ventricle) form the functional unit of the low-pressure compartment; 85% of the circulating blood volume is found in the low-pressure compartment, only 15% in the high-pressure compartment. The mean pressure in the low-pressure compartment is between 5 and 15 mmHg, and in the high-pressure compartment around 100 mmHg; the elasticity (compliance = change in volume per unit change in pressure) of the low-pressure compartment exceeds that of the high-pressure compartment by a factor of 100–200. For this reason, the low-pressure compartment is especially well equipped to accommodate variations in volume without significant changes in mean pressure levels. Regulation of blood pressure predominantly takes place through arterial receptors in the high-pressure compartment (baroreceptors in the region of the aortic arch and the carotid sinus), homeostasis of blood pressure being maintained by complementary changes in peripheral resistance and cardiac output. By contrast, receptors in the low-pressure compartment (for instance, in the right and

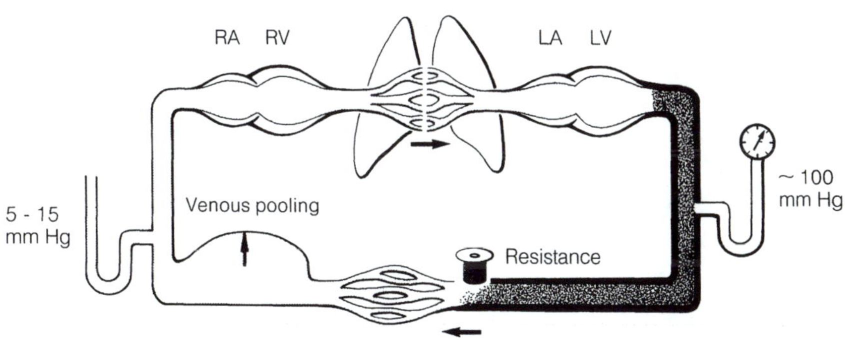

Fig. 10. Cardiovascular system, showing low-pressure and high-pressure compartments. Change in peripheral resistance in the arterioles of the high-pressure compartment, venous pooling in the low-pressure compartment. Mean pressure levels are indicated. *RA* right atrium, *RV* right ventricle, *LA* left atrium, *LV* left ventricle

left atria) are responsible for the regulation of blood volume and water balance, which occurs mainly as hormonal regulation via the renin–angiotensin–aldosterone system and the atrial natriuretic peptide (ANP), and possibly also the antidiuretic hormone (ADH).

Calcium antagonists act in the heart on the pacemaker and conduction system, on the myocardium and on the coronary arteries. In the vascular system, vasodilation, especially of the arterial vascular system, is prominent; the venous system is, in contrast to the effect of nitrates, only slightly affected (Stone et al. 1980; Braunwald 1982). Peripheral vasodilation and the subsequent drop in arterial pressure cause baroreceptor reflexes which activate the sympathetic nervous system and reduce vagal impulse.

The electrophysiological and haemodynamic actions of calcium antagonists on the cardiovascular system, as measured in volunteers or patients, are thus the result of the primary actions of these substances on the heart and vessels (as can be demonstrated in an isolated muscle preparation or in the heart after autonomic blockade) and also of the resultant autonomic counter-regulation effects. In the pages that follow, although their primary (i.e. direct) cardiac actions will be considered as necessary, discussion centers on the overall electrophysiological and haemodynamic effects of the calcium antagonists decisive in practical therapy.

Calcium antagonists form a chemically heterogeneous group of substances which differ among themselves, especially in terms of the ratio of vasodilation to cardiac effect. Thus, the extent of autonomic counter-regulation varies from substance to substance, and the various groups of substances: the verapamil group, diltiazem and the dihydropyridine group, are therefore discussed separately.

3.1.2 Pacemaker and Conduction System

In a spontaneously beating isolated atrial preparation or on direct injection into the artery of the sino-atrial (SA) node, calcium antagonists slow down the discharge frequency of the SA node, prolong the recovery time of the SA node and delay SA conduction (Antmann et al. 1980; Seipel and Breithardt 1982; Schulz

Table 6. *Relative* efficacy of calcium antagonists with administration of *equimolar* doses in vitro (test series with isolated preparation of cardiac muscle). Relative efficacy of nifedipine = 1 (After Stone et al. 1980)

	Nifedipine	Verapamil	Diltiazem
Negative chronotropy	1	1	1
Negative inotropy	1	1/13	1/40
Negative dromotropy	1	1/2	1/2
Vasodilation	1	1/12	1/26

et al. 1983 b; Singh et al. 1984). In respect to this direct negative chronotropic effect, nifedipine, verapamil and diltiazem show only minor differences (Table 6). In vivo the direct negative chronotropic effect of calcium antagonists is modified by autonomic counter-regulation (Sect. 3.1.1) and a nonspecific sympathetic antagonism (described in the case of verapamil and diltiazem; Haas 1964; Bender 1966; Singh et al. 1984) in such a manner that the discharge frequency of the SA node, the recovery time of the SA node and the SA conduction time show only negligible changes. The dihydropyridine derivatives such as nifedipine are more potent vasodilators than verapamil and diltiazem. Hence, they have to be given in lower equimolar doses to avoid excessive hypotension. The dihydropyridines are thus more vasoselective in their mechanism of action. In therapeutic doses their direct effects on the heart are usually negligible. Therefore autonomic counter-regulation results, mostly in the acute phase, in a rise in heart rate (Lydtin et al. 1975) as well as an occasional shortened SA node recovery time (Furlanello et al. 1980). Verapamil can either increase or decrease the heart rate slightly; diltiazem usually reduces it (Mitchell et al. 1982). Thus, during therapy with nifedipine, at least initially the heart rate is higher than in the control group, whereas with verapamil and diltiazem it remains within or is slightly below the control range.

The situation is entirely different in patients with disorders of the SA node (e.g. sick sinus syndrome). Here the SA node responds to changes of the autonomic tone only inadequately (e.g. a pathological atropine test). The direct depressant effect of calcium antagonists on the discharge frequency of the SA node can consequently be only partially compensated by the autonomic counter-regulatory reflex. The result is development of sinus bradycardia, even extending to sinus arrest, and an increase in the SA node recovery time. This increased tendency to bradycardia in disturbed SA node function is observed predominantly with substances of the verapamil group and diltiazem (Sugimoto et al. 1980; Seipel and Breithardt 1982), although individual cases have also been reported with nifedipine (Furlanello et al. 1980). With advancing age the sensitivity of the baroreceptor reflex and consequently autonomic counter-regulation decline; the direct effect of calcium antagonists on the SA node therefore predominates, which can, depending on whether the substance is "vasoselective" or "nonselective", lead to a slight increase in heart rate (nifedipine) or to an increase in bradycardia (verapamil, diltiazem) (Abernethy et al. 1986).

All calcium antagonists also have inhibitory effects on the AV node, and AV conduction is prolonged (a negative dromotropic effect), as demonstrated experimentally in isolated atrial preparation or with direct injection into the AV nodal artery. The relative efficacies in vitro of the various compounds are shown in Table 6. In vivo, owing to the role of autonomic counter-regulation (which depends on the dose given and the extent of individual vasoselectivity), individual substances produce different effects. Thus, while nifedipine either has no influence or even accelerates AV conduction and the effective refractory period of the AV node (Furlanello et al. 1980; Mitchell et al. 1982), verapamil (Mitchell et al. 1982; Seipel et al. 1983), diltiazem (Sugimoto et al. 1980; Seipel et al. 1983), gallopamil (Beck et al. 1978; Neuss et al. 1983) and tiapamil (Singh et al. 1984) retard

AV conduction with a concomitant rise in the A–H interval [2] and in the effective and functional refractory period of the AV node. The Wenckebach point, i.e. the maximum atrial stimulation frequency with 1:1 AV transmission, decreases. In supraventricular tachycardias and in atrial fibrillation (or atrial flutter), ventricular rate is reduced, provided that the conduction from the atrium to the ventricle occurs regularly through the AV node. In the ECG the PQ time is prolonged, depending on the dose. Inhibition of AV conduction through substances of the verapamil group is rate-dependent, i.e. it increases with increasing heart rate (so-called use dependence; Ellenbogen et al. 1985; Sect. 1.2.4).

While verapamil influences the vascular system and AV conduction to an approximately equal degree, nifedipine is about 10 times more potent in relaxing vessels than in its effects on the heart (Taira et al. 1980; Millard et al. 1982). Felodopine is even 100 times more potent in vessels than in the heart. Correspondingly, in vivo only with substances of the dihydropyridine group, with strong peripheral vasodilation, acceleration of AV conduction due to autonomic counter-regulation is sufficient to balance the directly negative dromotropic effect. This is in line with the results reported by Mancia et al. (1978), indicating that overall autonomic counter-regulation exercises less influence on the AV node than on the sinus node.

Owing to these differing influences of the various calcium antagonists on AV conduction and on the refractory time of the AV node, the verapamil group and diltiazem have been more precisely defined as calcium antagonists with antiarrhythmic action (class IV according to Singh and Vaughan-Williams; Sect. 4.3). Besides causing less autonomic counter-regulation, these compounds are reported in higher doses to reduce sodium permeability (action on the fast sodium channel) and to possess a nonspecific sympathetic antagonism (Singh et al. 1984). In vivo, verapamil and gallopamil are reported to produce a more powerful effect in this context than tiapamil and diltiazem (Seipel and Breithardt 1982).

β-Receptor blockers also inhibit AV conduction; thus, if a calcium antagonist/β-receptor blocker combination therapy is considered, calcium antagonists of the dihydropyridine type would seem to offer distinct advantages (Lydtin et al. 1975; Lynch et al. 1980; Jones et al. 1985), because clinically relevant AV block can thereby be avoided.

The negative chronotropic and dromotropic action of calcium antagonists can be demonstrated even after blocking of the autonomic nervous system with atropine and propranolol; the inhibition is therefore not the result of an effect on the autonomic nervous system, but rather due to direct blocking of slow calcium influx in the region of the sinus node and the AV node. In vitro, the effect of calcium antagonists on the sinus and AV nodes can be only partially compensated by increasing the calcium supply (Opie 1984a); β-receptor agonists, such as iso-

---

[2] Concept of the His bundle-ECG: here, by means of intracardiac or oesophageal electrodes, potentials are taken from the upper right atrium (A′), the lower right atrium (A), the bundle of His (H) and the ventricle (V). The corresponding conduction times are called A′–A (intraatrial) conduction time, A–H (AV node) conduction time (time required for conduction from the basal atrium to the bundle of His), and H–V interval (conduction time in the intraventricular conduction system; further details in Csapo 1980).

proterenol or adrenaline, can, however, reduce or balance the effect of calcium antagonists on impulse formation and conduction both in vivo and in vitro (Zipes and Fischer 1974).

In the normal atrial and ventricular myocardium, as well as in the His–Purkinje system, neither impulse conduction nor refractory period is significantly changed by calcium antagonists, with the exception of bepridil. Consequently, QRS duration and the H–V interval in the His bundle ECG remain unaltered. Only in the case of verapamil and, to a lesser extent, in the case of nifedipine have relative prolongations of the atrial refractory period in relation to the duration of the action potential been reported (Landmark and Amlie 1976; Seipel et al. 1983). Bepridil, a substance additionally exhibiting an effect on the fast sodium channel, extends the refractory period in the atrium and the ventricle (Rowland et al. 1983). If the ventricles and/or atria show pathological dilation, or if myocardial ischaemia is present, then "slow-channel" (slow-response) action potentials may ensue. These are characterized by a resting membrane potential of around −50 mV, as well as by the slow rise and very small amplitude of the action potential. In contrast to normal atrial and ventricular action potentials, which are predominantly dependent on the fast sodium influx, slow calcium influx is decisive in these "slow-channel" action potentials. Such "slow-channel" action potentials can generate reentry arrhythmias in the atrial or ventricular region; thus, under specific conditions, a direct antiarrhythmic action (exceeding the improvement of the myocardial ischaemia) of calcium antagonists in the region of the atrial or ventricular myocardium may be explained through their effect on, i.e. suppression of, these "slow-channel" action potentials (Naumann d'Alnoncourt et al. 1982; see also Sect. 4.3.1).

These isolated results, together with the positive findings in experimentally induced arrhythmias and in arrhythmias on the basis of a "triggered repetitive activity" (generation of pacemaker potentials by preceding repolarisation), may not, however, be extrapolated to a general effect of calcium antagonists on impulse conduction (and on ectopic impulse generation) in the atrial and ventricular myocardium.

A specific antiarrhythmic action through influence on the fast sodium channel (class I a or I b according to Singh and Vaughan-Williams) is not exhibited by the calcium antagonists, with the exception of bepridil (Cranefield 1977; Antmann et al. 1980; Seipel and Breithardt 1982).

In accessory conduction pathways (e.g. Kent's bundle in the Wolff–Parkinson–White syndrome), calcium antagonists (literature references are only available for verapamil) usually do not alter conduction time or the refractory period (Spurrell et al. 1984; Seipel and Breithardt 1982). In some cases (in Petri et al. 1980, for instance, 7 of 17 patients) verapamil accelerates, analogously to the action of digitalis and β-receptor blockers, *anterograde* conduction from the atrium to the ventricle (via the accessory bundle). The effective anterograde refractory period is thereby shortened (Spurrell et al. 1974; McGovern et al. 1986). Thus, either verapamil directly shortens the effective refractory period of the anterograde conduction, or the reflex activation of the sympathetic nervous system accelerates conduction (McGovern et al. 1986). On the basis of these electrophysiological changes, if atrial fibrillation and a WPW syndrome are present, a rapid ventricular response can lead to ventricular tachycardia, and the condition may even extend to ventricular fibrillation (Klein et al. 1979; McGovern et al. 1986). When there is atrial fibrillation and premature ventricular excitation, e.g. in the context of a WPW syndrome, calcium antagonists of the verapamil type and (probably also) diltiazem are thus contraindicated (Sect. 4.3.2.3

Table 7. Electrophysiological properties of calcium antagonists. (After Singh et al. 1984; Been et al. 1985; Amlie et al. 1989)

	Verapamil	Gallopamil	Tiapamil	Diltiazem	Nifedipine	Felodipine	Bepridil[a]
RR interval	↑↓	↑↑	↑↑	↑	↓	↓	↑
QRS duration	○	○	○	○	○	○	↑?
PR interval	↑	↑	↑	↑	○	○	↑
AH interval	↑↑	↑↑	↑	↑	○	↑ (?)	↑↑
HV interval	○	○	○	○	○	○	↑ (?)
Atrium: Effective refractory time	○	○	○	○	○		↑
AV node: Effective refractory time	↑↑	↑↑	↑	↑	±	↓	↑↑
AV node: Functional refractory time	↑↑	↑	↑	↑	±		↑↑
Ventricle: Effective refractory time	○	○	○	○	○		↑
His-Purkinje system: Effective refractory time	○	○	○	○	○	↓	↑
Bypass tract: Effective refractory time	±	○	○	?	○		↑
Sinus node: Recovery time	○[b]	○[b]	○[b]	○[b]	○	○	○[b]

[a] Provisional results (preparations not yet commercially available)
[b] Prolonged in sick sinus syndrome
○ No effect; ↓ decrease; ↑ increase; ↑↑ marked increase; ↑↓ variable effect; ± no change

and 4.3.2.6). Although complications (e.g. ventricular tachycardias, ventricular fibrillation) have so far been observed only with intravenous administration (Petri et al. 1982 observed no change in AV conduction with oral administration), in long-term oral therapy with verapamil in the presence of WPW syndrome, electrophysiological testing must nevertheless be undertaken.

Paroxysmal supraventricular reentry tachycardias with retrograde conduction via the accessory pathway can be positively influenced by verapamil through inhibition of AV conduction (Spurrell et al. 1974). Paroxysmal supraventricular tachycardia can, however, progress to atrial fibrillation (McGovern et al. 1986). The electrophysiological properties of calcium antagonists are summarized in Table 7.

3.1.3 Myocardium

As described in Sect. 1.2, calcium antagonists by definition lead in an isolated muscle preparation to excitation–contraction decoupling, i.e. with virtually unchanged action potential the development of muscular tension is weakened according to dose, or with higher doses even completely inhibited (Spurrell et al. 1974). This negative inotropic effect of calcium antagonists, which has been demonstrated in vitro in isolated muscular fibres, can also be demonstrated in an isolated myocardial preparation. If equimolar doses are employed, nifedipine exhibits a more powerful negative inotropic effect than verapamil, which in turn is more powerful than diltiazem (Table 6). The negative inotropic effect reveals itself in the isolated myocardial preparation as an overall reduced or asynchronous wall motion and in an elevated left ventricular end-diastolic pressure (Hugenholtz 1985). Parallel with the decline of contractility, there is a decrease in oxygen and ATP consumption in the cardiac muscle.

Employing the technique of intracoronary injection, Kaltenbach et al. (1979) and Serruys et al. (1981) attempted to measure the direct myocardial effect of calcium antagonists on the human organism in vivo. Both groups of workers selected a dose (0.1 mg nifedipine intracoronarily), which, in the authors' opinion, had not measurably altered blood pressure, heart rate or cardiac oxygen demand following intravenous administration. After intracoronary injection, reduced cardiac oxygen demand (measured as an increase in oxygen saturation in the coronary sinus) was observed. Contractility of the heart muscle was reduced (overall contractility and regional contractility in the myocardial areas supplied by the relevant coronary artery), and left ventricular end-diastolic pressure increased.

With systemic (oral or intravenous) administration in vivo myocardial contractility is influenced by the direct negative inotropic effect of calcium antagonists, by the autonomic counter-regulation and by the initial state of the myocardium (i.e. normal heart muscle, failing heart muscle, resting situation, exercise situation, adequate coronary blood flow, myocardial ischaemia).

On the basis of the strong autonomic counter-regulation (due to their relative vasoselectivity), substances of the dihydropyridine group (nifedipine, nitrendipine, felodipine, nisoldipine, etc.) exhibit, both in volunteers and in patients with coronary heart disease, no negative inotropic effect. Indeed, in most cases a positive inotropic effect can be demonstrated, resulting in increased cardiac output and decreased left ventricular end-diastolic pressure (Lydtin et al. 1975). Even in patients with restricted left ventricular function, nifedipine has been reported not to reduce contractility, or even to improve it (Bellocci et al. 1981; Ludbrook et al. 1982).

Unloading of the left ventricle can be demonstrated not only in the acute situation, but also during long-term therapy; Lydtin et al. found in 1975 that after 3 weeks of treatment, reduced pulmonary capillary wedge pressure was maintained, corresponding to a reduced left ventricular filling pressure (at rest and with exercise). In the presence of coronary heart disease, left ventricular function is positively influenced by nifedipine, on one hand, through reduction of afterload (Lydtin in panel discussion 1975) and, on the other, through improved perfusion of ischaemic myocardial areas.

Some authors (Mauser et al. 1985; White et al. 1985) emphasize the improvement in ventricular diastolic function with the use of nifedipine (normalization of increased ventricular stiffness with a decrease in left ventricular end-diastolic pressure); other authors (Ludbrook et al. 1982) stress an improved ventricular systolic function without significant change of the diastolic parameters. Recent investigations (Kurnik et al. 1986) have again found that improved left ventricular function through use of nifedipine is probably rather a result of reduction of afterload (peripheral vasodilation) than of direct effects on the myocardium, in particular on intrinsic myocardial stiffness.

If nifedipine and β-receptor blockers are combined, autonomic counter-regulation is weakened (Lydtin et al. 1975). Nevertheless, in most cases there is little or no impairment of myocardial contractility owing to the complementary actions of negative inotropy (β-receptor blockers) and the reduction of afterload (nifedipine). There are data reported in the literature favouring dihydropyridines with high vasoselectivity (e.g. felodipine) in combination with β-receptor blockers (Joshi et al. 1981; Culling et al. 1984).

With substances of the verapamil group and (to a lesser extent) with diltiazem, direct negative inotropic action on the cardiac muscle predominates owing to the reduced peripheral vasodilating effects and less powerful autonomic counter-regulation. When prescribing these drugs, it is not always possible to predict precisely whether myocardial contractility will remain unchanged or decrease. Overall, contractility seems to be decreased more rapidly with verapamil than with diltiazem (Millard et al. 1982). In certain situations of verapamil therapy, a more pronounced negative inotropic effect must be expected, possibly leading to congestive heart failure (Singh et al. 1982, 1984; Braunwald 1984): especially in cases where there is severe previous myocardial damage (ejection fraction less than 30%, left ventricular end-diastolic pressure greater than 20 mmHg); in combination with β-receptor blockers; with the administration of higher doses (e.g. 10 mg verapamil as an intravenous bolus); in severe hypotension; and where there is a radiologically and/or echocardiographically enlarged heart.

In patients with coronary heart disease and an ischaemic myocardium, verapamil (Hecht et al. 1981), gallopamil (Sesto et al. 1983) and diltiazem (Dash et al. 1985) are reported to reduce regional disturbances of ventricular wall motion and improve the ejection fraction in these areas. Vlitstra et al. (1983) reached a contrary conclusion (no improvement of regional wall motion in ischaemic hypokinetic areas) for verapamil; for diltiazem an improved ventricular diastolic function is additionally described (relaxation and filling) (Brugger et al. 1984; Dash et al. 1985). In these investigations there is no uniform mode of action, probably owing to varying methods.

Following pathophysiological theory, verapamil (and possibly also diltiazem) would appear, owing to its more pronounced negative inotropic action, less suitable for combination therapy with β-receptor blockers, particularly when the heart is considerably damaged. Various clinical studies have shown, however, that, provided careful attention is given to contraindications, this combination treatment is tolerated by a surprisingly large number of patients without side effects (these investigations were carried out predominantly with diltiazem; Rocha et al. 1986; Gibson et al. 1986; see also Sect. 4.1.2.3).

3.1.4 Cardioprotection

On the basis of experimental data presented by Fleckenstein (1968), according to which intracellular calcium overload plays a crucial role in the pathogenesis of myocardial necrosis (Fig. 11), a specific cardioprotective action of calcium antagonists has been repeatedly postulated. Calcium antagonists are reputed, in conditions of ischaemia, hypoxia, reperfusion after ischaemia (Nayler 1980) or exces-

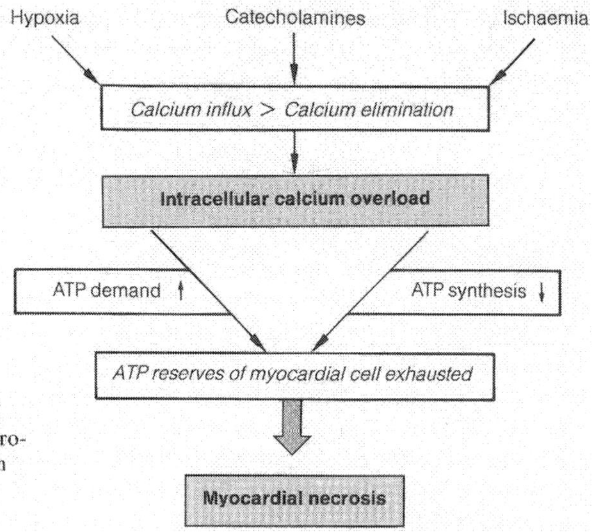

Fig. 11. Schematic summary of pathogenesis of myocardial necrosis through intracellular calcium overload. (After Fleckenstein 1984)

sive exposure to catecholamines, to prevent intracellular calcium overload by inhibiting calcium influx, thereby preventing development of myocardial necrosis or at least delaying its appearance. However, whether an increase in intracellular calcium concentration, in particular in free calcium concentration, is a cause or merely a consequence of myocardial necrosis remains to be definitively clarified (review by Cheung et al. 1986).

Various proposals can be made: the cardioprotective action of calcium antagonists in ischaemia may lie in their capacity to inhibit intracellular calcium overload (Fleckenstein 1980, 1982); in vivo they may also alleviate the ischaemia, thus prolonging survival time, by dilation of collateral coronary vessels, and/or by a decrease in energy and oxygen demand (afterload reduction, negative inotropic effect; Stone et al. 1980, Cheung et al. 1986).

Independent of such theoretical reflections on the mechanism of action, timing is the decisive factor in the use of calcium antagonists as cardioprotective agents. Only if given *before* a definitive ischaemia (e.g. coronary occlusion) has developed do they have a cardioprotective effect; but with that proviso they can prevent ischaemic damage to the myocardium, even in ischaemic contracture ("stone heart"; Henry et al. 1978). In the very early stages of manifested ischaemia, calcium antagonists can also have considerable therapeutic success (Nayler 1980). However, once the myocardial cell membrane has been destroyed and cytolysis has been initiated, no protective action can be expected (Hugenholtz et al. 1986).

In clinical practice use of calcium antagonists is encouraging in unstable angina pectoris or other conditions of threatening myocardial ischaemia, as well as prophylactically in open-heart surgery to improve ischaemic tolerance, i.e. in addition to hypothermia and cardioplegic solutions (Stone et al. 1980; Cheung et al. 1986; Hugenholtz et al. 1986). Following the onset of irreversible cytolysis, e.g. in acute myocardial infarction, no further cardioprotective effect can be expected (Chap. 4 indications, Sect. 4.1.3 unstable angina pectoris, 4.1.4 cardiac infarction, 4.5.3 cardiac surgery). For use of calcium antagonists (verapamil) as cardioprotective agents in tocolysis associated with β-sympathomimetics see Sect. 4.10.3.

3.1.5 Vascular System

By definition, calcium antagonists also produce excitation–contraction decoupling in vascular smooth muscle. They lower vascular muscle tone and thus act as vasodilators. This vasodilating effect manifests itself principally in the arteries and arterioles; the effect on the venous system is negligible and in haemodynamic terms can be largely ignored (Braunwald 1982; Singh et al. 1984). This differentiates the calcium antagonists from the nitrates with their marked vasodilating effect on the venous compartment of the cardiovascular system and permits discussion of synergistic effects of the two groups of substances in the treatment of angina pectoris.

Beside the coronary vessels (Sect. 3.1.5.1), calcium antagonists chiefly dilate the small arteries and arterioles, i.e. the precapillary resistance vessels, by inhibiting their myogenic activity. This lowers peripheral vascular resistance, and thereby blood pressure. The hypotensive action (above all in arterial hypertension) is thought to increase with age (Bühler et al. 1984; Abernethy et al. 1986), which is explained by the reduced autonomic counter-regulation (reduced baroreceptor sensitivity) and by the overall higher peripheral resistance (decrease in the compliance of the large vessels) associated with old age (see also Sect. 4.2.2 and 4.2.3). The more pronounced effect in the elderly may also be a consequence of higher drug plasma concentrations due to reduced drug metabolism (Landahl et al. 1988). In general, dihydropyridine derivatives overall exhibit a stronger vasodilating action than substances of the verapamil group or diltiazem; the ratios of the vasodilating efficacy of nifedipine : verapamil : diltiazem are reported as 26 : 10 : 1 (Kraemer and Kübler 1983); gallopamil has been found to have a more powerful action than verapamil in this context (Fleckenstein-Grün and Fleckenstein 1983), and the newer dihydropyridine derivatives nitrendipine and felodipine have been found to be more powerful than nifedipine.

Beside their vasodilating effect, calcium antagonists have the ability to reduce the reaction of vascular muscle to vasoconstricting stimuli (e.g. in vitro to electric stimulation, raised extracellular potassium, and stretching; and in vivo to acetylcholine, serotonin, histamine, ergotamine, cardiac glycosides, noradrenaline, angiotensin II, and thromboxane A_2). Calcium antagonists thus act as antivasoconstrictors (Hof 1985); they prevent vascular spasm or resolve existing spasm (Fleckenstein-Grün and Fleckenstein 1980). This spasmolytic effect can be demonstrated in almost all vascular systems (Hof 1985). It is of clinical relevance chiefly in the cardiovascular system (Prinzmetal's angina pectoris), in the extremities (Raynaud's syndrome), and in the cerebral vessels (migraine, subarachnoid haemorrhage). The potency of calcium antagonists in such situations is reported to exceed that of papaverine by a factor greater than 1000 (Stone et al. 1980; Fleckenstein 1982).

While the "classical" calcium antagonists (nifedipine, verapamil, diltiazem) differ only in respect of their relative potencies on the various regions of the vascular system, qualitative differences in respect of potency in causing vasodilation in various vascular beds are postulated for more recent calcium antagonists (e.g. nimodipine, flunarizine).

The vasodilation and resolution or prevention of spasm produced by calcium antagonists can lead to a new distribution of blood flow and oxygen supply in individual tissues ("redistribution of flow"). If sufficient collateral vessels or so-called dynamic stenoses (vascular stenoses consisting of a fixed organic and a variable, i.e. dilatable, muscular component) exist, improved circulation in ischaemic areas can result (Sect. 3.1.5.1).

3.1.5.1 Coronary Vessels

The description of a coronary dilating action for prenylamine (Lindner 1960) and for verapamil (Haas and Härtfelder 1962) initiated the development of the calcium antagonists, their common mechanism of action being subsequently analysed by Fleckenstein.

All calcium antagonists lower coronary arterial tone and thus have a coronary dilating effect. Their action centers upon both the large epicardial, extramural coronary arteries and upon the small arteriolar resistance vessels. With coronary dilation the resistance of coronary vessels is reduced and blood flow through the coronary arteries increases. This can, albeit with certain reservations, be detected in humans by means of electromagnetic flow probes in the coronary sinus, by isotopic techniques and, assuming a constant oxygen demand ($\dot{V}O_2$), by determining an increase in the oxygen content of the blood in the coronary sinus. Oxygen demand in the heart should either remain steady or fall slightly: the results appear to be influenced by structural arrangement of the experiments and by the mode of administration of the active substance (intracoronary versus intravenous; reviews by Singh et al. 1982; Hugenholtz 1985). In comparing efficacies of the various calcium antagonists, the same applies as in the smooth muscle of the peripheral resistance vessels: nifedipine is the most powerful coronary dilator, followed by verapamil and diltiazem. The coronary dilating effect of nitrates differs significantly both in strength and duration from that of the calcium antagonists (review by Fleckenstein 1982); combined administration of both groups of substances results in addition of individual effects (Raffenbeul and Lichtlen 1983). However, one may safely assume that here submaximal doses of the two substances were investigated.

In tissue ischaemia calcium antagonists increase blood flow in the collaterals (Stone et al. 1980), so long as tissue acidosis and endogenous coronary dilators (e.g. adenosine) have not already resulted in maximum dilation of these vessels.

Calcium antagonists can resolve and prevent coronary spasm (Bertrand et al. 1980; De Servi et al. 1980; Gunther et al. 1981; Hossack et al. 1984). Coronary spasm, by definition, plays a role in Prinzmetal's angina pectoris, and probably also in other forms of angina pectoris. Calcium antagonists are effective both in experimentally induced coronary spasm (cold stimuli in the cold pressure test, isometric stress test, and injections of ergonovine) and in spontaneous coronary spasm. Various factors have been considered as the possible cause of coronary spasm: sympathetic (α_2 receptors) and parasympathetic (acetylcholine) stimuli; an increase in blood pH (e.g. in hyperventilation), localized release of serotonin and thromboxane A_2 from microthrombi, and certain pharmaceutical agents (ergotamine derivatives, indomethacin, cocaine; Opie 1984; see also Sect. 4.1.3.1). In general, local factors appear to play a more important role than cardiac autonomic innervation, since coronary spasm is also observed after heart transplantation (review by Kupper 1986).

While the effectiveness of calcium antagonists in spasm of (coronary arteriographically) normal coronary arteries can be easily explained, their effect in atherosclerotic coronary arteries presupposes the presence of so-called dynamic stenoses. These consist of a fixed organic stenosis and a muscular component that can still react to vasoconstricting and vasodilating stimuli. In most cases an organic stenosis is eccentrically positioned, and the muscular component forms the so-called free wall segment. As Fig. 12 shows, when there is progression of organic stenosis, even the smallest changes in vasomotor tone are sufficient to induce an extreme degree of coronary artery stenosis or even complete occlusion. Conversely, where vasomotor tone is increased, calcium antagonists can reduce the degree of stenosis by dilating the free wall segment (MacAlpin 1980; Opie 1984a, b; Hugenholtz 1985).

By influencing vasomotor tone (coronary dilation) and by resolving or preventing coronary spasm, calcium antagonists can improve regional myocardial blood flow both in areas with normal vascular supply and in areas with coronary artery stenosis. With a more substantial increase of blood supply in poststenotic areas of the myocardium, the overall flow pattern in the heart is harmonized. Investigations in this area were carried out with nifedipine, diltiazem and verapamil, both under resting conditions and with rapid atrial stimulation (Engel

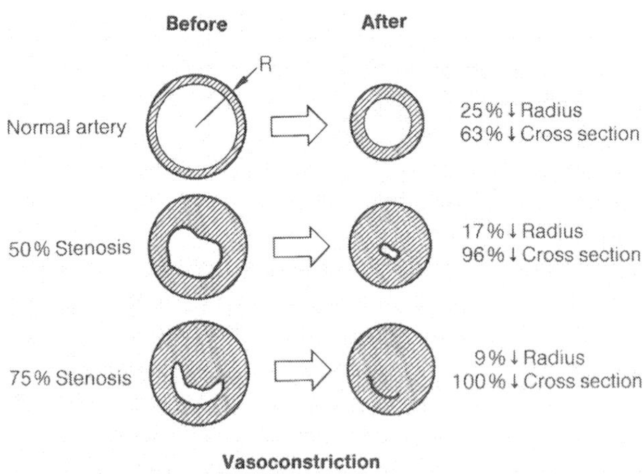

Fig. 12. Model of so-called dynamic coronary arterial stenosis. With increase of organic stenosis, even the smallest changes in vasomotor tone (shrinking radius) lead to an extreme degree of coronary artery stenosis or to occlusion (shrinking of cross section). (After MacAlpin 1981)

et al. 1980; Daniel et al. 1984). The improved regional wall movement in ischaemic areas of the myocardium (Sect. 5.1.3) can thus be explained by two vascular effects of calcium antagonists: improved blood supply in poststenotic myocardial areas, and unloading of the myocardium by lowering peripheral vascular resistance (i.e. reduction of afterload).

The importance of the action of calcium antagonists on peripheral vascular resistance and on arterial blood pressure, and thereby on the coronary blood flow was demonstrated in trials with nifedipine (Schulz et al. 1983a), in which in individual cases a reduction in arterial blood pressure reduces coronary perfusion pressure. This ultimately leads to passive vasoconstriction in regions of coronary artery stenosis, corresponding to an increase in coronary artery stenosis, with a resultant reduction in coronary blood flow.

The action of calcium antagonists on the coronary arteries can therefore be seen in isolation *only* in experimental situations or with strictly intracoronary injection of nonsystemically acting doses. The effects on the myocardium and on the peripheral resistance vessels must always be taken into account.

3.1.5.2 Other Vessels

The vasodilating action of calcium antagonists is not limited to coronary vessels and the peripheral resistance vessels; it can also be demonstrated in various other vascular systems. Furthermore, calcium antagonists can here, too, resolve or prevent vascular spasm.

The following brief review of the current situation must take into account that, all in all, systematic investigations are lacking in this area; most results rest on experimental studies, animal experiments or isolated observations, in which, moreover, human data were largely obtained from healthy volunteers.

As early as 1972, Lydtin and Lohmöller demonstrated an increase (100% – 150%) in peripheral muscle blood flow in the lower extremity after parenteral administration of nifedipine. Similar results were obtained by Nakamura et al. (1985), who found a decrease in vascular resistance in the forearm with nifedipine. With regard to verapamil, the evidence is contradictory. In animal studies Hof (1985) found no change in muscular blood flow; by contrast, Solti et al. (1978) and Nissen and Alexander (1975) reported an elevated blood flow in the lower leg, with corresponding decrease in vascular resistance, in patients with arterial-occlusive disease.

The evidence is also contradictory with regard to cutaneous blood flow (Ebner and Dünschede 1976; Hof 1985); however, the prevention of vascular spasm in the hand in Raynaud's syndrome is of clinical significance (Sect. 4.6.2).

Cerebral blood flow increases slightly with nifedipine (Schmitz et al. 1975); autoregulation of cerebral vessels, i.e. the capacity to react to a change in systemic blood pressure by adaptation of vascular resistance, remains unimpaired; a "steal effect" is not observed. The onset of action of calcium antagonists on the cerebral vessels is retarded in comparison with that in the peripheral vascular system, reflecting the necessary diffusion of the substance into the cerebrospinal fluid. The increased cerebral blood flow evoked by cerebral vasodilation accounts for the headache sometimes associated with the use of calcium antagonists (predominantly nifedipine; Sect. 5.1). The ability of calcium antagonists to prevent vasospasm in the cerebral vascular system is also of clinical relevance; it is therapeutically exploited in migraine and following subarachnoid haemorrhage (Sects. 4.7.1 and 4.7.2). Among the dihydropyridine group, nimodipine is thought to have a selective action on cerebral vessels. Without significantly influencing peripheral haemodynamics, this drug prevents intracerebral vasospasm (experimentally it has been shown to suppress depolarization- and receptor-operated muscle contraction of the cerebral vessels; Towart and Perzborn 1983; Schrör 1985). Flunarizine is also said to have a selective action on cerebral vessels (and cerebral metabolism; Sects. 1.2.5 and 4.7).

The renal vessels appear to be slightly dilated by calcium antagonists (Ebner and Dünschede 1976); there is thereby a reduction in renal vascular resistance. Since the arterial pressure and therewith the renal perfusion pressure are reduced concomitantly, the total renal blood supply is not significantly changed. Autoregulation of the renal vessels appears, in contrast to autoregulation of cerebral vessels, to be completely susceptible to blockade by nifedipine (Hashimoto 1980).

Blood flow to the liver and in the pulmonary vascular system appears to be largely unaffected in the healthy individual (Ebner and Dünschede 1976); it may even be slightly decreased (Hof 1985).

However, in the context of pulmonary hypertension, the increased pulmonary vascular resistance can be decreased, at least transiently, with calcium antagonists (Bucca et al. 1982; Gassner et al. 1983; Arnmann et al. 1984; Packer 1985; Bratel et al. 1988; Sect. 4.6.1). Results are available for both primary pulmonary hypertension and for the secondary forms, e.g. in chronic obstructive lung disease. Diltiazem and nifedipine appear to have a far more powerful action than verapamil. To some extent verapamil and nifedipine negatively influence right ventricular function, pulmonary gas exchange and the ventilation : perfusion ratio (reviews by Packer et al. 1984; Melot et al. 1984). These results and the data concerning diltiazem (no influence on gas exchange and ventilation : perfusion ratio) require further confirmation.

In the gastrointestinal tract the blood supply is slightly increased by nifedipine, diltiazem and verapamil in animal experiments. Furthermore, it has been shown that even in this region calcium antagonists can prevent vasospasm (Brobmann et al. 1980). This twofold action has prompted therapeutic trials in angina abdominalis, the results of which are not yet available (Sect. 4.6.3).

In Summary. Calcium antagonists act as vasodilators in various vascular systems. In addition, they have an antivasoconstricting action, i.e. they can resolve spasm (as expression of pathologically raised vasomotor tone), or, when used prophylactically, they can prevent it.

3.1.6 Integrated Haemodynamic Response

Against the background of this discussion of the action of calcium antagonists on individual organ systems, the integrated haemodynamic response of these agents can be summarized. As already emphasized, the primary actions of calcium antagonists on the pacemaker and conduction systems of the heart, the myocardium and vascular muscle (coronary arteries, peripheral vascular systems) are modified by means of autonomic counter-regulation. The relative vasoselectivity of the calcium antagonists determines whether or not direct inhibitory effects on cardiac contractility and conduction will appear when therapeutic doses are administered. With increasing vasoselectivity in the dihydropyridine class (e.g. felodipine and nitrendipine), there is a decreased potential for doses causing substantial reductions of blood pressure to have direct effects on the heart. The overall haemodynamic response to non-vasoselective calcium antagonists such as verapamil and diltiazem depends on the balance between direct inhibitory cardiac effects and the extent of autonomic counter-regulation. The extent of autonomic counter-regulation is, for instance, dependent on the peripheral vasodilation induced by a specific calcium antagonist, on the sensitivity of the baroreceptor reflex and on the initial autonomic conditions; i.e. there is a reduction of counter-regulation in the presence of increased sympathetic tone. Counter-regulation aimed at restoration of blood pressure is characterized by peripheral vasoconstriction, increased renin secretion and an increase in catecholamines (reflected by an increase in heart rate and cardiac contractility). Since the counter-regulation also influences the heart, the resultant action depends mainly on the functional state of the heart. Only a healthy pacemaker and conduction system and a sufficiently excitable heart muscle can react adequately to vagal withdrawal and adrenergic activation. In functional disturbances of the heart (sick sinus syndrome, congestive heart failure), the influence of autonomic counter-regulation is reduced, and the primary action of calcium antagonists on the sinus node, AV node and myocardium predominates. Autonomic counter-regulation also declines with advancing age (reduced sensitivity of the baroreceptor reflex). This explains, at least partly, why in the elderly hypertensive patient, calcium antagonist therapy results in a more marked reduction in blood pressure and a reflex tachycardia is more rarely observed than in younger patients (Sects. 3.1.2, 3.1.5 and 4.2).

In long-term therapy with dihydropyridines (data e.g. for felodipine, Smith et al. 1986) readjustment of the baroreceptor reflex (baroreceptor resetting) is postulated (i.e. a smaller increase in heart rate for the same degree of blood pressure reduction).

Beside autonomic counter-regulation, the primary action of calcium antagonists is also thought to be modified by a nonspecific antagonism towards the adrenergic system (Singh et al. 1984; see also Sect. 3.1.2) as well as by a vagolytic action, the latter being a subject of controversy regarding the human organism (Kirchheim and Gross 1975; Lydtin et al. 1975).

The following account restricts itself to verapamil, diltiazem and nifedipine. Investigations have shown that gallopamil behaves similarly to verapamil (overview by Kaltenbach and Hopf 1983); the newer dihydropyridines (nitrendipine, nisoldipine, felodipine) are, to judge by available results, comparable to nifedipine in their action (reviews by Tweddel et al. 1983; Johnsson et al. 1983; Scriabine et al. 1984), although further studies are required. Even the existing results relating to verapamil, diltiazem and nifedipine vary to a certain extent; this can be explained on the basis of the nonuniform composition of the test populations (on one hand, healthy volunteers, on the other hand, patients with arterial hypertension, coronary heart disease or congestive heart failure).

Verapamil and diltiazem exhibit a balanced action on the pacemaker and conduction system, on the myocardium and the vascular system, which contrasts to the strong peripheral vasodilating action associated with the dihydropyridines, the prototype being nifedipine, which is said to be 10 times more potent, and felodipine as much as 100 times more potent (Ljung et al. 1987), on the peripheral vascular system than on the heart (so-called vascular selectivity).

The total peripheral resistance, corresponding roughly to cardiac afterload, is reduced to the greatest extent by nifedipine and its successor substances (nitrendipine, felodipine, etc.). Krämer and Kübler (1983) give a ratio of 26:9:1 for the comparative strengths of nifedipine, verapamil and diltiazem in this context. Correspondingly, the greatest reduction in blood pressure (systolic and diastolic pressure) is observed following the administration of nifedipine (Lydtin et al. 1972, 1975), although dose-dependent lowering of blood pressure is also seen with verapamil and diltiazem.

Reduction of coronary vascular resistance is also greatest for nifedipine, although the difference to verapamil and diltiazem is considerably less than with respect to reduction of total peripheral resistance. In respect of pulmonary vascular resistance, the data so far available are incomplete. At least in the acute situation, calcium antagonists are thought to reduce pulmonary artery pressure. The results obtained to date relate to nifedipine, felodipine, verapamil and diltiazem (Gassner et al. 1983; Packer 1985; see also Sect. 4.6.1).

On account of its powerful vasodilating action, and corresponding autonomic counter-regulation, nifedipine accelerates the heart rate, both in the resting state and with exercise (Lydtin et al. 1972, 1975). Parallel to this, and depending on initial conditions, there is either an increase in contractility (recognizable by a shortening of the systolic time intervals) or contractility of the left ventricle remains unchanged. Cardiac output, stroke volume and the ejection fraction of the left ventricle increase correspondingly or remain unchanged (Lydtin et al. 1975). Because of a decrease in afterload and of the positive or neutral effect on contractility, left ventricular filling pressure (left ventricular end-diastolic pressure, LVEDP) falls. Individual cases, however, have been described in which left ventricular filling pressure remains unchanged or even increases (review by Stone et al. 1980). This increase in left ventricular filling pressure (i.e. pulmonary capillary wedge pressure, reflecting left atrial pressure), which is observed with exercise, even in healthy subjects, but most markedly in patients with coronary heart disease, is significantly reduced by nifedipine (Lydtin et al. 1980). Here, in addition to the reduction of afterload, improved coronary blood flow may possibly also play a role (Sect. 3.1.5.1).

Table 8. Summary of haemodynamic actions of calcium antagonists compared with nitrates and β-receptor blockers. (Details and literature references in the text)

	Nifedipine	Verapamil	Diltiazem	Nitrates	β-Receptor blockers
Heart rate	↑	=↓	↓	↑=	↓
Contractility (dP/dt_{max})	↑	↓=	=↓	=	↓
Cardiac output (CO)	↑	=↓	↑=↓	=	↓
Peripheral vascular resistance	↓↓	↓	↓	=↓	↑=
Coronary blood flow	↑↑	↑	↑	↑	↓=
Arterial blood pressure	↓↓	↓	↓	=↓	↓
Oxygen demand	=	=↓	=↓	↓	↓
Left ventricular filling pressure (LVEDP)	↓=	↓=↑	=↓	↓	↑
Rise in LVEDP (or pulmonary capillary pressure) under stress	↓	↓	–	↓	↓
Pulmonary artery wedge pressure	↓	↓=	↓=	↓	↑

↑ increase; ↑↑ marked increase; ↓ decrease; ↓↓ marked decrease; = no change; – no data available

In contrast to nifedipine, which may be regarded as relatively vasoselective, the direct negative chronotropic and negative inotropic actions of verapamil and diltiazem are more prominent despite some degree of autonomic counter-regulation. In most instances diltiazem slows the heart rate to a greater extent than does verapamil, whereas verapamil exhibits the stronger negative inotropic action. According to the initial state of the left ventricle, and to the degree of afterload reduction, cardiac output remains either unchanged or decreases. Correspondingly, left ventricular filling pressure (left ventricular end-diastolic pressure) increases with administration of both verapamil (Braunwald 1984; Amende et al. 1985) and diltiazem (Biamino et al. 1982). During stress tests in patients with coronary heart disease (e.g. by atrial stimulation), verapamil prevents an increase in pulmonary capillary wedge pressure (Ferlinz and Turbow 1980). This finding matches the results obtained with nifedipine (Lydtin et al. 1980).

Although calcium antagonists in general reduce afterload – and verapamil and diltiazem additionally exhibit a negative inotropic action – the oxygen demand of the heart is only negligibly reduced during treatment with calcium antagonists. Tauchert et al. (1980) found an unchanged oxygen demand with nifedipine and a slightly reduced oxygen demand with verapamil; Amende et al. (1985) found no change in oxygen demand when verapamil was administered; Biamino et al. (1982) observed with diltiazem a decrease in myocardial oxygen demand. All in all, the values for reduction of oxygen consumption determined in vivo were lower than those expected on theoretical pathophysiological reasoning. Both β-receptor blockers and nitrates are superior to the calcium antagonists in this respect (Tauchert et al. 1980).

The principal haemodynamic data pertaining to calcium antagonist therapy are summarized in Table 8, together with corresponding data, as far as available, on nitrates and β-receptor blockers. As the literature contains partly contradictory statements concerning some parameters, those readers wishing to pursue this topic in detail are referred to recent detailed reviews (Singh et al. 1984; Hugenholtz 1985).

Since calcium antagonists are combined with nitrates or β-receptor blockers in various indications (e.g. arterial hypertension, coronary heart disease), the haemodynamic effects to be expected from such combined treatment are dealt with briefly here. In combination with nitrates the heart rate increases more sharply than with nifedipine alone, and in comparison with verapamil or diltiazem alone, the heart rate also increases. This is due to the additional decrease in afterload and to the known preload reduction of nitrates. It should also be borne in mind that alongside the additive reduction of blood pressure, there is a concomitant reduction in coronary perfusion pressure. By combining nifedipine and β-receptor blockers, the increase in heart rate observed in healthy volunteers taking nifedipine is diminished; in fact, the rate may even fall. Peripheral vascular resistance and arterial blood pressure decrease; cardiac output and left ventricular filling pressure usually remain unchanged (Lydtin et al. 1975; Nelson et al. 1984; Opie 1984a). If in congestive heart failure the use of a β-receptor blocker with or without a calcium antagonist is considered at all, then preference should be given to combination therapy with dihydropyridines (Nelson et al. 1984), in which, as a rule, no negative haemodynamic changes occur. In individual cases, however, heart function may further deteriorate as autonomic counter-regulation (reflex activation of the sympathetic nervous system) is inhibited by β-receptor blockade, and the negative inotropic primary action of calcium antagonists prevails (Koch 1980; Sheridan and Thomas 1987).

Combinations of verapamil or diltiazem with a β-receptor blocker are not absolutely contraindicated, but need very close supervision, especially during the initial stages of treatment. By combining verapamil or diltiazem with a β-receptor blocker, the negative chronotropic (heart rate), the negative dromotropic (AV conduction) and the negative inotropic (contractility) effects are additive. Consequently, a clear decrease in heart rate is to be expected; in individual cases AV block may occur, and a deterioration of myocardial function is possible. However, Rocha et al. (1986) observed, in a combination of propranolol and diltiazem, no decrease in cardiac output.

Overall, combinations of calcium antagonists with nitrates or β-receptor blockers markedly reduce cardiac oxygen demand. Combinations of nifedipine and a β-receptor blocker appear most suited to the requirements of a pharmacologically useful therapy. By combination, the undesirable side effects of both substances (rise in heart rate with nifedipine, increase in peripheral vascular resistance and negative inotropic action with β-receptor blockers) substantially counteract one another. Undesirable additive effects, as seen in the combination of a calcium antagonist with a nitrate (decrease in blood pressure, excessive increase in heart rate), or of verapamil (or diltiazem) with a β-receptor blocker (negative chronotropic, negative dromotropic, negative inotropic action), are generally avoided. Further details on combination therapy are to be found in Chap. 4. The important data on combination therapy with nifedipine and β-receptor blockers are summarized in Table 9.

Table 9. Haemodynamic actions of a combination therapy with nifedipine and β-receptor blockers in comparison with monotherapy. (After Lydtin et al. 1975; Koch 1980; Opie 1980; Krämer and Kübler 1983; Nelson et al. 1984)

	Nifedipine	β-Receptor-blockers	Nifedipine + β-receptor blocker
Heart rate	↑	↓	↓=
Arterial blood pressure	↓	↓	↓↓
Cardiac output	↑	↓	=
Peripheral vascular resistance	↓	↑=	↓=
Left ventricular filling pressure (LVEDP)	↓=	↑	=

↑ increase; ↓ decrease; ↓↓ marked decrease; = no change

References

Abernethy DR, Schwartz JB, Todd EL, Luchi R, Snow E (1986) Verapamil pharmacodynamics and disposition in young and elderly hypertensive patients Ann Intern Med 105: 329–336

Amende I, Simon R, Lichtlen PR (1985) Die Wirkung von intravenösem Verapamil auf die Ventrikelfunktion und den Myokardstoffwechsel Z Kardiol 74 (Suppl 3): 9

Amlie JP, Endresen K, Sire S (1989) The effect of felodipine on the sinus and atrioventricular node in patients with ischaemic heart disease J Cardiovasc Pharmacol (in press)

Antman EM, Stone PH, Muller JE, Braunwald E (1980) Calcium channel blocking agents in the treatment of cardiovascular disorders. Part I: Basic and clinical electrophysiologic effects Ann Intern Med 93: 875–885

Arnman K, Ryden L, Smedgärd P, Thorhallsson E (1984) Felodipine in primary pulmonary hypertension Acta Med Scand 215: 275–280

Beck OA, Witt E, Lehmann HU, Hochrein H (1978) Die Wirkung von Gallopamil (D 600) auf die intrakardiale Erregungsleitung und Sinusknotenautomatie beim Menschen Z Kardiol 67: 522–526

Been M, MacFarlane PW, Hillis WS (1985) Electrophysiological effects of felodipine Drugs 29 (Suppl 2): 76–80

Bellocci F, Ansalone G, Scabbia E (1981) Sustained benefical effect of nifedipine in chronic refractory heart failure Am J Cardiol 47: 407 (abstract)

Bender F, Koijama N, Replok HD, Oelmann G (1966) Behandlung tachykarder Rhythmusstörungen des Herzens durch Beta-Rezeptorenblockade des Atrioventrikulargewebes Med Welt 17: 1120–1123

Bertrand ME, Lablanche IM, Tilmant PY (1980) Treatment of spasm of the coronary artery with nifedipine Eur Heart J (Suppl B)1: 65–69

Biamino G, Oeff M, Prokein E, Schröder R (1982) Verhalten von Hämodynamik und Koronardurchblutung nach intravenöser Gabe von Diltiazem bei koronarer Herzerkrankung In: Bender F, Greeff K (eds) Calciumantagonisten zur Behandlung der Angina pectoris, Hypertonie und Arrhythmie Excerpta Medica Amsterdam pp 94–105

Bratel T, Hedenstierna G, Lundquist H, Nyquist O, Ripe E (1988) Cardiac function and central haemodynamics in severe chronic obstructive lung disease – acute and long term effects of felodipine Eur Respir J 1: 262–268

Braunwald E (1982) Mechanism of action of calcium channel blocking agents N Engl J Med 307: 1618–1627

Braunwald E (1984) Heart disease – a textbook of cardiovascular medicine Saunders Philadelphia

Brobmann GF, Mayer M, Grimm W, Safer A (1980) Aufhebung Herzglycosid-induzierter Spasmen der Mesenterialgefäße durch Calcium-Antagonisten im Hundeexperiment In: Fleckenstein A, Roskamm H (eds) Calcium-Antagonismus Springer Berlin pp 230–242

Brugger P, Lässer W, Klein G (1984) Koronare Herzkrankheit / Wirkung von Diltiazem auf die diastolische linksventrikuläre Funktion Münch Med Wschr 35: 996–998

Bucca C, Scalvo M, Rolla G, Borello G, Polizzi S (1982) Verapamil versus oxygen in pulmonary hypertension complicating chronic obstructive lung diseases Bull Physio-path resp 18 (Suppl 4): 87–93

Bühler FR, Erne P, Müller FB, Hulthén UL, Kiowski W, Block LH Bolli P (1984) Calcium-Antagonisten bei Hypertonie – ein neues therapeutisches Konzept In: Althaus U, Burckhardt D, Vogt E (eds) Calcium-Antagonismus International Symposium on Calcium-Antagonism Universimed Frankfurt/Main pp 123–132

Cheung JY, Bonventre JV, Malis CD, Leaf A (1986) Calcium and ischemic injury N Engl J Med 314: 1670–1675

Cranefield DF (1977) Action potentials, after potentials and arrhythmias Circ Res 41: 415–423

Csapo G (1980) Konventionelle und intracardiale Elektrokardiographie Ciba-Geigy Wehr/Baden pp 47–53

Culling W, Ruttley MSM, Sheridan DJ (1984) Acute haemodynamic effects of felodipine during beta blockade in patients with coronary artery disease Br Heart J 52: 431–434

Daniel WG, Engel HJ, Lichtlen PR (1984) Effects of calcium antagonists on regional myocardial blood flow In: Althaus U, Burckhardt D, Vogt E (eds) Calcium-Antagonismus International Symposium on Calcium-Antagonism Universimed Frankfurt/Main pp 152–161

Dash H, Copenhaver GL, Ensinger S (1985) Improvement in regional wall motion and left ventricular relaxation after administration of diltiazem in patients with coronary artery disease Circulation 72: 353–363

DeServi S, Mussini A, Specchia G, Bramucci E Gavazzi A, Falcone C, Ardissino D, Guagliumi G, Bobba P (1980) Effects of nifedipine on coronary blood flow and coronary resistance during cold pressor test and isometric exercise in patients with coronary artery disease Eur Heart J 1 (Suppl B): 43–47

Ebner F, Dünschede HB (1976) Haemodynamics, therapeutic mechanism of action and clinical findings of adalat use based on worldwide clinical trials In: Jatene AD, Lichtlen PR (eds) 3rd International Adalat Symposium Excerpta Medica Amsterdam pp 283–300

Ellenbogen KA, German LD, O'Callaghan WG, Colavita PG, Marchese AC, Gilbert MR, Strauss MC (1985) Frequency-dependent effects of verapamil on atrioventricular nodal conduction in man Circulation 72: 344–352

Engel HJ, Wolf R, Hundeshagen H, Lichtlen PR (1980) Different effects of nitroglycerin and nifedipine on regional myocardial blood flow during pacing induced angina pectoris Eur Heart J 1 (Suppl B): 53–58

Ferlinz J, Turbow ME (1980) Anti-anginal and myocardial metabolic properties of verapamil in coronary artery disease Am J Cardiol 46: 1019–1026

Fleckenstein A (1968) Myocardstoffwechsel und Nekrose In: Heilmeyer L, Holtmeier HJ (eds) VI Symposium der Deutschen Ges. für Fortschritte auf dem Gebiet der inneren Medizin über „Herzinfarkt und Schock" Thieme Stuttgart pp 94–109

Fleckenstein A, Fleckenstein-Grün G (1980) Cardiovascular protection by Ca antagonists Eur Heart J 1 (Suppl B): 15–21

Fleckenstein A (1982) Calciumantagonismus – ein neues Prinzip in der kardiovasculären Therapie In: Roskamm H, Reindell H (eds) Herzkrankheiten Springer Berlin pp 777–796

Fleckenstein-Grün G, Fleckenstein A (1975) Ca-dependent changes in coronary smooth muscle tone and the action of Ca-antagonistic compounds with special reference to adalat In: Lochner W, Braasch W, Kroneberg G (eds) 2nd International Adalat Symposium Springer Berlin pp 68–76

Fleckenstein-Grün G, Fleckenstein A (1980) Calcium-Antagonismus, ein Grundprinzip der Vasodilatation In: Fleckenstein A, Roskamm H (eds) Calcium-Antagonismus Springer Berlin pp 191–207

Fleckenstein-Grün G, Fleckenstein A (1983) Blockierungen der Ca+ +-abhängigen bioelektrischen Automatie und elektromechanischen Kopplung glatter Muskelzellen durch Gallopamil In: Kaltenbach M, Hopf R (eds) Gallopamil Springer Berlin pp 35–49

Furlanello F, Disertori M, Vergara G, Del Favero A (1980) Study on the electrophysiological effects of nifedipine in man In: Puech P, Krebs R (eds) 4th International Adalat Symposium Excerpta Medica Amsterdam pp 227–235

Gaßner A, Pichler M, Fridrich L Sykora J, Tizek H, Lenz K (1983) Einfluß von Nifedipin auf die Hämodynamik bei präkapillärer pulmonaler Hypertension in Ruhe und unter Belastung Dtsch Med Wschr 108: 1790–1794

Gauer OH, Henry JP (1963) Circulatory basis of fluid volume control Physiol Rev 43: 423–481

Gauer OH (1972) Kreislauf des Blutes In: Trautwein W, Gauer OH, Koepchen HP (eds) Herz und Kreislauf Urban & Schwarzenberg München pp 81–326

Gibson RS, Boden WE, Theroux P, Strauss HD, Pratt CM, Gheorghiade M, Capone RJ, Crawford MH, Schlant RC, Kleiger RE, Young MH, Schechtman K, Perryman MB, Roberts R (1986) Diltiazem and reinfarction in patients with non-Q-wave myocardial infarction: results of a double-blind, randomized, multicenter trial N Engl J Med 315: 423–429

Gunther S, Green L, Muller JE, Mudge GH, Grossman W (1981) Prevention by nifedipine of abnormal coronary vasoconstriction in patients with coronary artery disease Circulation 63: 849–855

Haas H, Härtfelder G (1962) Alpha-Isopropyl(N-methylhomoveratryl)-aminopropyl-3,4-dimethoxy-phenylacetonitrol, eine Substanz mit coronargefäßerweiternden Eigenschaften Arzneimittelforschung 12: 549

Haas H (1964) Selektive Sympathikolyse und Myokardfunktion Dtsch Med Wschr 89: 2117–2121

Hashimoto K, Ono H O'Hara N (1980) Blockade of renal autoregulatory vasoconstriction by calcium antagonists In: Fleckenstein A, Roskamm H (eds) Calcium-Antagonismus Springer Berlin pp 221–229

Hecht HS, Chew CYC, Burnman MH, Hopkins J, Schnugg S, Singh BN (1981) Verapamil in chronic stable angina: Amelioration of pacing induced abnormalities of left ventricular

ejection fraction, regional valve motion, lactate metabolism and hemodynamics Am J Cardiol 48: 536–544
Henry PD, Shuchleib R, Borda LJ, Roberts R, Williamson JR, Sobel BE (1978) Effects of nifedipine on myocardial perfusion and ischemic injury in dogs Circ Res 43: 372–380
Hof RP (1985) Pharmakologische Charakterisierung der Kalziumantagonisten: Wirkungsweise und Wirkungsorte Triangel 24 (3/4): 143–156
Hossack KF, Brown BG, Stewart DK, Dodge HT (1984) Diltiazem-induced blockade of sympathetically mediated constriction of normal and diseased coronary arteries: Lack of epicardial coronary dilatory effect in humans Circulation 70: 465–471
Hugenholtz PG (1985) Calcium Antagonists In: Abshagen U (ed) Clinical pharmacology of antiangial drugs Springer Berlin pp 459–538
Hugenholtz PG, Serruys PW, Fleckenstein A, Nayler W (1986) Why calciumantagonists will be most useful before or during early myocardial ischemia and not after infarction has been established Eur Heart J 7: 270–278
Johnsson G, Murray G, Tweddel A, Hutton L (1983) Haemodynamic effects of a new vasodilator drug felodipine in healthy subjects Eur J Clin Pharmacol 24: 49–53
Jones CR, Rae AP, Been M, deVane PJ, Jamieson RR, Hornung RS, Hillis WS (1985) Electrophysiological effects of felodipine in combination with metoprolol Drugs 29 (Suppl 2): 81–86
Joshi PI, Dalal JJ, Ruttley SJ, Sheridan DJ, Henderson AH (1981) Nifedipine and left ventricular function in beta-blocked patients Br Heart J 45: 457–459
Kaltenbach M, Schulz W, Kober G (1979) Effects of nifedipine after intravenous and intracoronary administration Am J Cardiol 44: 832–838
Kaltenbach M, Hopf R (1983) Gallopamil Springer Berlin
Kirchheim H, Gross R (1975) Hemodynamic effects of Adalat in the unanesthetized dog (and Lydtin H. Discussion remarks) In: Lochner W, Braasch W, Kroneberg G (eds) 2nd International Adalat Symposium Springer Berlin pp 88–91
Klein GJ, Bashore TM, Sellers TD, Pritchett ELC, Smith WM, Gallagher JJ (1979) Ventricular fibrillation in the Wolff-Parkinson-White Syndrome N Engl J Med 301: 1080–1085
Koch G (1980) Beta-receptor and calcium blockade in ischemic heart disease: effects on systemic and pulmonary hemodynamics and on plasma catecholamines at rest and during exercise In: Puech P, Krebs R (eds) 4th International Adalat Symposium Excerpta Medica Amsterdam pp 131–142
Krämer B, Kübler W (1983) Calcium-Antagonisten bei koronarer Herzerkrankung: Wirkmechanismen und therapeutische Möglichkeiten Deutsches Ärzteblatt 29: 21–29
Kupper W (1986) Koronararterienspasmus In: Astra Chemicals Herz-Kreislauf Information pp 1–35
Kurnik PB, Courtois MR, Ludbrook PA (1986) Effects of nifedipine on intrinsic myocardial stiffness in man Circulation 74: 126–134
Landahl S, Edgar B, Gabrielsson M, Larsson M, Lernfeldt B, Lundborg P, Regardh CG (1988) Pharmacokinetics and blood pressure effects of felodipine in elderly hypertensive patients Clin Pharmacokinetics 14: 374–383
Landmark K, Amlie JP (1976) A study of the verapamil-induced changes in conductivity and refractoriness and monophasic action potentials of the dog heart in situ Eur J Cardiol 4: 419–427
Lindner E (1960) Phenyl-propyl-dihphenyl-propyl-amin, eine neue Substanz mit coronargefäßerweiternder Wirkung Arzneimittelforschungung 10: 569
Ljung B, Kjellstedt A, Orebäck B (1987) Vascular versus myocardial selectivity of calcium antagonists studied by concentration–time–effect relations J Cardiovasc Pharmacol 10 (Suppl 1): 34–39
Ludbrook PA, Tiefenbrunn AJ, Reed FR, Sobel BE (1982) Acute hemodynamik response to sublingual nifedipine: Dependence on left ventricular function Circulation 65: 489–498
Lydtin H, Lohmöller R (1972) Hämodynamische Wirkungen von Bay a 1040 vor und nach Atropingabe In: Schlegel B (ed) Verhandlungen der Deutschen Gesellschaft für Medizin Bergmann München pp 1554–1559
Lydtin H, Lohmöller G, Lohmöller R, Walter I (1975) Hemodynamic studies on nifedipine in man In: Hashimoto K, Kimura E, Kobayashi T (eds) 1st International Nifedipine Adalat Symposium University of Tokyo Press Tokyo pp 97–105

Lydtin H, Schierl W, Lohmöller G (1980) Exercise pulmonary wedge pressure after acute and chronic administration of nifedipine in ischemic heart disease In: Puech P, Krebs R (eds) 4th International Adalat Symposium Excerpta Medica Amsterdam pp 249–254

Lynch P, Dargie H, Krikler S, Krikler D (1980) Objective assessment of anti-anginal treatment: a double blind comparison of propranolol, nifedipine and their combination Br Med J 281: 184–187

MacAlpin RN (1980) Contribution of dynamic vascular wall thickening to luminal narrowing during coronary arterial constriction Circulation 61: 296–301

Mancia G, Bonazzi O, Ferrari A, Gardumi M, Gregorini L, Perondi R, Pozzoni (1978) Baroreceptor control of atrioventricular conduction system in man In: Schwartz PI, Brown AM, Mallini A, Zanchetti A (eds) Neural mechanisms in cardiac arrhythmias Raven Press New York pp 339–343

Mauser M, Karsch KR, Wagner S, Seipel L (1985) Änderung der diastolischen Ventrikeleigenschaften durch intravenöse Nifedipininfusion bei Patienten mit instabiler Angina pectoris Z Kardiol 74: 590–597

McGovern B, Garan H, Ruskin JN (1986) Precipitation of cardiac arrest by verapamil in patients with Wolff-Parkinson-White syndrome Ann Intern Med 104: 791–794

Millard RW, Lathrop DA, Grupp G, Ashraf M, Grupp I, Schwartz A (1982) Differential cardiovascular effects of calcium channel blocking agents: Potential mechanisms Am J Cardiol 49: 499–505

Mitchell LB, Schroeder JS, Mason JW (1982) Comparative clinical electrophysiologic effects of diltiazem, verapamil and nifedipine: a review Am J Cardiol 49: 629–634

Nakamura Y, Ikeda T, Takata S, Yamamoto M, Kitamura T, Hattori N (1985) Wirkung von Nifedipin auf den Gefäßwiderstand des Unterarms und die venöse Kapazität bei Gesunden und Patienten mit Herzinsuffizienz Int J Cardiol 9: 27

Naumann d'Alnoncourt C, Zierhut W, Lüderitz B (1982) Wirkung von Diltiazem auf abnorme Reizbildung in Purkinje-Fasern bei Ischämie In: Bender F, Greeff K (eds) Calciumantagonisten zur Behandlung der Angina pectoris, Hypertonie und Arrhythmie Excerpta Medica Amsterdam pp 270–277

Nayler WG (1980) The pharmacological protection of the ischaemic heart: the use of calcium and beta-adrenoceptor antagonists Eur Heart J 1 (Suppl B): 5–13

Nelson GIC, Silke B, Ahuja RC, Hussain M, Forsyth D, Taylor SH (1984) The effect on left ventricular performance of nifedipine and metoprolol singly and together in exercise-induced angina pectoris Eur Heart J 5: 67–79

Neuss H, Mitrovic I, Stieren B, Bühler V, Schlepper M (1983) Pharmakodynamik und Elektrophysiologie von Gallopamil In: Kaltenbach M, Hopf R (eds) Gallopamil Springer Berlin pp 101–108

Nissen P, Alexander K (1975) Unterschenkeldurchblutung und Sauerstoffverbrauch des Beines bei Patienten mit chronischem Gliedmaßenarterienverschluß unter intraarterieller Verapamilinfusion Folia Angiologica 23: 318–322

Opie LH (1984a) Calcium antagonists (Slow channel blockers) In: Opie LH (ed) The Heart Grune & Stratton London pp 246–260

Opie LH (1984b) Vasodilators and vascular smooth muscle In: Opie LH (ed) The Heart Grune & Stratton London pp 230–245

Packer M, Medina N, Yushak M (1984) Adverse hemodynamic and clinical effects of calcium channel blockade in pulmonary hypertension secondary to obliterative pulmonary vascular disease J Am Coll Cardiol 4: 890–901

Packer M (1985) Vasodilator therapy for primary pulmonary hypertension Ann Intern Med 103: 258–270

Panel discussion (1975) In: Hashimoto K, Kimura E, Kobayashi T (eds) 1st International Nifedipine „Adalat" Symposium University of Tokyo Press Tokyo pp 299–304

Petri H, Kafka W, Hall D, Rudolph W (1980) Potential acceleration of the ventricular rate in WPW with atrial fibrillation after verapamil Circulation 62 (Suppl 3): III–262

Petri M, Kafka W, Hansen W, Rudolph W (1982) Discrepant effect of intravenous and oral verapamil on the ventricular rate in WPW syndrome during atrial fibrillation Circulation 66 (4 pt 2): II–270 (abstract)

Rafflenbeul W, Lichtlen PR (1983) Release of residual vascular tone in coronary artery stenoses with nifedipine and glyceryl trinitrate In: Kaltenbach M, Neufeld HN (eds) 5th International Adalat Symposium Excerpta Medica Amsterdam pp 300–308

Rocha P, Baron B, Delestrain A, Pathe M, Cazor IL, Kahn JC (1986) Hemodynamic effects of intravenous diltiazem in patients treated chronically with propranolol Am Heart J 111: 62–68

Rowland E, McKenna W; Krikler D (1983) Electrophysiological and antiarrhythmic effects of bepridil in reentry AV tachycardia – comparison with verapamil and ajmaline Circulation 86 (Suppl III): III 311

Schmitz H, Schierl W, Beck O, Lydtin H (1975) Über die Wirkung von Nifedipine (Adalat) auf regionale Hirndurchblutung und Unterschenkeldurchblutung In: Schlegel B (ed) Verhandlungen der Deutschen Gesellschaft für Innere Medizin Bergmann München pp 888–890

Schrör K (1985) Physiologische Grundlagen, Pharmakologie und Prinzipien der klinischen Anwendung In: Hossmann V, Grötz I, Schrör K (eds) Kalziumantagonisten und zerebrale Erkrankungen Springer Berlin pp 1–39

Schulz W, Wendt T, Kaltenbach M, Kober G (1983a) Active and passive changes in coronary artery diameters after vasodilatation In: Kaltenbach M, Neufeld HN (eds) 5th International Adalat Symposium Excerpta Medica Amsterdam pp 309–321

Schulz W, Kaltenbach M, Kober G (1983b) Chronotropic response after injection of nifedipine into the sinus node artery in man In:Kaltenbach M, Neufeld HN (eds) 5th International Adalat Symposium Excerpta Medica Amsterdam pp 322–327

Scriabine A, Vanov S, Deck K (eds) (1984) Nitrendipine Urban & Schwarzenberg Baltimore

Seipel L, Breithardt G (1982) Electrophysiological actions of calcium antagonists in the heart Cardiology 69 (Suppl 1): 105–116

Seipel L, Breithardt G, Borggrefe M (1983) Antiarrhythmische Therapie mit Calciumantagonisten Internist 24: 415–421

Serruys PW, Brower RW, Ten Kate HJ, Bom AH, Hugenholtz PG (1981) Regional wall motion from radiopaque markers after intravenous and intracoronary injections of nifedipine Circulation 63: 584–591

Sesto M, Ivancic'R, Custovic F (1983) Die Wirkung von Gallopamil auf die Hämodynamik bei Patienten mit KHK In: Kaltenbach M, Hopf R (eds) Gallopamil Springer New York pp 97–100

Sheridan DJ, Thomas P (1987) Vascular versus myocardial selectivity of calcium antagonists J Cardiovasc Pharmacol 10 (Suppl 1): 165–168

Singh BN, Vaughan-Williams EM (1972) A fourth class of antidysrhythmic action? Effect of verapamil on ouabain toxicity, on atrial and ventricular intracellular potentials and on other features of cardiac function Cardiovasc Res 6; 109–119

Singh BN, Hecht HS, Nademanee K, Chew CYC (1982) Electrophysiologic and hemodynamic effects of slow channel blocking drugs Prog Cardiovasc Drugs 15: 103–132

Singh BN, Josephson MA, Nademanee K (1984) Elektrophysiologische und hämodynamische Wirkungen von Calcium-Antagonisten: klinischer Überblick In: Althaus U, Burckhardt D, Vogt E (eds) CalciumAntagonismus Internat. Symposium Calcium-Antagonism Universimed Frankfurt/Main pp 57–85

Smith SA, Mace PJE, Littler WA (1986) Felodipine, blood pressure and cardiovascular responses in hypertensive humans Hypertension 8: 1172–1178

Solti F, Iskum M, Syabó Z, Czakó E (1978) Die Wirkung des Ca^{++}-Antagonisten Isoptin auf die periphere Durchblutung bei oblitierenden Angiopathien Therapiewoche 28: 5993–5994

Spurrell RAJ, Krikler DM, Sowton GE (1974) The effect of verapamil on the electrophysiologic properties of the anomalous atrio-ventricular connections in Wolff-Parkinson-White syndroms Br Heart J 36:256–264

Stone PH, Antman EM, Muller JE, Braunwald E (1980) Calcium channel blocking agents in the treatment of cardiovascular disorders. Part II Hemodynamic effects and clinical applicationes Ann Intern Med 93: 886–904

Sugimoto T, Ishikawa T, Kaseno K, Nakase S (1980) Electrophysiologic effects of diltiazem, a calcium antagonist in patients with impaired sinus or atrioventricular node function Angiology 31: 700–709

Taira N, Motomura S, Narimatsu A, Satoh K, Yanagisawa T (1980) The effect of calcium antagonists on atrioventricular conduction In: Fleckenstein A, Roskamm H (eds) Calcium-Antagonismus Springer Berlin pp 42–43

Tauchert M, Behrenbeck DW, Niehues B, Hilger HH (1980) Therapeutische Wirkung von Calcium-Antagonisten auf den globalen myokardialen Blutfluß In: Fleckenstein A, Roskamm H (eds) Calcium-Antagonisten Springer Berlin pp 265–269

Towart R, Perzborn E (1983) Differential effects of nifedipine on vascular smooth muscle: selective inhibition of agonist- and thromboxane-induced contractions of coronary vessels In: Kaltenbach M, Neufeld HN (eds) 5th International Adalat Symposium Excerpta Medica Amsterdam pp 27–33

Tweddel AC, Johnsson G, Pringle TH, Murray RG, Hutton I (1983) The systemic and coronary haemodynamic effects of felodipine in patients with coronary heart disease Eur Heart J 4: 699–705

Vliestra RE, Farias MAC, Frye RL, Smith HC, Ritman E (1983) Effect of verapamil on left ventricular function: a randomized, placebo-controlled study Am J Cardiol 51: 1213–1217

White HD, Polak CBJ, Wymne J, Hohman BL, Antman EM, Nesto RW (1985) Addition of nifedipine to maximal nitrate and betaadrenoreceptor blocker therapy in coronary artery disease Am J Cardiol 55: 1303–1307

Zipes DP, Fischer JC (1974) Effects of agents which inhibit the slow channel on sinus node automaticity and atrioventricular conduction in the dog Circ Res 34: 184–192

3.2 Smooth Muscle

Besides the vascular smooth muscle (Sect. 3.1.5), calcium antagonists also act on the gastrointestinal tract, the bronchi and the urogenital tract.

3.2.1 Gastrointestinal Tract

Calcium antagonists inhibit the smooth muscle of the digestive tract under experimental conditions in both animals (review by Richter et al. 1982) and humans. They reduce the tone of smooth muscle and prevent or resolve spasm. Most investigations concern verapamil, diltiazem and nifedipine; other calcium antagonists have not yet been systematically investigated with regard to their action on the gastrointestinal tract. A comprehensive summary of available experimental and clinical results in this area is provided by Mössner (1987).

In the oesophagus, calcium antagonists reduce the amplitude and duration of peristaltic waves, and the pressure at the lower oesophageal sphincter. This action is limited to oesophageal smooth muscle; striated muscle is unaffected (Richter et al. 1982). Calcium antagonists have been found successful in treating disturbances of oesophageal motility (e.g. achalasia, oesophageal spasm; Sect. 4.8).

Also in the small and large intestines calcium antagonists appear to inhibit peristalsis and smooth muscle tone to some degree; this may lead to constipation. Excessive or uncoordinated intestinal motility (e.g. as in irritable bowel syndrome or proctalgia fugax) is positively affected (Sect. 4.8), and calcium antagonists would also seem to lend themselves to the treatment of diarrhoea due to acceleration of stool passage (Mössner 1987). In contrast to these findings, gastric acid secretion, gastric motility and gastric emptying are not affected by calcium antagonists (Mössner 1987).

Theoretically, calcium antagonists should have a relaxant effect on the smooth muscle of the biliary tract; however, to date no systematic studies, e.g. in biliary colic, have been carried out. Schmidt (1986, personal communication) observed in 15 patients a reduced (cholecystokinin-induced) contraction of the gallbladder when nifedipine was given, thereby raising the possibility of gallstone formation, as a result of reduced gallbladder motility, with long-term calcium antagonist therapy. Porschen et al. (1987), however, in a study of nine healthy volunteers, with ceruletide stimulation, were unable to confirm such findings (no significant difference in gallbladder contraction in comparison with placebo).

3.2.2 Respiratory Tract

The basal tone of bronchial smooth muscle, i.e. bronchial diameter in the resting state, is not affected by calcium antagonists (McFadden 1981; Editorial 1983; Ritchie et al. 1984). This applies as much to healthy subjects as to patients with coronary heart disease, asthma or chronic obstructive pulmonary disease. Only Emslander et al. (1984, 1986) have reported, in patients with chronic obstructive pulmonary disease, a brief period of bronchodilation after administration of nifedipine, corresponding to the effect of a β_2 sympathomimetic (salbutamol). Both in animal experiments and in patients with asthma, calcium antagonists inhibit bronchoconstriction induced by exercise (Cerrina et al. 1981; Patel 1983; Butchers et al. 1981; Patel and Peers 1988), hyperventilation, exposure to an antigen, or by administration of histamine or metacholine (Editorial 1983). The inhibitory effect on bronchoconstriction was in most cases incomplete, and there was considerable variation among individuals. These results indicate that calcium antagonists probably reduce reactivity of the bronchial system to bronchoconstricting stimuli, without additional bronchodilation.

This is an important consideration in clinical practice since calcium antagonists, in contrast to β-receptor blockers, can be prescribed in angina pectoris with asthma or chronic obstructive pulmonary disease without a negative effect on the bronchial system, and that in combination with β-sympathomimetics (salbutamol, fenoterol, terbutaline) bronchospasmolysis will not be reduced (Editorial 1983).

It should also be borne in mind in asthma that calcium antagonists, in addition to their effect on the bronchial smooth muscle, also possibly affect mucus secretion, degranulation of mast cells and activation of inflammatory cells (Sect. 3.3.3; Editorial 1983).

3.2.3 Urogenital Tract

Calcium antagonists lower the activity of the myometrium and thus have a relaxant effect on the uterus (Forman et al. 1979; Ulsten et al. 1980). This effect can be demonstrated in the normal uterus, in patients with dysmenorrhoea and in the gravid uterus (inhibition of labour).

Calcium antagonists also have a relaxant effect on the muscles of the ureter and the bladder (Maggi and Meli 1984; Forman et al. 1978; Carrol 1985). Positive therapeutic effects can be anticipated in conditions with spasm (e.g. ureteral colic) and with raised muscle tone (detrusor hyperflexia in the bladder). Bladder emptying (e.g. in prostatic adenoma), on the other hand, may be disturbed. Furthermore, in vitro high concentrations of calcium antagonists are thought to inhibit the smooth muscle of the penis and the vas deferens (Fovaeus et al. 1987, personal communication).

References

Butchers PR, Skidmore IF, Vardey CJ, Wheeldon A (1981) Calcium antagonists in exercise-induced asthma Br Med J 282: 1792
Carrol W (1985) Nifedipine and ureteral colic Ann Intern Med 102: 864
Cerrina J, Denjean A, Alexandre G, Lockhart A, Duroux P (1981) Inhibition of exercise-induced asthma by a calcium antagonist, nifedipine Am Rev Respir Dis 123: 156–160
Editorial (1983) Calcium-channel blockers and asthma Thorax 38: 481–485
Emslander HP, Sauer E, Freisinger H, Sebening H, Daum S (1984) Akute Bronchospasmolyse mit dem Kalzium-Antagonisten Nifedipin Atemw.-Lungenkrkh. 10: 379–382
Emslander HP, Sauer E, Munteanu J, Sebening H, Daum S (1986) The acute effect of nifedipine on airways resistance in patients with chronic obstructive lung disease In: Lichtlen PR (ed) 6th International Adalat Symposium Excerpta Medica Amsterdam pp 497–503
Forman A, Andersson KE, Henricksson L, Rud T, Ulmsten U (1978) Effects of nifedipine on the smooth muscle of human urinary tract in vitro and in vivo Acta Pharmacol Toxicol 43: 111–118
Forman A, Andersson KE, Persson CGA, Ulmsten U (1979) Relaxant effects of nifedipine on isolated human myometrium Acta Pharmacol Toxicol 45: 81–86
Maggi CA, Meli A (1984) The effects of nifedipine and verapamil on KCl-induced rhythmic contractions of guinea pig ureter in vitro Experientia 40: 681
McFadden JR (1981) Calcium-channel blocking agents and asthma Ann Intern Med 95: 232–233
Mössner J (1987) Calciumantagonisten in der experimentellen und klinischen Gastroenterologie Dtsch Med Wschr 112: 1092–1099
Patel KR (1983) Sodium cromoglycate and verapamil alone and in combination in exercise induced asthma Br Med J 286: 606
Patel KR, Peers EM (1988) Felodipine modifies exercise-induced asthma J Cardiovasc Pharmacol 12 (Suppl 6): 182
Porschen R, Pieper S, Bernhardt L, Schade B, Wienbeck M (1987) Beeinflussen Kalziumantagonisten die Gallenblase in vivo? Klin Wochenschr 65 (Suppl IX): 30
Richter J E,Sinar DR, Cordova CM, Castell DO (1982) Verapamil – a potent inhibitor of esophageal contractions in the baboon Gastroenterology 82: 882–886
Ritchie DM, Sierchio JN, Bishop CM, Hedli CC, Levinson SL, Capetola RJ (1984) Evaluation of calcium entry blockers in several models of immediate hypersensitivity J Pharmacol Exp Ther 229:690–695
Ulmsten U, Andersson KE, Wingerup L (1980) Treatment of premature labor with the calcium antagonist nifedipine Arch Gynecol 229: 1–5

3.3 Other Organ Systems

3.3.1 Kidney

During treatment with calcium antagonists, previously normal renal function is not impaired, even in long-term therapy (Ebner and Dünschede 1976). A precondition is adequate renal perfusion pressure. (For discussion concerning the use of calcium antagonists in impaired renal function, see Sect. 5.1.)

For substances of the dihydropyridine group (nifedipine, nitrendipine, felodipine), a diuretic and natriuretic effect at the onset of therapy has been described. An increase in sodium excretion and in urinary volume occurs, although renal blood flow and glomerular filtration rate apparently remain unchanged (Ene et al. 1985; Di Bona 1985). Causally, a direct renal tubular action is postulated; at higher doses the effect recedes, which is explained (Ene et al. 1985) on the basis of a haemodynamic adaptation process (sodium retention through activation of the renin–angiotensin–aldosterone system). In long-term therapy this diuretic effect cannot be demonstrated (Hulthen et al. 1987); sodium and water balance is well maintained, and body weight remains constant.

In kidney transplantation (cadaver kidneys) Wagner and his co-workers (1985, 1986) investigated the influence of diltiazem on the development of acute renal failure following transplantation. Diltiazem reduced the incidence of rejection reactions (primary transplant failure) and improved glomerular filtration rate in transplants with primary graft function. In addition to a cytoprotective effect on the kidney transplant, a reduced cyclosporin A nephrotoxicity with diltiazem therapy is under discussion.

A general protective effect of calcium antagonists against postischaemic renal failure has, however, not yet been proven. Hull and Hasbargen (1985) could not, even with administration of a calcium antagonist before the ischaemic episode, observe any differences between the treated and the untreated control groups (although this was admittedly a retrospective, nonrandomized study).

3.3.2 Metabolism – Endocrinology

In animal studies the calcium-dependent insulin secretion of the pancreas is inhibited by calcium antagonists dose-dependently (review by Trost and Weidmann 1984). In the human organism, however, glucose metabolism is scarcely influenced by calcium antagonists, even during long-term therapy. Fasting and postprandial blood glucose levels, the insulin level, and HbA_1 remain unchanged (Ebner et al. 1983; Ferrara et al. 1985; Kindermann et al. 1985; Hedner et al. 1987). In a very few individual cases of nifedipine therapy, however, disturbed glucose tolerance with an increase in blood glucose has been observed (Ebner et al. 1983; Arzneitelegramm 1985). This diabetogenic effect is postulated to be due to inhibition of insulin secretion and activation of the sympathetic nervous system. To judge by the numerous therapeutic studies with nifedipine that have revealed no influence on glucose tolerance (review by Ebner et al. 1983), this seems to be a relatively rare effect observed with nifedipine. Calcium antagonists have been used on a trial basis to inhibit insulin secretion in insulinoma with recurrent episodes of hypoglycaemia (diltiazem 180 mg daily; Taniguchi et al. 1987; two further recent case reports are cited by Mössner 1987).

On the basis of the unanimity of results recorded in the literature, it can be concluded that calcium antagonists do not affect lipid metabolism; even during long-term therapy, levels of triglycerides, LDL-cholesterol and HDL-cholesterol remain unchanged (Ferrara et al. 1985; Kindermann et al. 1985; Trost 1985; Nihlsson-Ehle 1988). In contrast to thiazide diuretics, calcium antagonists cause no increase in uric acid levels.

As a result of their peripheral vasodilatory action, calcium antagonists activate, as a counter-regulatory autonomic reflex, the sympathetic nervous system; this manifests itself by an increase in noradrenaline and renin levels. Levels of noradrenaline and renin are higher in short-term therapy and with nifedipine than in long-term therapy (Bühler et al. 1984) and with verapamil (corresponding to autonomic adaptation and the more powerfully vasodilatory action of nifedipine). In contrast, nifedipine was observed to lower an excessively raised noradrenaline secretion in phaeochromocytoma (Serfas et al. 1983); the entire question of calcium dependence of noradrenaline secretion in the adrenal medulla is open to discussion. Aldosterone, ADH, cortisol and angiotensin show, even long-term, no essential change, and secondary hyperaldosteronism does not occur during calcium antagonist therapy (Laederach et al. 1983; Sluiter et al. 1985). This has given rise to the hypothesis that calcium antagonists can reduce the aldosterone-releasing effect of renin.

Hormone secretion by the pituitary gland also appears to be calcium-dependent; in animal studies secretion of ACTH, TSH, LH and prolactin is inhibited in isolated cases (review by Triggle and Swamy 1983; Merritt et al. 1984). Systematic investigations in humans are limited (e.g. felodipine does not affect the glycoregulatory hormone response to insulin-induced hypoglycaemia in hypertensives; Katzman et al. 1987; Elmfeld 1987, personal communication).

3.3.3 Erythrocytes – Thrombocytes – Mast Cells

Under physiological conditions erythrocytes contain only small amounts of calcium; membrane permeability to calcium ions is low. In tissue ischaemia, accompanied by an increase in CO_2 and a decrease in pH, an increased influx of calcium ions into the erythrocytes occurs. The composition of the red cell membrane is thereby changed and the intracellular ATP concentration lowered, resulting in a decrease in erythrocyte deformability and flexibility. A similarly disturbed deformability of erythrocytes is reported to occur in patients with severe hypertension, coronary heart disease, other forms of arterial occlusive disease and in sickle cell anaemia (de Clerck 1981; Hopkins and Hill 1985; Sowemimo et al. 1985).

Calcium antagonists improve the disturbed erythrocyte deformability (increase in erythrocyte flexibility, measured as raised filterability) and thereby decrease blood viscosity. Most studies concern flunarizine (Scott et al. 1980, de Clerck 1981); a few studies report analogous results with nitrendipine, felodipine and verapamil (de Clerck 1981, Hopkins and Hill 1985, Sowemimo et al. 1985).

In sickle cell anaemia bepridil and flunarizine reduce the percentage of sickle cells by reducing mean corpuscular haemoglobin concentration (MCHC) and improving erythrocyte deformability (Reilly and Asakura 1986).

Calcium ions also participate in platelet activation (Lüscher 1984). The calcium antagonists nifedipine, diltiazem, verapamil and perhexiline inhibit platelet aggregation in vitro and dissolve existing platelet aggregates (McAlpine et al. 1985). Furthermore, an inhibition of thromboxane A_2 synthesis is postulated for nifedipine. This was recently confirmed for the first time in vivo (a study of 20 hypertensive patients, dose 30 mg/day; details in Uehara et al. 1986). In the case of tissue ischaemia, calcium antagonists can thus increase blood supply through vasodilation and simultaneously improve the rheological properties of blood (erythrocyte deformability, platelet aggregation). Overall, the clinical significance of the effects of calcium antagonists on formed blood components remains an open question.

In allergic reactions of the immediate type calcium plays a role in the release of mediators from mast cells and basophils. The theoretical assumption that administration of calcium antagonists prevents release of histamine (mast cell degranulation) awaits confirmation despite several in vitro studies (Middleton 1981; Ritchie et al. 1984; Sects. 3.2.2 and 4.9). The effect of calcium antagonists on histamine- and antigen-induced asthma probably reflects a direct action on bronchial smooth muscle (Sect. 3.2.2).

3.3.4 Eye

Intraocular pressure remains unaffected or slightly decreases with administration of calcium antagonists (Schnell 1975; Ebner and Dünschede 1976). This applies to both healthy subjects and patients with glaucoma.

In animal trials it has been possible to delay or prevent the formation of cataract in rats with diabetes mellitus (Fleckenstein et al. 1984). To date corresponding investigations have not been carried out in humans.

3.3.5 Bone–Mineral Metabolism

Although calcium plays a decisive role in bone metabolism, and a change in calcium concentration can result in bone disease (e.g. osteomalacia), bone metabolism is not affected by calcium antagonists. Even in long-term therapy no negative consequences for bone formation and bone metabolism are to be expected from treatment with calcium antagonists (Lehmann 1986).

Calcium metabolism remains unchanged by normal dosages (e.g. 360 mg verapamil per day). Intestinal calcium uptake, serum calcium level and renal excretion are also unaffected (Sjöden et al. 1987, personal communication). Neither does the serum potassium level change significantly during therapy with calcium antagonists.

3.3.6 Central Nervous System

Calcium antagonists can pass the blood–brain barrier (concentrations of nifedipine and of verapamil in cerebrospinal fluid at about 5% of plasma concentration; Doran et al. 1985) and act principally on the vessels of the central nervous system (vasodilation, prevention of vasospasm; cf. Sect. 3.1.5.2). A direct influence on brain cells (cell membrane, metabolism) is postulated for some clinical syndromes (migraine, epilepsy, cerebral ischaemia, Menière's syndrome). Further details are discussed in Sect. 4.7, concerning calcium antagonists in neurological disease.

References

Arzneitelegramm (1985) Weitere Stellungnahmen zur Bewertung des Kalzium-Antagonisten Nifedipin (Adalat) Arzneitelegramm 1: 3–4

Bühler FR, Erne P, Müller FB, Hulthén UL, Kiowski W, Block LH, Bolli P (1984) Calcium-Antagonisten bei Hypertonie – ein neues therapeutisches Konzept In: Althaus U, Burckhardt D, Vogt E (eds) CalciumAntagonismus-Internat. Symposium on Calcium-Antagonism Universimed Frankfurt/Main pp 123–132

De Clerk F (1981) Zur Beeinflussung der Blutviskosität durch Flunarizin Fortschr Med 99: 592

Di Bona GF (1985) Effects of felodipine on renal function in animals Drugs 29 (Suppl 2): 168–175

Doran AR, Narang PK, Meigs CY, Wolkowitz OM, Roy A, Breier A, Pickar D (1985) Verapamil concentrations in cerebrospinal fluid after oral administration N Engl J Med 312: 1261

Ebner F, Dünschede HB (1976) Haemodynamics, therapeutic mechanism of action and clinical findings of Adalat use based on worldwide clinical trials In: Jatene AD, Lichtlen PR (eds) 3rd International Adalat Symposium Excerpta Medica Amsterdam pp 283–300

Ebner F, Leisten L, Lejeune P, Donath M, Rämsch KD (1983) Administration of nifedipine to patients treated with digitalis-glycosides, anti-diabetic agents and betablockers In:

Kaltenbach M, Neufeld HN (eds) 5th International Adalat Symposium Excerpta Medica Amsterdam pp 328–338

Ene MD, Williamson PJ, Roberts CJC, Waddell G (1985) The natriuresis following oral administration of the calcium antagonists nifedipine and nitrendipine Br J Clin Pharmac 19: 423–427

Ferrara LA, Soro S, Fasano ML (1985) Effects of nitrendipine on glucose and lipid serum concentrations Curr Ther Res 37: 614–618

Fleckenstein A, Frey M, Fleckenstein-Grün G (1984) Zwanzig Jahre Calcium-Antagonismus aus physiologischer und pathophysiologischer Sicht, Rückblick und Ausblick In: Althaus U, Burckhardt D, Vogt E (eds) CalciumAntagonismus- Internat. Symposium on Calcium-Antagonism Universimed Frankfurt/Main pp 4–29

Hedner T, Elmfeldt D, Von Schenck H, Sjögren E, Smith U (1987) Glucose tolerance in hypertensive patients during treatment with the calcium antagonist, felodipine Br J Clin Pharmacol 24: 145–149

Hopkins RJ, Hill TWK (1985) Effects of felodipine on red blood cell deformability Drugs 29 (Suppl 2): 42–43

Hull RW, Hasbargen JA (1985) No clinical evidence for protective effects of calcium channel blockers against acute renal failure N Engl J Med 313: 1477–1478

Hulthén UL, Katzman PL, Hökfelt B (1987) Effect of long term felodipine treatment on renal vascular tone, glomerular filtration rate and renal tubular function in essential hypertension Drugs 34 (Suppl 3): 67–68

Katzman PL, Hulthén UL, Hökfelt B (1987) Glucoregulatory hormone response to insulin-induced hypoglycaemia following long-term calcium antagonism with felodipine in patients with essential hypertension Acta Endocrinol (Copenh) 116: 473–478

Kindermann W, Schmitt W, Stengele E (1985) Einfluß von Calcium-Antagonisten auf die körperliche Leistungsfähigkeit und den Metabolismus Dtsch Med Wschr 110: 1657–1661

Laederach K, Gerber A, Weidmann PU (1984) Wirkung von Calcium-Antagonisten auf Blutdruck, Katecholamin-, Renin- und Aldosteronspiegel sowie auf kardiovaskuläre Pressorfaktoren bei Normotonikern und Hypertonikern In: Althaus U, Burckhart D, Vogt E (eds) Calcium-Antagonismus – Internat.Symposium on Calcium-Antagonism Universimed Frankfurt/Main pp 98–122

Lehmann HU (1986) Knochenschäden durch Calcium-Antagonisten? Dtsch Med Wschr 111: 476

Lüscher EF (1984) Calciumantagonisten und Thrombozytenaggregation In: Althaus U, Burckhart D, Vogt E (eds) CalciumAntagonismus-Internat.Symposium on Calcium-Antagonism Universimed Frankfurt/Main pp 254–265

McAlpine HM, Walker ID, Davidson JF, Lawrie TDV, Lorimer AR (1985) Acute haemodynamic and platelet effects of felodipine in hypertensive patients Drugs 29 (Suppl 2): 161–164

Merritt JE, Tomlinson S, Brown BL (1984) Flunarizine, a calcium influx blocker, inhibits TRH-but not potassium-stimulated prolactin secretion Acta Endocrinologica 107: 31–35

Mössner J (1987) Calciumantagonisten in der experimentellen und klinischen Gastroenterologie Dtsch Med Wschr 112: 1092–1099

Nilsson-Ehle P (1988) Felodipine does not affect plasma lipoprotein metabolism J Cardiovasc Pharmacol 12 (Suppl 6): S183

Reilly MP, Asakura T (1986) Antisickling effect of bepridil Lancet 1: 848

Schnell D (1975) Response of intraocular pressure in normal subjects and glaucoma patients to single and repeated doses of the coronary drug adalat In: Lochner W, Braasch W, Kroneberg G (eds) 2nd International Adalat Symposium Springer Berlin pp 290–302

Scott CK, Persico FJ, Carpenter K, Chasin M (1980) The effects of flunarazine, a new calcium antagonist, on human red blood cells in vitro Angiology 31: 320–330

Serfas D, Shoback DM, Lorell BH (1983) Phaeochromocytoma and hypertrophic cardiomyopathy: apparent suppression of symptoms and noradrenaline secretion by calcium-channel blockade Lancet 2: 711–713

Sluiter HE, Huysmans FTM, Thien TA, van Lier HJJ, Koene AP (1985) Haemodynamic, hormonal and diuretic effects of felodipine in healthy normotensive volunteers Drugs 29 (Suppl 2): 26–35

Sowemimo-Coker SO, Debbas NMG, Kovacs IB, Turner P (1985) Ex vivo effects of nifedip-

ine,nisoldipine and nitrendipine on filterability of red blood cells from healthy volunteers Br J Clin Pharmacol 20: 152–154
Taniguchi H, Murakami K, Morita S, Kazumi T, Yoshino G, Maeda M, Baba S (1977) Calcium antagonist (diltiazem) for reversal of hypoglycaemic symptoms in insulinoma Lancet 2: 501
Triggle DJ, Swamy VC (1983) Calcium antagonists – some chemical-pharmacologic aspects Circ Res 52 (Suppl 1): 17–28
Trost BN, Weidmann P (1984) Effects of nitrendipine and other calcium antagonists on glucose metabolism in man J Cardiovasc Pharmacol 6 (Suppl 7): 986–995
Trost BN (1985) Nitrendipin in der Klinik Fortschr Med 103 (Suppl 7): 13
Uehara S, Handa H, Hirayama A (1986) Effects of the calcium antagonist nifedipine on thromboxane B2 level and aggregation in hypertensive patients Drug Res 36 (II): 1687–1689
Wagner K, Neumayer HH (1985) Prevention of delayed graft function in cadaver kidney transplants by diltiazem Lancet 1:1355–1356
Wagner K, Albrecht S, Neumayer HH (1986) Protektiver Einfluß des Calciumantagonisten Diltiazem auf das akute Nierenversagen nach Nierentransplantation Dtsch med Wschr 111: 1363–1367

3.4 Atherosclerosis – Antiatherogenic Effects

Atherosclerotic vessels contain calcium carbonate, and thus calcium. The role of calcium in the pathogenesis of atherosclerosis thus called for intensive further investigation. Fleckenstein et al. (1984) showed that the calcium content of the coronary arteries and aorta continuously increases with advancing age, and that in octo- and nonagenarians is 20–100 times higher than in the first decade of life. Among smokers and diabetics higher calcium levels were found in the vessels than in groups of the same age without diabetes or nicotine abuse. A conceptual pathophysiological model (Henry 1985; Hugenholtz et al. 1986) proceeds from the proposition that with advancing age and a raised serum cholesterol level, the composition of the cell membrane changes (including an increase in cholesterol) with a consequent cellular uptake of calcium. The raised intracellular calcium concentration produces, as a secondary effect, an increased vasomotor tone, thereby narrowing the vascular lumen. Further factors also favour the progression of atherosclerosis: possible cell damage through high intracellular concentrations of calcium; endothelial defects through nicotine abuse or hypertension; local clot formation.

Numerous animal trials have been undertaken to determine the role of calcium antagonists in preventing atherosclerotic wall changes. In such studies rapid progression of atherosclerosis was induced by either a high-cholesterol diet (Henry and Bentley 1981) or vitamin D_3 or dihydrotachysterol (AT 10) administration (Frey et al. 1980). Both these investigations showed that prophylactically administered calcium antagonists reduced progression of atherosclerosis, thus supporting the hypothesis of an antiatherogenic effect for these drugs. Nifedipine, verapamil, diltiazem and flunarizine were used in this context (review by Henry 1985). Not all animal trials have shown a positive effect with calcium antagonists; Tilton (1985) found, in Watanabe rabbits on a high-cholesterol diet and verapamil therapy, unchanged formation of atherosclerotic plaques in the aorta, compared with the control group. In all the investigations cited here, calcium antagonists showed no influence on the serum cholesterol level (Henry 1985).

Although a thoroughly plausible pathogenetic model and the predominantly positive outcome of the animal trials support the hypothesis of an antiatherogenic effect for calcium antagonists, there has as yet been no definitive demonstration of their efficacy in humans. First results of the INTACT study (International Nifedipine Trial on Antiatherosclerotic Therapy) have recently been presented (Lichtlen 1989, personal communication). In this study the effect of 80 mg nifedipine daily on atherosclerotic changes in the coronary arteries was analysed (two coronary arteriographic studies; when patients were randomised between active treatment and placebo, and after 3 years of treatment – computer controlled evaluation technique, Hugenholtz et al. 1986). After 3 years a significant reduction of new lesions was reported for the nifedipine group, whereas progression of preexisting stenoses was not affected. A final analysis of these data is pending. A similar study is being carried out with verapamil (360 mg daily) in patients who have undergone coronary artery bypass surgery. In this study control arteriography is projected after 1 and 3 years. In an open, nonrandomized study involving 26 patients, Kober and Schneider (1988) found, with administration of 120–480 mg verapamil, as determined by coronary angiography, "slower progression, a higher regression rate, and rarer development of new stenoses".

In Summary. At the present time, there is still no definitive proof of a specific antiatherogenic action for calcium antagonists in humans. Calcium antagonists can, however, favourably influence subcomponents of atherogenesis, for instance by lowering an elevated blood pressure, preventing platelet aggregation, and resolving or preventing vasospasm.

References

Fleckenstein A, Frey M, Fleckenstein-Grün G (1984) Zwanzig Jahre Calcium-Antagonismus aus physiologischer und pathophysiologischer Sicht, Rückblick und Ausblick In: Althaus U, Burckhardt D, Vogt E (eds) Calcium-Antagonismus Internat. Symposium on Calcium-Antagonism Universimed Frankfurt/Main pp 4–29

Frey M, Keidel J, Fleckenstein A (1980) Verhütung experimenteller Gefäßverkalkungen (Mönckebergs Typ der Arteriosklerose) durch Calcium-Antagonisten bei Ratten In: Fleckenstein A, Roskam H (eds) Calcium-Antagonismus Springer Berlin pp 258–264

Henry PD, Bentley KL (1981) Suppression of atherogenesis in cholesterol-fed rabbit treated with nifedipine J Clin Invest 68: 1366–1369

Henry PD (1985) Atherosclerosis, calcium and calcium antagonists Circulation 72: 456–459

Hugenholtz PG, Lichtlen P, van der Giessen W, Becker AE, Nayler WG, Fleckenstein A, Hülsmann WC (1986) On a possible role for calcium antagonists in atherosclerosis. A personal view Eur Heart J 7: 546–559

Kober G, Schneider W (1988) Kann Verapamil die Progredienz der Koronarsklerose hemmen? In: Bender F, Fleckenstein A (eds) Therapie und Prävention mit Kalziumantagonisten Steinkopff Darmstadt pp 53–59

Tilton GD (1985) Failure of a slow channel calcium antagonist, verapamil to retard atherosclerosis in the Watanabe heritable hyperlipidemic rabbit: An animal model of familial hypercholesterolemia J Am Coll Cardiol 6: 141–144

4 Indications for Calcium Antagonists

Originally developed as "coronary dilators", calcium antagonists have proven themselves to be reliable drugs in the treatment of stable angina pectoris, and they are now also widely used in other forms of coronary heart disease. Although their antihypertensive action was first described as long ago as the early 1960s, it is only in the last few years that calcium antagonists have established themselves worldwide in the treatment of hypertension. They have also been successfully and regularly employed in the treatment of cardiac arrhythmias (mainly supraventricular tachycardia, atrial fibrillation and atrial flutter) and hypertrophic cardiomyopathy. However, in contrast, the use of calcium antagonists as vasodilators in congestive heart failure and aortic regurgitation, and their use in heart surgery, coronary arteriography and coronary angioplasty remain relatively new indications.

The rationale for the employment of this class of drugs in vascular, neurological and other extracardial indications is based mainly on their inhibitory effect on smooth muscle, especially on vascular smooth muscle. Finally, an additional cytoprotective action for calcium antagonists, particularly in the central nervous system, is currently under discussion (see also Chap. 6).

The various indications for calcium antagonists and their status in overall treatment schemes are discussed in subsequent sections. The issues raised by situations in which there is at present inadequate experience or contradictory results are frankly addressed. It is our strong conviction that for many of the newer indications, the initial positive results cannot yet be extrapolated to permit any definitive recommendations for therapy.

Further indications, for which there are in most cases only isolated observations from case histories, are discussed in Sect. 4.11. Possible future indications, suggested at present only by experimental data, are considered in Chap. 6, Perspective.

Naturally, as with any medication, the therapeutic use of calcium antagonists must be based on sound principles of economic and rational medical care. The goals of therapy have to be clearly defined, and its course must be critically reviewed at regular intervals to ensure prompt recognition of therapeutic failure and/or severe side effects; the rationale for any treatment plan must be reconsidered constantly.

Treatment with calcium antagonists aims at an improvement in the quality of life and/or, ideally, at prolongation of life. While the realization of the first of these two aims, i.e. a symptomatic effect, has been sufficiently well established for a range of indications, there are as yet no studies which unequivocally document a life-prolonging effect due to therapy with calcium antagonists (e.g. in primary and secondary prophylaxis of coronary heart disease or in the treatment of hypertension). If such intervention studies aimed at prolonging life (and/or, for example, preventing myocardial infarction) are carried out in the treatment of mainly asymptomatic patients with mild hypertension or those with "silent myocardial ischaemia", the medication administered must have a particularly high safety margin with respect to uncommon, substance-specific side effects; the incidence of severe, potentially lethal side effects should be less than 1 in 1000 patient years. Furthermore, this type of study requires randomized and double-blind trials with sufficiently large numbers of patients (as highlighted, for instance, by the so-called Clofibrate Study; WHO 1980).

4.1 Coronary Heart Disease

4.1.1 Presenting Forms: Coronary Physiology

The degree of coronary heart disease is determined by the extent of stenosing or occluding atherosclerotic changes in the coronary arteries. In some cases, stenoses or occlusions can also be caused by inflammation (panarteritis nodosa, stenoses of coronary ostia in lues), embolism, the formation of muscular bridges, or by vasospasm.

The spectrum of the clinical manifestations of coronary heart disease embraces the various forms of angina pectoris, acute myocardial infarction, cardiac arrhythmias, sudden cardiac death, and slow progressive congestive heart failure without angina pectoris (Fig. 13). To these different symptomatic forms of coronary heart disease can now be added the recently recognized asymptomatic "silent myocardial ischaemia" (Sect. 4.1.6). The important risk factors in coronary heart disease are also shown in Fig. 13.

The natural history of coronary heart disease varies considerably. Thus, a patient may develop, as an initial manifestation of the disease, a stable, exercise-

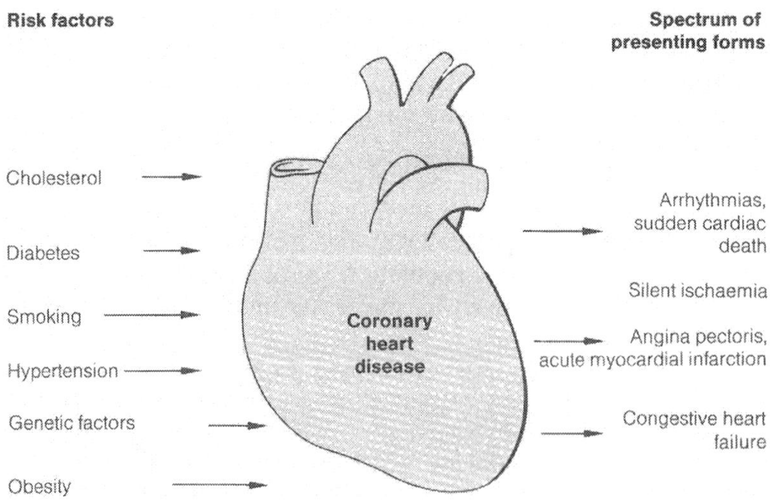

Fig. 13. Risk factors (*left*) and spectrum of presenting forms (*right*) in coronary heart disease

dependent angina pectoris, or may experience without warning acute myocardial infarction and/or sudden cardiac death. It has recently become possible by the technique of Holter ECG monitoring (computer-controlled ST segment analysis) to diagnose "silent myocardial ischaemia". This supplements ergometry and thallium scintigraphy, and may perhaps in future permit treatment of asymptomatic patients with a high risk of myocardial infarction or of sudden cardiac death (Sect. 4.1.6).

In angina pectoris, the most common manifestation of coronary heart disease, three forms can be distinguished: stable angina pectoris, unstable angina pectoris, and, as a special case, Prinzmetal's angina pectoris (vasospastic angina pectoris, also called variant angina pectoris).

Stable angina pectoris, known also as chronic or effort angina, is characterized by an absence of symptoms when the patient is at rest, and by the manifestation of angina as soon as a certain, relatively constant level of exercise is attained, reflecting the patient's coronary reserve. Small fluctuations, especially those of a circadian nature, in this maximal exercise level are possible (Waters et al. 1985).

Angina pectoris is defined as unstable if the intensity and/or frequency of attacks increases (crescendo angina pectoris), i.e. when the symptom-free maximal exercise level decreases, or when the patient experiences pain even at rest. The first attack of angina pectoris, i.e. a "new case", is also labelled as unstable during the first 4 weeks (Cohn and Braunwald 1984). This condition can change into stable angina pectoris, or it can be a harbinger of myocardial infarction. While ECG changes (usually depression of the ST segment) may be present in stable and unstable angina, serum levels of myocardial enzymes (creatine phosphokinase, CK; aspartate aminotransferase, SGOT; lactate dehydrogenase, LDH) are, by definition, not raised.

Prinzmetal's angina pectoris occurs chiefly at rest, without warning, and is explained by spasm in the large epicardial coronary vessels. At least two-thirds of spastically reacting vessels reveal atherosclerotic changes; in the remaining cases the coronary angiogram reveals normal coronary arteries, although flat plaques may, of course, escape detection. It is doubtful whether in entirely "normal" coronary arteries spasm plays a significant role under normal conditions, i.e. without the influence of any vasoconstricting medication (MacAlpin 1980). Attacks of Prinzmetal's angina pectoris are characterized by transmural ischaemia, and consequently elevations of the ST segment can be recorded. ECG changes are, in contrast to acute myocardial infarction, spontaneous or, with appropriate therapy, reversible within a short time. Myocardial enzymes again remain normal.

The vascular lesion present in angina pectoris can vary between two extreme forms, from an advanced fixed concentric stenosis to a normal coronary artery with intermittent spasm, all transitional forms being possible (Fig. 14). In most cases there is a dynamic coronary artery stenosis, i.e. the affected coronary artery shows partial obstruction of the lumen (often eccentric) and a free wall segment which can still react to vasoconstricting and vasodilating changes in vascular smooth muscle tone. As Fig. 15 shows, a fixed, eccentrically situated coronary artery stenosis, depending on the degree of vascular smooth muscle tone of the

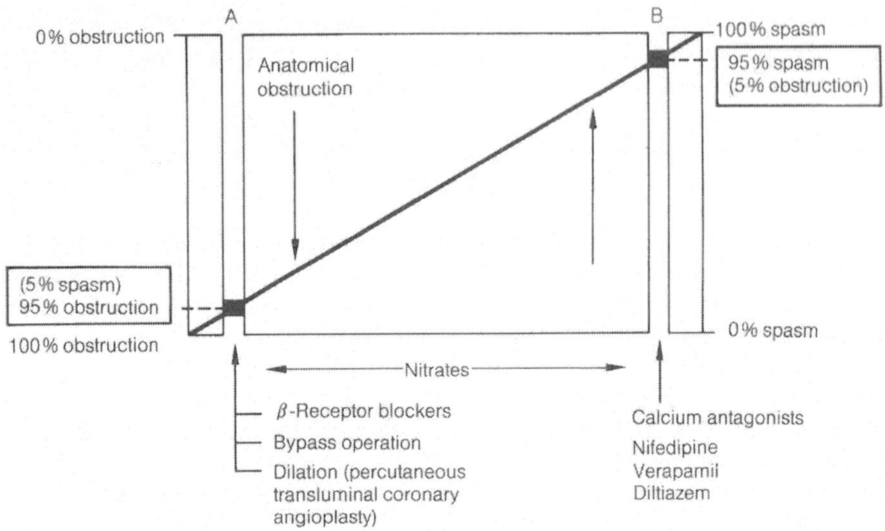

Fig. 14. Transitional forms of angina pectoris between the two extremes of "anatomically fixed coronary artery stenosis" and "coronary spasm in an anatomically normal coronary artery", together with therapeutic principles. (After Hugenholtz et al. 1984)

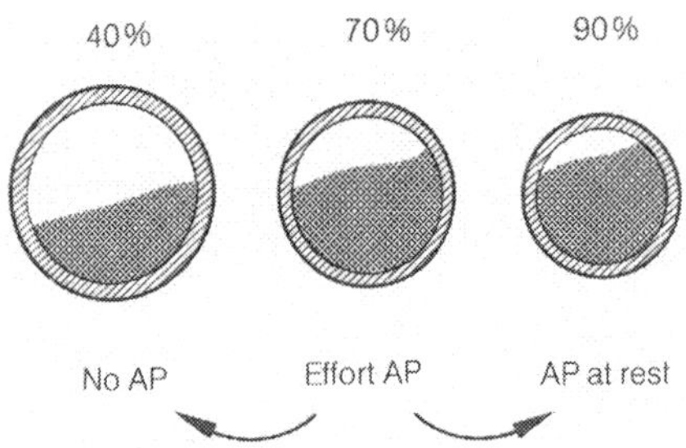

Fig. 15. Significance of the free wall segment for clinical symptomatology (asymptomatic, effort angina, angina at rest). Concept of "dynamic coronary artery stenosis" (anatomic component constant)

free wall segment, can remain asymptomatic in one case, while in another it evokes effort angina pectoris, or even angina pectoris at rest. With growth of the fixed stenosis, very small variations in vascular smooth muscle tone are enough to induce severe coronary artery stenosis, even to the extent of complete occlusion (Sect. 3.1.5).

The three essential factors in myocardial ischaemia influencing the coronary artery – fixed, i.e. atherosclerotic coronary artery stenosis; coronary vasomotor tone or coronary spasm; and platelet aggregation – are summarized in Fig. 16, in relation to the various presenting forms of angina pectoris. Unstable angina pectoris due to reversible myocardial ischaemia is juxtaposed to myocardial infarction with irreversible myocardial ischaemia. Each of the three components is capable of producing myocardial infarction: for example, coronary artery stenosis can, through haemorrhage or thrombosis in some part of an atherosclerotic plaque, lead to complete coronary occlusion; a coronary spasm can persist until irreversible damage develops, and a condition of hypercoagulability (e.g. due to an unspecific infection or to oestrogen therapy) can initiate spontaneous platelet

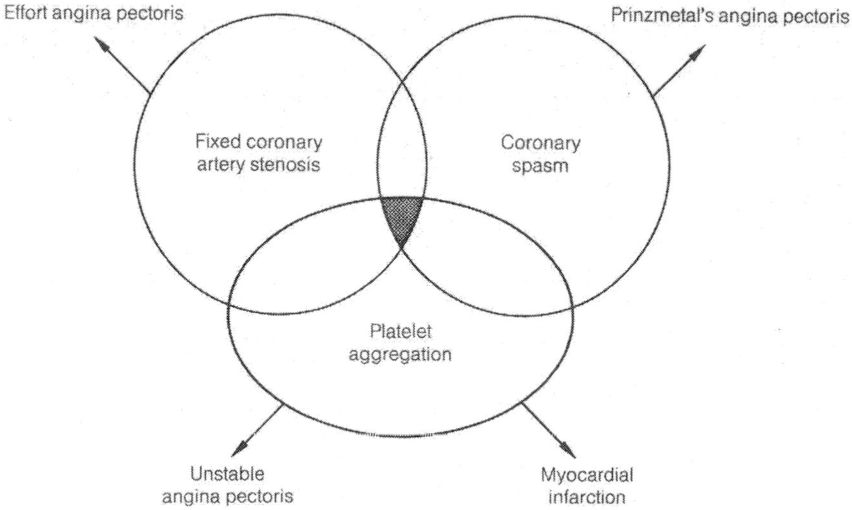

Fig. 16. Presenting forms of coronary heart disease and their determinants

aggregation. This simplified account highlights that all forms of angina pectoris, even myocardial infarction, are the expression of a multifactorial dynamic process. Consequently, any therapeutic approach must aim to influence all three pathogenetic mechanisms.

An appreciation of the aetiology of angina pectoris and the action of calcium antagonists and other antianginal[3] drugs in this condition is based on a few basic principles.

Myocardial oxygen demand is dependent on heart rate, arterial blood pressure, left ventricular pressure, heart size, and contractility (measurable as maximum rate of increase in left ventricular pressure dP/dt_{max}; Lochner 1975). Myocardial oxygen supply is determined by coronary flow and blood oxygen concentration; coronary flow is in turn determined by perfusion pressure (diastolic aortic pressure minus pressure in the right atrium) and by coronary vascular resistance.

Coronary insufficiency is defined as a disturbed balance between myocardial oxygen supply and demand. The reversible form of coronary insufficiency leads to angina pectoris, the irreversible form to acute myocardial infarction. Myocardial oxygen demand increases with exercise owing to increased heart rate and blood pressure. This mechanism underlies stable, effort-dependent angina pectoris in its classical form, since a fixed coronary artery stenosis does not permit an appropriate increase in oxygen supply. If angina pectoris occurs without a change in heart rate or blood pressure, that is, the oxygen demand is unchanged, then the oxygen supply must be disturbed. This results mainly from an increase in coronary vascular resistance with a reduction in coronary blood flow. The raised vascular resistance can be caused by coronary spasm, vasoconstricting stimuli (e.g. release of thromboxane A_2 from platelets), or by inadequate local vasodilation. Angina pectoris at rest and Prinzmetal's angina pectoris are currently

[3] Since it has recently become a practice to distinguish between the symptomatic forms of myocardial ischaemia (angina pectoris, myocardial infarction) and asymptomatic "silent" myocardial ischaemia, the concept of antianginal therapy will in future probably be replaced by that of antiischaemic therapy. The aim of all treatment is to prevent ischaemic episodes

explained mainly in terms of transient elevations in coronary vasomotor tone. As recent results have demonstrated (Chierchia et al. 1984), however, chronic stable angina pectoris can also be characterized by attacks in which there is no rise in heart rate and which are explained in terms of a changed vasomotor tone (Maseri and Chierchia 1981).

In rare cases myocardial oxygen supply may be restricted by arterial hypoxia (e.g. due to disturbances in pulmonary diffusion or to carboxyhaemoglobin formation) or by fairly severe anaemia (reduced oxygen-transporting capacity of the blood); this is known as "secondary coronary insufficiency".

The essential therapeutic aim in angina pectoris is to reduce the frequency of attacks by lowering oxygen demand or by increasing the oxygen supply.

Myocardial oxygen demand diminishes when the heart rate, preload (corresponding to left ventricular end-diastolic volume), afterload (dependent on peripheral vascular resistance or arterial blood pressure, and on the size of the heart), and contractility are reduced. An excessive depression of blood pressure must be avoided, for in this situation coronary perfusion pressure and coronary blood flow are also reduced.

Improvement of myocardial oxygen supply is theoretically possible by elimination of arterial hypoxia or severe anaemia, and by increasing a depressed diastolic aortic pressure, e.g. in aortic regurgitation. More important, however, is to increase coronary blood flow by reducing coronary vascular resistance. The latter can be attained by reduction of vascular smooth muscle tone (coronary vascular resistance; Fig. 17), or of heart rate, contractility and left ventricular end-diastolic pressure (together forming the extravascular, "myocardial" component of coronary vascular resistance; Fig. 17). Changes in vascular smooth muscle

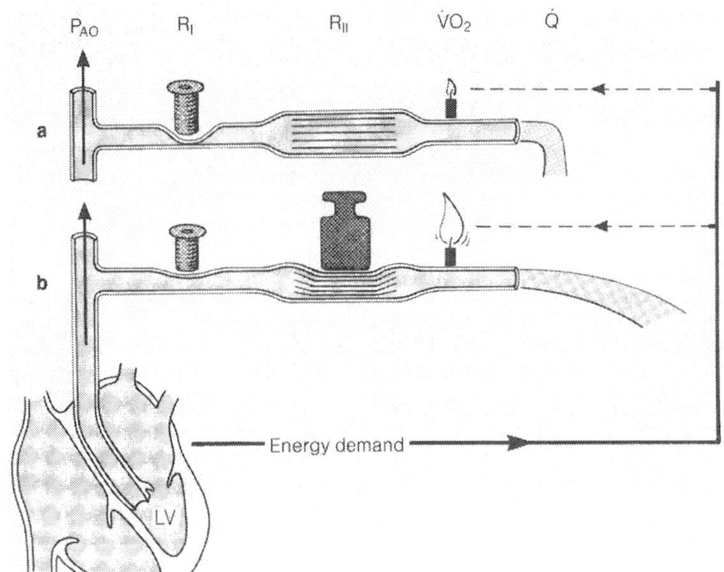

Fig. 17 a, b. Schematic diagram of relationship between coronary vascular resistance (R_1), coronary extravascular ("myocardial") resistance (R_{II}), oxygen demand ($\dot{V}O_2$) and coronary blood flow ($\dot{Q}$) in a normal coronary system at rest **(a)** and during exercise **(b)**. *LV* left ventricle; P_{AO} aortic pressure

Table 10. Haemodynamic effects of antianginal drugs. (After Hugenholtz et al. 1984)

	Heart rate	Preload	After-load	Contrac-tility	Coronary tone	Coronary blood flow
Nitrates	↑	↓	↓=	=	↓=[a]	↑=
β-Receptor blockers	↓	↑=	↑	↓=	↑	=
Calcium antagonists	↑=↓[c]	=	↓	↓=	↓[b]	↑
β-Receptor blockers and calcium anta-gonists	↓=	↓	↓	↓=	↓	↑

[a] Large epicardial coronary arteries only
[b] Large epicardial coronary arteries and coronary arterioles (resistance vessels)
[c] Increase (particularly) with dihydropyridines, decrease with diltiazem
↑ increase; ↑=slight increase; ↓ decrease; ↓=slight decrease; =no change

tone can only occur in the presence of normal coronary arteries or of dynamic coronary artery stenoses. The actions of various antianginal drugs on the determinants of myocardial oxygen demand and supply are summarized in Table 10.

4.1.2 Stable Angina Pectoris

Stable angina pectoris (chronic angina pectoris or effort angina pectoris) is characterized by the patient experiencing typical symptoms under somatic or psychic stress. Further precipitating factors include cold, humid and/or hot environments, heavy meals, fever, hyperthyreosis, anaemia and treatment with drugs that increase cardiac contractility. Until a few years ago, it was accepted doctrine that in stable angina pectoris a fixed coronary arterial stenosis was always present, that maximum oxygen supply was thereby restricted and that onset of pain could develop only with an increase in oxygen consumption (via an increase in heart rate and/or blood pressure). Only more recently has the basic pathogenetic principle of stable angina pectoris been replaced by a multifactorial model (Maseri and Chierchia 1981; Braunwald 1982; Hugenholtz et al. 1984). According to current concepts, coronary blood flow, and thereby myocardial oxygen supply, is determined by fixed coronary artery stenoses, by fluctuations in vasomotor tone, possibly including intercurrent spasm, and also by myocardial wall stress (depending on filling pressure, aortic pressure, ventricular diameter and wall thickness). If the heart rate, as determined in Holter ECGs, is correlated with anginal episodes in patients with stable angina pectoris, some attacks are seen to coincide with an increase in heart rate (Quyyumi et al. 1984), whereas others occur without a change in heart rate, and the triggering mechanism could well be a rise in coronary vascular resistance (due to increased vasomotor tone), resulting in a diminished oxygen supply (Chierchia et al. 1984). This pathogenetic model of stable angina pectoris permits antianginal therapy with drugs which restrict an increase in blood pressure and heart rate and/or reduce coronary vasomotor tone and arterial (peripheral) vascular resistance.

4.1.2.1 Monotherapy with Calcium Antagonists

Today the efficacy of calcium antagonists as monotherapy for stable angina pectoris is unequivocally well documented. Without claiming to be complete, Table 11 provides a summary of worldwide studies using nifedipine, verapamil and diltiazem. Gallopamil (featured mainly in German studies) acts similarly to verapamil. Although available studies, on account of differences in study organization, dosages, trial end-points and patient selection, are only partially comparable, and although not all studies meet the statistical criteria (double-blind approach, randomization, placebo groups, etc.), the results, nevertheless, permit a consensus in evaluation. Calcium antagonists reduce the number of attacks of angina pectoris and thereby also the consumption of nitroglycerin (or of other fast-acting nitrates). In exercise tests (e.g. ergometry) the maximum exercise level rises with calcium antagonists, the occurrence of anginal symptoms is distinctly delayed, and the depression of the ST segment (duration, extent and frequency) is reduced (reviews by Stone et al. 1980; Ellrodt et al. 1980; Subramanian et al. 1982a, 1983; Hopf et al. 1982; Fox et al. 1983; Eichstädt 1984; Hugenholtz 1985; Fach and Becker 1985). The increases in blood pressure and in pulmonary capillary wedge pressure, as measured during exercise, are also reduced (Lydtin et al. 1980).

Although there is considerable variation in the percentages quoted in the studies (e.g. relating to the increase of maximum exercise tolerance, the regression of depression of the ST segment, the decrease in the number of attacks of angina pectoris, etc.), the conclusion can nevertheless be drawn that a dose of 30–60 mg nifedipine, 240–360 mg verapamil, 180–360 mg diltiazem, or 150 mg gallopamil (ideally given as three subdoses) will achieve a good antianginal effect. Obviously, in some cases even smaller doses may produce adequate results (Sect. 2.2, Pharmacokinetics).

Numerous studies have compared calcium antagonists with β-receptor blockers (in most cases propranolol or metoprolol) in stable angina pectoris. No significant differences between the two therapies in terms of symptoms, anginal attacks, exercise tolerance or depression of the ST segment have been reported (Kaltenbach 1975; Stone et al. 1980; Mueller and Chahine 1981; Arnmann and Rydén 1982; Subramanian et al. 1983; Schroeder et al. 1985). Thus, in the selection of a calcium antagonist or another antianginal drug in the treatment of angina pectoris, supplementary indications, contraindications and individual patient tolerance are of crucial importance (Sect. 4.1.2.4, Practical Considerations).

Calcium antagonists are effective, both immediately following a single dose and after administration over days, weeks or even months; in contrast to nitrates (Abrams 1986), neither tolerance nor tachyphylaxis develop (Becker et al. 1975; Stone et al. 1980). A specific withdrawal syndrome, as has been described for β-receptor blockers, has not been observed following discontinuation of calcium antagonists (Frishman et al. 1982; Gottlieb et al. 1984; Wilcox et al. 1986). Only in isolated cases has the abrupt withdrawal of calcium antagonists been associated with the occurrence of acute myocardial infarction (Kozeny et al. 1986; diltiazem) or a clear increase in myocardial ischaemia (symptoms, alterations of the ST segment; Subramanian et al. 1983; diltiazem and verapamil).

Table 11. Calcium antagonists in stable angina pectoris. Monotherapy with verapamil, gallopamil, diltiazem and nifedipine. (A selection of studies on acute effect and long-term therapy. Oral dose unless otherwise stated)

Studies with verapamil		
Acute effect		*Dose*
Hopf et al.	1982	5 mg i.v.
		160 mg
		240 mg
		320 mg
Rouleau et al.	1983	80 mg
		120 mg
Fach and Becker	1985	160 mg
		320 mg
Long-term therapy		
Sandler et al.	1968	3 × 120
Livesley et al.	1973	3 × 120
Fagher et al.	1977	3 × 120
Arnman and Rydén	1982	3 × 120
Frishman et al.	1982	3 × 160
Pine et al.	1982	3 × 120
		3 × 160
Tan et al.	1982	3 × 120
Weiner et al.	1984	4 × 120
Leon	1984	3 × 160
Subramanian	1984	3 × 120
Studies with gallopamil		
Acute effect		
Hopf et al.	1983 a	50 mg
Theisen and Jahrmärker	1983	50 mg
Rettig and Sen	1983	50 mg
Long-term therapy		
Bouzo	1983	3 × 50 mg
Khurmi et al.	1984 a	3 × 50 mg
Kroesen	1983	3 × 50 mg
Mitrovic et al.	1983	3 × 50 mg
Rettig and Sen	1983	3 × 50 mg
Subramanian	1984	3 × 50 mg
Studies with diltiazem		
Acute effect		
Kober et al.	1981	60 mg
Hopf et al.	1982	60 mg
Hossack et al.	1982 a	120 mg
Anderson et al.	1984	120 mg
Long-term therapy		
Hossack et al.	1982 b	4 × 60 mg
Klein et al.	1982	3 × 90 mg
Pool and Seagren	1982	4 × 60 mg
Strauss et al.	1982	3 × 120 mg
Go and Hollenberg	1984	3 × 120 mg
		Dose
Heppe and Sturn	1984	2 × 90 mg
Hossack et al.	1984	3 × 120 mg
Khurmi et al.	1984 b	3 × 120 mg
Subramanian	1984	3 × 120 mg
Weiner et al.	1984	4 × 90 mg
Schroeder et al.	1985	4 × 90 mg
Studies with nifedipine		
Acute effect		
Cherchi et al.	1975	1 mg i.v.
Ekelund et al.	1975	10 mg
Broustet et al.	1980 a	20 mg
Brügmann et al.	1982 a	10 mg
		20 mg
		30 mg
Hopf et al.	1982	10 mg
		20 mg
		1 mg i.v.
		0.1 mg intra-cor.
Brügmann et al.	1983	20 mg SR
Emanuelsson and Holmberg	1983	10 mg
Speccia et al.	1983	20 mg
Fach and Becker	1985	20 mg
Shen et al.	1985	20 mg
Long-term therapy		
Bobba	1975	4 × 10 mg
Alvarado and Pineros	1976	3 × 10 mg
Folle et al.	1976	3 × 10 mg
Dargie et al.	1981	3 × 10 mg
		3 × 20 mg
Mueller and Chahine	1981	3 × 20 mg
Brügmann et al.	1982 b	3 × 10 mg
Fox et al.	1983	3 × 10 mg to 3 × 40 mg
Sherman and Liang	1983	3 × 10 mg
		3 × 20 mg
Deanfield et al.	1984	3 × 20 mg
Subramanian	1984	3 × 20 mg
Study with felodipine		
Acute effect		
Detri et al.	1983	5–10 mg

Further surveys on nifedipine by Ebner and Dünschede (1976); Eichstädt (1984); Hugenholtz (1985); Rudolph et al. (1985). SR, Slow-release

There have as yet been no large-scale comparative studies on the efficacy of the various calcium antagonists (in the same patient groups). Subramanian (1982a, 1984) has attempted direct comparison of various calcium antagonists by adopting identical inclusion criteria, uniform investigation conditions and computer-aided evaluation. With respect to maximum exercise tolerance, this study indicates an advantage for calcium antagonists that reduce the heart rate at rest (verapamil, gallopamil, diltiazem). These substances would thus appear well suited to first-line treatment (as monotherapy) of stable angina pectoris. The respective contraindications (Sect. 5.6) must, of course, be borne in mind. Calcium antagonists of the dihydropyridine group (e.g. nifedipine) are also often used as monotherapy of stable angina pectoris (their positive effect has been shown in numerous studies; Table 11); however, the effect of nifedipine (as monotherapy) appears in general to be weaker, and in individual cases paradoxical reactions (an increase in angina pectoris) have been described (Hopf et al. 1982; Schulz et al. 1983; Fach and Becker 1985). A reflex tachycardia resulting from vasodilation and a significant reduction in coronary perfusion pressure together with a particular coronary morphology (Schulz et al. 1983) have been proposed as a possible explanation for this paradoxical action. In their investigations into the effectiveness of nifedipine, both Rudolph et al. (1985) and Brügmann et al. (1982) recognized a subgroup of so-called nonresponders in whom a dilation of coronary stenoses under nifedipine was not possible, and who were therefore assumed not to respond to nifedipine; in this subgroup even higher doses (up to 30 mg) in the acute test remained without effect. To summarize, in treatment with nifedipine, a failure of antianginal action or an increase in angina pectoris symptoms may occur in between 10% (Subramanian) and 30% (Rudolph) of patients; the treatment regimen for these individuals can either be changed to a calcium antagonist of the verapamil type or to diltiazem, or to combination therapy of a calcium antagonist with a β-receptor blocker.

The authors claim no personal experience concerning the use of other calcium antagonists (prenylamine, fenidiline, perhexiline, bepridil, tiapamil and more recent substances of the dihydropyridine group, such as felodipine, nicardipine and nisoldipine) in the treatment of stable angina pectoris. According to available data, the findings relating to nifedipine may also be extended to the entire dihydropyridine group. There is no consensus in the literature regarding the effects of fenidiline, perhexiline and prenylamine (Stone et al. 1980; Hopf et al. 1982; Eichstädt 1984; Fach and Becker 1985), so that it is not possible to give clear recommendations concerning the use of these substances in the treatment of stable angina pectoris. Bepridil, tiapamil and the newer dihydropyridine derivatives appear to be effective (Detri et al. 1983; Hill et al. 1984; Emanuelson et al. 1984; Schneider et al. 1985; Silke et al. 1985; Keefe et al. 1986).

4.1.2.2 Mechanism of Action

In contrast to the clinical effectiveness of calcium antagonists in stable angina pectoris, the precise mechanism of their action has yet to be unequivocally elucidated (Hugenholtz 1985). In considering the possibilities of positively influencing coronary insufficiency, i.e. the imbalance between oxygen supply and demand,

Table 12. Mechanisms of action of calcium antagonists in stable angina pectoris. (After Braunwald 1982)

1. Improvement of oxygen supply through increase in coronary blood flow a) Dilation of coronary arteries (free wall segment, poststenotic) b) Improved subendocardial perfusion	Coronary action
2. Reduction in oxygen demand a) Decrease in afterload (decrease in peripheral vascular resistance)	Peripheral action
b) Decrease in contractility c) Decrease in heart rate[a]	Myocardial action

[a] Only verapamil group and diltiazem

Braunwald (1982) postulates a multifactorial mechanism of action (Table 12). As a simplification, one may distinguish between a coronary, a peripheral and a myocardial component (Lydtin 1975, in panel discussion).

Calcium antagonists improve the oxygen supply in ischaemic myocardial areas by lowering vascular smooth muscle tone in the area of the free wall segment of atherosclerotic coronary artery stenoses and, parallel to this, by dilating collaterals and submaximally dilated poststenotic vascular regions. Furthermore, the ability of the coronary vascular muscle to react to vasoconstricting stimuli (e.g. exposure to cold) is reduced. The end-result is an increased poststenotic blood flow (Engel and Lichtlen 1981); a diffuse dilation of the coronary arteries is avoided, so that a "steal phenomenon" need not be anticipated. Since fixed coronary artery stenoses cannot be dilated by drugs, and, furthermore, collaterals as well as poststenotic vascular regions under ischaemic conditions are in most cases dilated to the maximum extent, calcium antagonists are assumed to affect coronary arteries only in the presence of dilatable free wall segments in the stenotic region. Rudolph et al. (1985) observed an association between the dilation of stenosis, the increase in poststenotic blood flow and the effectiveness of antianginal treatment with nifedipine. In contrast to these findings, many clinical studies have been unable to confirm any increase in oxygen supply due to calcium antagonist therapy; i.e. there is no increase in the rate–pressure product (blood pressure times heart rate) at the onset of anginal symptoms under exercise conditions (Stone et al. 1980; Mueller and Chahine 1981). If oxygen supply is improved, however, an increase in the attainable ischaemia-free rate-pressure product should be observed, as has been reported by Detri et al. (1983) for felodipine. Calcium antagonists lower peripheral vascular resistance and consequently myocardial afterload, and by this means can reduce the myocardial oxygen consumption. This effect is particularly marked with substances of the dihydropyridine group, but in principle it applies to all calcium antagonists. Simultaneously, systolic wall tension falls, and with it also a myocardial (extravascular) component of coronary vascular resistance (which is independent of coronary artery diameter); thus, for a given level of exercise, not only can oxygen demand be reduced, but oxygen supply can also be increased in an area where supply is critical, and the threshold of angina pectoris can be raised, even without an alteration in the coronary artery diameter. Lowering of peripheral resistance can, in a reflex manner, induce an increase in heart rate and in contractility which diminishes the oxygen-sparing effect. If the direct and indirect effects of a reduction in afterload play a less prominent role compared with direct myocardial effects, as may be assumed for verapamil, gallopamil and diltiazem, oxygen demand is nevertheless still reduced, because these substances either reduce heart rate and contractility or produce only a slighter increase in both under exercise conditions.

Although, for a given substance and for an individual patient, it is impossible to quantify the various components of the antianginal effect, the fact remains that all calcium antagonists lower the rate–pressure product for a given level of exercise. This has been demonstrated in several studies with calcium antagonists in which the patient is subjected to increasing exercise levels, and whereby it has been possible to show a shift to higher degrees of maximum tolerated exercise (i.e. up to the appearance of anginal symptoms and/or ischaemia-typical ECG changes; Stone et al. 1980; Ellrodt et al. 1980; Mueller and Chahine 1981; Rouleau et al. 1983).

Independent of their actions on heart rate and contractility, a direct "antiischaemic effect of calcium antagonists on the heart muscle" due to an effect on myocardial metabolism has also been proposed (Kober et al. 1986).

In Summary. Calcium antagonists, as β-receptor blockers and nitrates, lower oxygen consumption of the (ischaemia-threatened) heart; results to date also indicate an improvement in oxygen supply. An important prerequisite for the latter, however, is the presence of "dynamic coronary artery stenoses". Details concerning the mechanism of action are dealt with in special study designs (Emanuelson and Holmberg 1983; Specchia et al. 1983; Rouleau et al. 1983) and in review studies (Stone et al. 1980; Rudolph et al. 1985; Hugenholtz 1985).

4.1.2.3 Combination Therapy with Calcium Antagonists

β-Receptor blockers, calcium antagonists and nitrates influence heart rate, contractility, wall tension, afterload and coronary blood flow in differing ways (Fig. 18). In many cases, therefore, combination therapy allows further improvement of myocardial oxygen balance or the reduction or elimination of undesirable effects of a particular group of substances. The essential haemodynamic effects of combination therapy are described in detail in Sect. 3.1.6.

Combination therapy with calcium antagonists and β-receptor blockers has been successfully applied in numerous studies (a selection is summarized in Table 13). In most cases a more powerful antianginal effect was seen than in monotherapy involving only one of the two groups of substances. Patients showed a better response to exercise, experienced fewer episodes of angina pectoris, and the extent and duration of the ST segment depression during exercise were reduced (Lydtin et al. 1975; Ekelund and Orö 1976; Winniford and Hillis 1982; Opie 1984; Rudolph et al. 1985). Left ventricular hypokinesia and left ventricular end-systolic volume were reduced equally during exercise (Johnston et al. 1985). In these

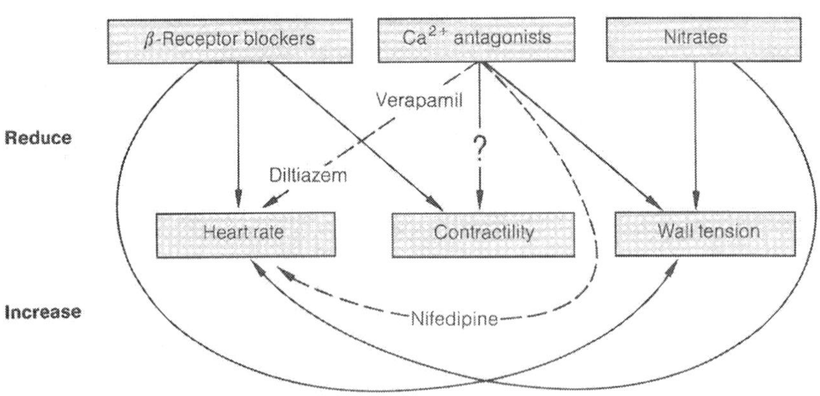

Fig. 18. Synergism and antagonism of various antianginal drugs in combination therapy of coronary heart disease

Table 13. Combination therapy in stable angina pectoris using calcium antagonists and β-receptor blockers. (Dose oral unless otherwise stated; for long-term therapy daily doses are indicated)

	Calcium antagonist	β-Receptor blocker
Acute effect		
Lydtin et al. 1975	Nifedipine i. v.	Practolol i. v.
Broustet et al. 1980 b	Nifedipine 20 mg	Atenolol 200 mg
Pfisterer et al. 1982	Nifedipine 10 mg	Acebutolol 100 mg
Choong et al. 1983	Nifedipine 20 mg	Metroprolol 100 mg
Bassan et al. 1983	Nifedipine 10 mg	Propranolol 60–360 mg daily
	Verapamil 120 mg	Propranolol 60–360 mg daily
Rizzon et al. 1983	Nifedipine 10 mg	Acebutolol 100 mg
Silke et al. 1985	Nicardipine 10 mg i. v.	Metoprolol 10 mg i. v.
Long-term therapy		
Ekelund and Orö 1976	Nifedipine 30 mg	Metoprolol 150 mg or Atenolol 300–400 mg
Kenmure and Scruton 1986	Nifedipine 30 mg	Propranolol 120 mg
Fox et al. 1980	Nifedipine 50 mg	Propranolol 120–480 mg
Lynch et al. 1980	Nifedipine 30–60 mg	Propranolol 240–480 mg
Tweddel et al. 1980	Nifedipine 30 mg	Propranolol individualized
Dargie et al. 1981	Nifedipine 30–60 mg	Propranolol 240–480 mg
Subramanian et al. 1982 b	Verapamil 360 mg	Propranolol 240 mg
Hung et al. 1983	Diltiazem 360 mg	Propranolol 240 mg
Winniford et al. 1983	Verapamil 320–480 mg	Propranolol 160–480 mg
Leon 1984	Verapamil 480 mg	Propranolol 160–320 mg
Johnston et al. 1985	Nifedipine 60 mg	Propranolol 160–480 mg
	Verapamil 360 mg	Propranolol 160–480 mg
	Diltiazem 240 mg	Propranolol 160–480 mg
Kenny et al. 1985	Diltiazem 180 mg	individualized
Strauss et al. 1985	Diltiazem 360 mg	Propranolol individualized
Humen et al. 1986	Diltiazem 240 mg	Propranolol individualized
	Diltiazem 360 mg	

Further surveys by Winniford and Hillis 1982; Ebner et al. 1983; Jee and Opie 1984; Julian 1985; Leon et al. 1985; Johnston et al. 1985; Subramanian 1986; Packer 1989

studies, 30–60 mg nifedipine, 180–360 mg diltiazem, and 240–480 mg verapamil were combined with various doses of a β-receptor blocker (usually propranolol, metoprolol, acebutolol or atenolol; Table 13). While monotherapy with calcium antagonists probably positively influences both myocardial oxygen consumption and oxygen supply, β-receptor blockers produce only a reduction in myocardial oxygen demand. From a pathophysiological standpoint, a combination of nifedipine with a β-receptor blocker appears to offer advantages; the increase in the heart rate at rest under nifedipine (due to autonomic counter-regulation) is weakened by β-receptor blockers, while cardiac output remains virtually constant, in spite of the negative inotropic effect of β-receptor blockers, owing to a reduction in afterload. Coronary blood flow remains unchanged or may even increase slightly. Consequently, despite the wide use of combinations of nifedipine and β-receptor blockers, only isolated negative haemodynamic effects have been reported in the literature, namely hypotension and/or congestive heart failure (reviews by Winniford and Hillis 1982; Packer et al. 1982; Oesterle and Schroeder 1982).

In contrast, however, a combination of verapamil or diltiazem with a β-receptor blocker results in an additive effect of the negative inotropic and negative chronotropic actions of both groups of substances. Especially after administration of verapamil, therefore, one frequently observes signs of congestive heart failure and bradycardiac arrhythmias (AV block II or III, bradyarrhythmia and in isolated cases cardiac arrest; review by Packer 1982; Johnston et al. 1985; Arzneitelegramm 1985). These negative effects are observed principally in patients with reduced left ventricular function and/or disturbances of the pacemaker–conduction system. Subramanian (1986) therefore recommends that, in the presence of congestive heart failure, of an ejection fraction less than 30% and/or of a systolic blood pressure below 100 mmHg, a combination of verapamil and a β-receptor blocker should not be considered. In general, the literature advises against parenteral administration of verapamil to patients who have been previously treated with propranolol or another β-receptor blocker, or vice versa.

Because of the shorter period of observation and the smaller number of patients investigated, results with diltiazem do not as yet permit any definitive assessment; but severe side effects seem to occur less frequently than with verapamil (Strauss et al. 1985; Gibson et al. 1986).

To date only a few researchers have directly compared the effectiveness of the various combinations (verapamil or diltiazem or nifedipine plus a β-receptor blocker; Bassan et al. 1983; Johnston et al. 1985). Such comparisons have revealed no significant differences among the various patient groups. Individual patients, however, seem by both subjective and objective criteria to show a better response with a particular combination. In so far as any trend is discernible from the few studies and their short periods of investigation, a combination of verapamil and a β-receptor blocker would appear to be most effective, although this therapy is most frequently associated with side effects (Leon et al. 1985; Johnston et al. 1985). The combinations nifedipine plus a β-receptor blocker and diltiazem plus a β-receptor blocker are also effective and, furthermore, better tolerated (Leon et al. 1985; Johnston et al. 1985; Julian 1985). In patients with reduced left ventricular function or bradycardiac arrhythmias, the combination nifedipine plus a β-receptor blocker is to be preferred; caution is, however, in this case already necessary in monotherapy with a β-receptor blocker. Although the importance of β-receptor blocker induced coronary spasm is discussed controversially in the literature (Kupper 1984; Cieslinski et al. 1987), it stimulates additional support for combination therapy in angina pectoris.

In Summary. The effectiveness of combination therapy with calcium antagonists and β-receptor blockers in stable angina pectoris is well established (Table 13). In general, treatment is well tolerated, although in patients with reduced left ventricular function, with a history of arrhythmias and, especially in cases where verapamil is prescribed, possible side effects can be anticipated (congestive heart failure, bradycardia, hypotension).

In pathophysiological terms the combination of calcium antagonists and nitrates would appear theoretically well founded owing to the combined effects on preload (nitrates) and afterload (calcium antagonists); but adequate clinical studies on relatively large groups of patients are not yet available (Julian 1985; Rudolph et al. 1985). The few studies available (Schnellbacher et al. 1980; Hopf et al. 1983b; Rizzon et al. 1983) fail to show, even in the acute situation, any significant advantage for combination therapy over monotherapy. On the whole, nitrates appear most suited as an adjunct to basic therapy with calcium antagonists, for prophylactic administration to reduce the occurrence of angina pectoris in special stress situations. In individual cases combination therapy can lead to a more marked decrease in blood pressure and to tachycardia. Verapamil and diltiazem are the agents of choice, because they have a stronger tendency to lower the heart rate, and not, as nifedipine, to reinforce the increased heart rate observed with nitrates (Julian 1985).

Combination therapy (in stable angina pectoris) with calcium antagonists, β-receptor blockers and nitrates has occasionally been discussed in the literature. If calcium antagonists are given in addition to the maximum β-receptor blocker and nitrate therapy (Boden et al. 1985; Nesto et al. 1985; White et al. 1985), an additional antianginal effect with an increase in exercise tolerance and a decrease both in the frequency of angina pectoris and in ST segment depression may occur. In direct comparison (Tolins et al. 1984), however, a "triple combination" of propranolol, nifedipine and isosorbide dinitrate was found to be inferior to a "dual combination" of propranolol and nifedipine in respect to maximum exercise tolerance (Fig. 19). In a similar study, Katz et al. (1982) found no advantage for a combination of nifedipine, propranolol and nitrates as against maximum tolerated doses of propranolol and nitrates. We, too, have in individual cases used a triple combination of a calcium antagonist (usually nifedipine), a β-receptor blocker and isosorbide dinitrate. Such "maximum therapy" is mostly employed when coronary angioplasty or bypass surgery are not possible on account of coronary arteriography findings, or when a patient is too old for, or categorically rejects such diagnostic and therapeutic measures (Sect. 4.1.2.4, Practical Considerations).

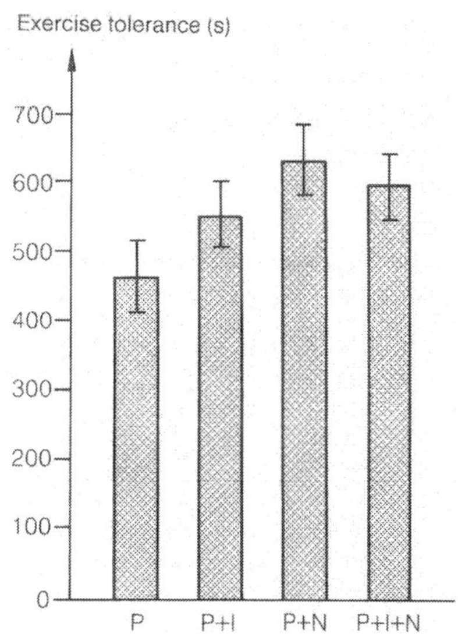

Fig. 19. Exercise tolerance under antianginal combination therapy. "Maximal therapy" (triple therapy) does not coincide with optimal therapy. *P* propranolol, *N* nifedipine, *I* isosorbide dinitrate. (After Tolins et al. 1984)

4.1.2.4 Practical Considerations

In the *prophylactic* treatment of angina pectoris, β-receptor blockers, calcium antagonists, and long-acting nitrates currently compete with one another as first-line drugs; a decision in favour of a particular group of substances is often determined by consideration of possible side effects (Table 14). Therapy of the *acute* angina pectoris *attack* continues to be the domain of the fast-acting nitro compounds (nitroglycerin, isosorbide dinitrate and isosorbide mononitrate in the form of a spray, chewable capsule, or sublingual tablet).

As basic therapy in angina pectoris, we recommend a β-receptor blocker; since the antianginal effect depends upon blockade of β_1 receptors, we favour cardioselective (β_1-receptor-selective) β-receptor blockers such as metoprolol, atenolol or acebutolol. The contraindications (asthma or chronic obstructive pulmonary disease, congestive heart failure, advanced arterial occlusive disease, bradycardiac arrhythmias) are, of course, to be borne in mind. We consider β-receptor blockers to be agents of first choice because of their proven antianginal effect, their additional antiarrhythmic (and perhaps antifibrillatory) action, and their possible primary prophylactic effect. A primary prophylactic effect has, admittedly, been hitherto assured neither for β-receptor blockers nor for any other antianginal drug, but extrapolation is possible from the results attained with β-receptor blockers in secondary prevention following myocardial infarction (reduction of reinfarction, of sudden cardiac death, and of total mortality).

If there are clear contraindications to the use of β-receptor blockers, or if their use produces severe side effects, or if the antianginal effect is insufficient, then we initiate therapy with a calcium antagonist. In these situations, monotherapy with verapamil, gallopamil, or diltiazem appears to provide minor advantages over nifedipine (Subramanian 1984). Nifedipine must be given in individually adjusted doses in order to avoid a sharp decrease in blood pressure or an undesirable reflex increase in heart rate (Deanfield et al. 1983). On combining nifedipine with a β-receptor blocker, doses of both substances can usually be reduced, and, as a rule, the undesirable effects of nifedipine as monotherapy (drop in blood pressure,

Table 14. Main side effects of antianginal drugs. (After Braunwald)

	Hypotension, flushing, headache	Negative inotropy	Bradycardia AV block[a]	Gastro-intestinal symptoms	Broncho-constriction[b]
β-Receptor blockers	○	++	+++	+	+++
Nitrates	+++	○	○	○	○
Molsidomine	+++	○	○	○	○
Nifedipine	+++	○	○	○	○
Verapamil	+	+	++	++	○
Diltiazem	+	+	+	○	○

[a] Mainly in sick sinus syndrome or in conduction disturbances
[b] In obstructive pulmonary disease
○ Appears never or only extremely rarely; + rare, weakly expressed; ++ occasional, medium intensity; +++ frequent, considerable intensity

increase in heart rate) can be avoided. In our experience such a combination is effective and well tolerated in the great majority of patients with stable angina pectoris; it satisfactorily reduces the frequency of attacks and attains an adequate exercise tolerance. With regard to potential side effects (development of congestive heart failure, bradycardiac arrhythmias), we hold this combination to be safer than that of verapamil, gallopamil or diltiazem with a β-receptor blocker. In individual cases, however, we have also used these combinations with success. Our general recommendation of an initial nifedipine plus β-receptor blocker combination does not, of course, exclude the possibility that single groups of patients may respond as well, or even better, to one of the other combinations. This standpoint is supported by the comparative studies so far available on combination therapy for angina pectoris (Sect. 4.1.2.3).

In the case that adequate doses of β-receptor blockers and calcium antagonists in combination do not provide satisfactory control of angina pectoris, one should first ascertain whether the diagnosis of chronic stable angina pectoris still holds or whether an unstable angina pectoris (angina pectoris at rest, crescendo angina pectoris) has developed (Sect. 4.1.3). If stable angina pectoris is confirmed, supplementary prophylactic nitrate therapy can be prescribed. In this situation the dose of nitrates should not be too high (maximum: 60 mg isosorbide dinitrate non-slow-release or 40 mg isosorbide 5-mononitrate daily), and a nitrate-free interval should ideally be built into the regimen (e.g. administration in the morning and at midday, or at midday and in the evening) in order to avoid nitrate tolerance. A so-called triple therapy for stable angina pectoris, consisting of a β-receptor blocker, a calcium antagonist and a nitrate, should, however, remain an exception. On the one hand, patient compliance and physical performance often diminish under this therapy; on the other, in this group of patients coronary arteriography often reveals findings (severe proximally situated coronary artery stenoses) indicating the need for percutaneous transluminal coronary angioplasty (PTCA) or bypass surgery.

In contrast to our recommended and practised procedure, long-term nitrates are still used in many other hospitals and in wider medical practice as primary therapy in stable angina pectoris. The effectiveness of such therapy is without doubt well attested, albeit with the proviso that nitrate tolerance is to be avoided by interposing a nitrate-free interval (or a period of low levels of isosorbide 5-mononitrate). This can be achieved by giving the nitrate in the morning and at midday, or at midday and in the evening, or alternatively administering a morning dose of a slow-release nitrate (e.g. 120 mg isosorbide dinitrate or 40–60 mg isosorbide-5-mononitrate in slow-release form). If transdermal nitrate patches are used, they should be worn no longer than about 12 h (either during the night or during the day).

There is as yet no unequivocal evidence that treatment with β-receptor blockers or calcium antagonists is superior to nitrate therapy. All forms of therapy show an effect on the symptom "angina pectoris", but a clear positive influence on the natural history of coronary heart disease (progression of atherosclerosis – see Sect. 3.4) or a primary protective effect on acute myocardial infarction, i.e. a life-prolonging action, has not to date been demonstrated for any of the three forms of therapy.

Factors favouring treatment with β-receptor blockers or calcium antagonists include the absence of the development of tolerance (hence the possibility of 24-h protection), its excellent patient acceptance, and its actions in concomitant diseases such as arterial hypertension or arrhythmias (in the latter indication dihydropyridine derivatives are ineffective).

Further details, in particular those relating to therapy with β-receptor blockers and nitrates, are to be found in recent reviews (Lohmöller and Lydtin 1984; Lydtin and Trenkwalder 1985; Yusuf et al. 1985; Abrams 1986; Silber 1986; Lydtin 1989).

References

Abrams J (1986) Tolerance to organic nitrates Circulation 74: 1181–1185

Alvarado E, Pineros J (1976) The efficacy of Adalat in angina pectoris patients. Results of a 6-month trial In: Jatene AD, Lichtlen PR (eds) 3rd International Adalat Symposium Excerpta Medica Amsterdam pp 261–267

Anderson JL, Wagner JM, Datz FL, Christian PE, Bray BE, Taylor AT (1984) Comparative effects of diltiazem, propranolol and placebo on exercise performance using radionuclide ventriculography in patients with symptomatic coronary artery disease: results of a double-blind, randomized, crossover study Am Heart J 107: 698–706

Arnman K, Rydén L (1982) Comparison of metoprolol and verapamil in the treatment of angina pectoris Am J Cardiol 49: 821–827

Arznei-Telegramm (1985) Betablocker plus Verapamil Arzneitelegramm 6: 48

Bassan MM, Weiler-Ravell D, Shalev O (1983) Comparison of the antiangial effectiveness of nifedipine, verapamil and isosorbide dinitrate in patients receiving propranolol: a double-blind study Circulation 68: 568–575

Becker HJ, Kaltenbach M, Kober G (1975) Comparison of the effects of Adalat with other substances on myocardial ischemia under loading conditions In: Lochner W, Braasch W, Kroneberg G (eds) 2nd International Adalat Symposium Springer Berlin pp 156–163

Bobba P (1975) Nifedipine, long-acting nitrates and betablockers in the treatment of angina pectoris – a clinical comparative study In: Hashimoto K, Kimura E, Kobayashi T (eds) 1st International Nifedipine „Adalat" Symposium University of Tokyo Press Tokyo pp 190–191

Boden WE, Bough EW, Reichmann MJ, Rich VB, Young PM, Korr KS, Shulman RS (1985) Benefical effects of high-dose diltiazem in patients with persistent effort angina on β-blockers and nitrates: a randomized, double-blind, placebocontrolled cross-over study Circulation 71: 1197–1205

Bouzo H (1983) Gallopamil und Propranolol bei stabiler belastungsinduzierter Angina pectoris Therapiewoche 33: 6465–6468

Braunwald E (1982) Mechanism of action of calcium channel blocking agents N Engl J Med 307: 1618–1627

Broustet JP, Guern P, Cherrier JF, Pic A (1980a) Nifedipine and glyceryl trinitrate in the prevention of exertional angina In: Puech P, Krebs R (eds) 4th International Adalat Symposium Excerpta Medica Asterdam pp 64–72

Broustet JP, Rumeau P, Guern P, Cherrier JF, Picand A, Bonnet J (1980b) Comparison of the combination of nifedipine and atenolol with the combination of nitroglycerine and atenolol in patients with angina pectoris Eur Heart J 1 (Suppl B): 59–64

Brügmann U, Dirschinger J, Blasini R, Rudolph W (1982a) Antiischämische Wirksamkeit unterschiedlicher Nifedipin Dosen Herz 7: 235–242

Brügmann U, Blasini R, Goebel G, Mannes A, Dirschinger J, Rudolph W (1982b) Unterschiedliches Ansprechen von Patienten mit stabiler Belastungs-Angina-pectoris auf Nifedipin Herz 7: 117–125

Brügmann U, Blasini R, Rudolph W (1983) Antiischämische Wirkung von Nifedipin in Retard-Form Herz 8: 206–210

Cherchi A, Fonzo R, Bina M (1975) Influence of nifedipine on the effort tolerance test in angina patients In: Hashimoto K, Kimura E, Kobayashi T (eds) 1st International Nifedipine „Adalat" Symposium University of Tokyo Press Tokyo pp 85–95

Chierchia S, Gallino A, Smith G, Deanfield J, Morgan M, Croom M, Maseri A (1984) Role of heart rate in pathophysiology of chronic stable angina Lancet 2: 1353–1357

Choong CYP, Roubin GS, Shen WF, Harris JP, Kelly DT, (1983) Hemodynamic basis of improved exercise capacity with addition of metoprolol to nifedipine in patients with exercise angina Circulation 68 (Suppl III) III–53

Cieslinski G, Hopf R, Kober G, Kaltenbach M (1987) Können β-Rezeptorenblocker Koronarspasmen auslösen? Dtsch Med Wschr 112: 593–596

Cohn PF, Braunwald E (1984) Chronic ischemic heart disease In: Braunwald E (ed) Heart disease – a textbook of cardiovascular medicine Saunders Philadelphia pp 1334–1383

Dargie HJ, Lynch PG, Krikler DM, Harris L, Krikler S (1981) Nifedipine and propranolol: a beneficial drug interaction Am J Med 71: 676–682

Deanfield J, Wright C, Fox K (1983) Treatment of angina pectoris with nifedipine: importance of dose titration Br Med J 286: 1467

Deanfield J, Wright C, Krikler S, Ribeiro P, Fox K (1984) Cigarette smoking and the treatment of angina with propranolol, atenolol, and nifedipine N Engl J Med 310: 951–954

Detry JR, DeCoster PM, Renkin J (1983) Hemodynamic effects of felodipine at rest and during exercise in exertional angina pectoris Am J Cardiol 52: 453–457

Ebner F, Dünschede HB (1976) Heamodynamics, therapeutic mechanism of action and clinical findings of Adalat use based on worldwide clinical trials In: Jatene AD, Lichtlen PR (eds) 3rd International Adalat Symposium Excerpta Medica Amsterdam pp 283–300

Ebner F, Leisten L, Lejeune P, Donath M, Rämsch KD (1983) Administration of nifedipine to patients treated with digitalis-glycosides, anti-diabetic agents and betablockers In: Kaltenbach M, Neufeld HN (eds) 5th International Adalat Symposium Excerpta Medica Amsterdam pp 328–338

Eichstädt H (1984) Calcium Antagonisten In: Roskamm H (ed) Handbuch der inneren Medizin Bd IX/3 Springer Berlin pp 1069–1113

Ekelund LG, Atterhög JH, Melin AL (1975) Effect of nifedipine on exercise tolerance in patients with angina pectoris In: Hashimoto K, Kimura E, Kobayashi T (eds) 1st International Nifedipine „Adalat" Symposium University of Tokyo Press Tokyo pp 144–149

Ekelund LG, Orö L (1976) Antianginal efficiency of Adalat with and without a beta-blocker. A subacute study with exercise tests In: Jatene AD, Lichtlen PR (eds) 3rd International Adalat Symposium Excerpta Medica Amsterdam pp 218–225

Ellrodt G, Chew CYC, Singh BN (1980) Therapeutic implications of slow-channel blockade in cardiocirculatory disorders Circulation 62: 669–679

Emanuelsson H, Holmberg S (1983) Mechanisms of angina relief after nifedipine: a hemodynamic and myocardial metabolic study Circulation 68: 124–130

Emanuelsson H, Hjalmarson A, Holmberg S, Waagstein F (1984) Effects of felodipine on systemic and coronary haemodynamics in patients with angina pectoris Eur Heart J 5: 308–316

Engel HJ, Lichtlen PR (1981) Benefical enhancement of coronary blood flow by nifedipine Am J Med 71: 658–666

Fach WA, Becker HJ (1985) Kalzium-Antagonisten in der Therapie der stabilen Angina pectoris Herz und Gefäße 5: 347–358

Fagher B, Svenson SE, Persson S (1977) Double-blind comparison of verapamil and propranolol in the treatment of angina pectoris Postgrad Med J 53: 61–65

Folle LE, Ortiz A, Artucio R, Dighiero J (1976) Efficacy of Adalat in angina pectoris patients in a controlled clinical trial compared with placebo In: Jatene AD, Lichtlen PR (eds) 3rd International Adalat Symposium Excerpta Medica Amsterdam pp 200–205

Fox K, Jonathan A, Selwyn AP (1980) Combined high-dosage administration of nifedipine and propranolol in patients with angina pectoris In: Puech P, Krebs R (eds) 4th International Adalat Symposium Excerpta Medica Amsterdam pp 147–153

Fox KM, Deanfield J, Selwyn A, Krikler S, Wright C (1983) Treatment of chronic stable angina pectoris with nifedipine In: Kaltenbach M, Neufeld HN (eds) 5th International Adalat Symposium Excerpta Medica Amsterdam pp 197–204

Frishman WH, Klein NA, Strom JA, Willens H, LeJemtel TH, Jentzer J, Siegel L, Klein P, Kirschen N, Silverman R, Pollack S, Doyle R, Kirsten E, Sonnenblick EH (1982) Superiority of verapamil to propranolol in stable angina pectoris: A double-blind, randomized crossover trial Circulation 65 (Suppl I): 51–59

Gibson RS, Boden WE, Theroux P, Strauss HD, Pratt CM, Gheorghiade M, Capone RJ, Crawford MH, Schlant RC, Kleiger RE, Young PM, Schechtman K, Perryman B, Roberts R, Diltiazem Reinfarction Study Group (1986) Diltiazem and reinfarction in patients with non-Q-wave myocardial infarction N Engl J Med 315: 423–429

Go Jr M, Hollenberg M (1984) Improved efficacy of high-dose versus medium and low-dose diltiazem therapy for chronic stable angina pectoris Am J Cardiol 53: 669

Gottlieb SO, Ouyang P, Achuff SC, Baughman KL, Trail TA, Mellits ED, Weisfeldt ML, Gerstenblith G (1984) Acute nifedipine withdrawal: consequences of preoperative and late cessation of therapy in patients with prior unstable angina J Am Coll Cardiol 4: 382–388

Heepe W, Sturn B (1984) Stabile Belastungs-Angina Münch med Wschr 126: 908–911

Hill JA, O'Brian JT, Alpert JS, Gore JM, Zusman RM, Christensen D, Boucher CA, Vetrovec G, Borer JS, Friedman C, Mack R, Conti CR, Pepine CJ (1985) Effect of bepridil in patients with chronic stable angina: results of a multicenter trial Circulation 71: 98–103

Hopf R, Becker HJ, Kober G, Dowinsky S, Kaltenbach M (1982) Therapie der Angina pectoris mit Calcium-Antagonisten Herz 7: 221–234

Hopf R, Drews H, Kaltenbach M (1983a) Die antianginöse Wirkung von Gallopamil im Vergleich zu Nifedipin In: Kaltenbach M, Hopf R (eds) Gallopamil Springer Berlin pp 127–136

Hopf R, Pietruska M, Downisky S, Kaltenbach M (1983b) Combined administration of various doses of nifedipine and isosorbide dinitrate in patients with angina pectoris In: Kaltenbach M, Neufeld HN (eds) 5th International Adalat Symposium Excerpta Medica Amsterdam pp 209–216

Hopf R, Drews H, Kaltenbach M (1984) Die antianginöse Wirkung von Gallopamil im Vergleich mit einem anderen Calciumantagonisten und Placebo Z Kardiol 73:578–585

Hossack KF, Bruce RA, Ritterman JB, Kusumi F, Trimble S (1982a) Divergent effects of diltiazem in patients with exertional angina Am J Cardiol 49: 538–546

Hossack KF, Pool PE, Steele P, Crawford MH, DeMaria AN, Cohen LS, Ports TA (1982b) Efficacy of diltiazem in angina on effort: A multicenter trial Am J Cardiol 49: 567–572

Hossack KF, Kannagi T, Day B, Bruce RA (1984) Long-term study of high-dose diltiazem in chronic stable exertional angina Am Heart J 107: 1215–1220

Hugenholtz PG, Michels R, Serruys PW, Simoons ML (1984) Behandlung der instabilen Angina pectoris mit CalciumAntagonisten In: Althaus U, Burckhardt D, Vogt E (eds) Calcium-Antagonismus Internat. Symposium on Calcium-Antagonism Universimed Frankfurt/Main pp 138–156

Hugenholtz PG (1985) Calcium Antagonists In: Abshagen U (ed) Clinical pharmacology of antianginal drugs. Springer Berlin pp 459–548

Humen DP, O'Brien P, Purves P, Johnson D, Kostuk WJ, (1986) Effort angina with adequate beta-receptor blockade: comparison with diltiazem alone and in combination J Am Coll Cardiol 7: 329–335

Hung J, Lamb IH, Conally SJ, Jutzy KR, Goris ML, Schroeder JS (1983) The effect of diltiazem and propranolol, alone and in combination, on exercise performance and left ventricular function in patients with stable affort angina: a double-blind, randomized, and placebo-controlled study Circulation 68: 560–567

Jee LD, Opie LH (1984) Nifedipine for hypertension and angina pectoris: interactions during combination therapy In: Opie LH (ed) Calcium antagonists and cardiovascular disease Raven Press New York pp 339–346

Johnston DL, Lesoway R, Humen DP, Kostuk WJ (1985) Clinical and hemodynamic evaluation of propranolol in combination with verapamil, nifedipine and diltiazem in exertional angina pectoris: a placebo-controlled, double blind, randomized, crossover study Am J Cardiol 55: 680–687

Julian DG (1985) Comparisons and combinations in anti-anginal therapy Eur Heart J 6 (Suppl A): 37–45

Kaltenbach M (1975) Assessment of antianginal substances by means of ST depression in the exercise EKG In: Hashimoto K, Kimura E, Kobayashi T (eds) 1st International Nifedipine "Adalat" Symposium University of Tokyo Press Tokyo pp 126–135
Katz RI, Weintraub WS, Bodenheimer MM (1982) Failure of low dose nifedipine to improve exercise tolerance in stable angina Am J Cardiol 49: 895
O'Keefe JC, Maltz MB, Butrous GS, Camm AJ (1986) Efficacy of incremental doses of tiapamil on exercise performance in patients with chronic stable angina pectoris Eur Heart J 7: 843–847
Kenmure ACF, Scruton JH (1976) A double-blind controlled trial of the antianginal efficacy of nifedipine compared with propranolol In: Jatene AD, Lichtlen PR (eds) 3rd International Adalat Symposium Excerpta Medica Amsterdam pp 268–271
Kenny J, Daly K, Bergman G, Kerkez S, Jewitt DE (1985) Beneficial effects of diltiazem combined with beta blockade in angina pectoris Eur Heart J 6: 418–423
Khurmi NS, O'Hara MJ, Bowles J, Subramanian VB, Raftery EB (1984a) Randomized double-blind comparison of gallopamil and propranolol in stable angina pectoris Am J Cardiol 53: 684–688
Khurmi NS, Bowles MJ, O'Hara MJ, Subramanian B, Raftery EB (1984b) Long-term efficacy of diltiazem assessed with multistage graded exercise tests in patients with chronic stable angina pectoris Am J Cardiol 54: 738–743
Klein W, Brandt D, Fluch N, Maurer E (1982) Therapie der stabilen Belastungsangina mit Calcium-Antagonisten – ein Vergleich von Diltiazem und Nifedipin Z Kardiol 71: 398–405
Kober G, Berlad T, Hopf R, Kaltenbach M (1981) Die Wirkung von Diltiazem und Nifedipin auf ST-Senkung und Herzfrequenz im Belastungs-EKG bei Patienten mit koronarer Herzerkrankung Z Kardiol 70: 59–65
Kober G, Kästner R, Hopf R, Kaltenbach M (1986) Die direkte myokardiale antiischämische Wirkung von Diltiazem beim Menschen Z Kardiol 75: 386–393
Kozeny GA, Ragona BP, Bansal VK, Hurley RM, Dixon DW, Vertuno LL, Hano JE (1986) Myocardial infarction with normal results of coronary angiography following diltiazem withdrawal Am J Med 80: 1184–1186
Kroesen P (1983) Vergleich der antianginösen Wirksamkeit von Gallopamil zu Isosorbiddinitrat bei Langzeitmedikation Med Welt 34: 1437–1440
Kupper W (1984) Betarezeptorenblocker und Koronarspasmus In: Lohmann FW (ed) Die klinische Bedeutung der Beta-1 – Selektivität Verlag für angewandte Wissenschaften pp 99–112
Leon MB (1984) Verapamil alone and combined with propranolol in treating patients with chronic stable angina pectoris In: Althaus U, Burckhardt D, Vogt E (eds) Calcium-Antagonismus – Internat. Symposium on Calcium-Antagonism Universimed Frankfurt/Main pp 181–189
Leon MB, Rosing DR, Bonow RO, Epstein SE (1985) Combination therapy with calcium-channel blockers and beta blockers for chronic stable angina pectoris Am J Cardiol 55: 69 B–80 B
Livesley B, Catley PF, Campbell RC, Oram S (1973) Doubleblind evaluation of verapamil, propranolol, and isosorbide dinitrate against a placebo in the treatment of angina pectoris Br Med J 1: 375–378
Lochner W (1975) Present basis of coronary therapy In: Lochner W, Braasch W, Kroneberg G (eds) 2nd International Adalat Symposium Springer Berlin pp 2–10
Lohmöller G, Lydtin H (1984) Betarezeptorenblocker In: Roskamm H (ed) Handbuch der inneren Medizin, Bd IX/3 Koronarerkrankungen. Springer Berlin pp 1007–1067
Lydtin H, Lohmöller G, Lohmöller R, Walter I (1975) Hemodynamic studies on nifedipine in man In: Hashimoto K, Kimura E, Kobayashi T (eds) 1st International Nifedipine „Adalat" Symposium University of Tokyo Press Tokyo pp 97–106
Lydtin H, Schierl W, Lohmöller G (1980) Exercise pulmonary wedge pressure after acute and chronic administration of nifedipine in ischemic heart disease In: Puech P, Krebs R (eds) Proceedings of the 4th International Adalat Symposium Excerpta Medica Amsterdam pp 249–254
Lydtin H, Trenkwalder P (1985) Gibt es eine Differentialtherapie mit Beta-Rezeptorenblockern? Herz und Gefäße 5: 68–76
Lydtin H (1989) Konservative Therapie der stabilen und instabilen Angina pectoris In:

Jahrmärker H, Theisen K, Holzgartner H (eds) Das kardiale Risiko AMV München pp 75–86
Lynch P, Dargie H, Krikler S, Krikler D (1980) Objective assessment of antianginal treatment: a double-blind comparsion of propranolol, nifedipine and their combination Br Med J 281: 184–187
MacAlpin RN (1980) Contribution of dynamic vascular wall thickening to luminal narrowing during coronary arterial constriction Circulation 61: 296–301
Maseri A, Chierchia S (1981) A new rationale for the clinical approach to the patient with angina pectoris Am J Med 71: 639–644
Meyer J (1983) Der Einsatz von Calciumantagonisten in der Therapie der koronaren Herzkrankheit Internist 24: 408–414
Mitrovíc V, Niemelä L, Neuss H, Schlepper M (1983) Zur antianginösen Wirkung des Kalziumantagonisten Gallopamil In: Kaltenbach M, Hopf R (eds) Gallopamil Springer Berlin pp 109–115
Mueller HS, Chahine RA (1981) Interim report of multicenter double-blind, placebo-controlled studies of nifedipine in chronic stable angina Am J Med 71: 645–657
Nesto RW, White HD, Ganz P, Koslowski J, Wynne J, Holman BL, Antman E (1985) Addition of nifedipine to maximal betablocker-nitrate-therapy: effects on exercise capacity and global left ventricular performance at rest and during exercise Am J Cardiol 55: 3E
Oesterle SN, Schroeder JS (1982) Editoral: calcium-entry blockade, beta-adrenergic blockade and the reflex control of circulation Circulation 65: 669–670
Opie LH (1984) Angina pectoris In: Opie LH (ed) The Heart Grune & Stratton London pp 338–350
Packer M (1989) Combined beta-adrenergic and calcium-entry blockade in angina pectoris N Engl J Med 320: 709–718
Packer M, Meller J, Medina N, Yushak M, Smith H, Holt J, Guererro J, Todd GD, McAllister G, Gorlin R (1982) Hemodynamic consequences of combined beta-adrenergic and slow calcium channel blockade in man Circulation 65: 660–668
Panel discussion (1975) In: Hashimoto K, Kimura E, Kobayashi T (eds) 1st International Nifedipine „Adalat" Symposium University of Tokyo Press Tokyo pp 299–304
Pfisterer M, Müller-Brand J, Burkart F (1982) Combined acebutolol/nifedipine therapy in patients with chronic coronary artery disease: additional improvement of ischemia induced left ventricular dysfunction Am J Cardiol 49: 1259–1265
Pine MB, Citron PD, Bailly DJ, Butman S, Plasencia GD, Landa DW, Wong RK (1982) Verapamil versus placebo in relieving stable angina pectoris Circulation 65: 17–22
Pool PE, Seagren SC (1982) Long-term efficacy of diltiazem in chronic stable angina associated with atherosclerosis: effect on treadmill exercise Am J Cardiol 49: 573–577
Quyyumi AA, Wright CA, Mockus LJ, Fox KM (1984) Mechanisms of nocturnal angina pectoris: importance of increased myocardial oxygen demand in patients with severe coronary artery disease Lancet 1: 1207–1209
Rettig G, Sen S (1983) Akut- und Langzeiteffekte von Gallopamil bei Patienten mit stabiler Angina pectoris In: Kaltenbach M, Hopf R (eds) Gallopamil Springer Berlin pp 141–147
Rizzon P, Scrutinio D, De Toma L, Mangini S, Lagioia R, Accettura D, Mitrotti S (1983) Effects of nifedipine, isosorbide dinitrate, and acebutolol alone and in combination on exercise tolerance in patients with angina pectoris In: Kaltenbach M, Neufeld HN (eds) 5th International Adalat Symposium Excerpta Medica Amsterdam pp 218–231
Rouleau JL, Chatterjee K, Ports TA, Doyle MB, Hiramatsu B, Parmley WW (1983) Mechanism of relief of pacing-induced angina with oral verapamil: reduced oxygen demand Circulation 67: 94–100
Rudolph W, Dirschinger J, Kraus F, Hall D (1985) Behandlung der belastungsinduzierten Angina pectoris mit Nifedipin Med Klin 80: 548–554
Sandler G, Clayton GA, Thornicroft SG (1968) Clinical evaluation of verapamil in angina pectoris Br Med J 3: 224–227
Schneider W, Mehlhorn C, Kaltenbach M, Bussmann WD (1985) Die antianginöse Wirksamkeit des Calciumantagonisten Bepridil bei der stabilen Angina pectoris Z Kardiol 74: 341–347
Schnellbacher K, Kalusche D, Roskamm H (1980) Hämodynamik während belastungsinduzierter Angina pectoris nach einmaliger Gabe von Fendilin und Nifedipin In: Fleckenstein A, Roskamm H (eds) Calcium-Antagonismus Springer Berlin pp 314–317

Schroeder JS, Hung J, Lamb IH, Connolly SJ, Jutzy KR, Goris ML (1985) The effect of diltiazem and propranolol, alone and in combination, on exercise performance and left ventricular function in patients with stable exertional angina In: Just H, Schroeder JS (eds) Advances in clinical applications of calcium antagonist drugs Excerpta Medica Amsterdam pp 115–123

Schulz W, Jost S, Kaltenbach M, Kober G (1983) Antianginöse Wirksamkeit des Kalziumantagonisten Nifedipin in Abhängigkeit vom Koronarbefall Z Kardiol 72: 657–664

Shen WF, Roubin GS, Choong CYP, Hutton BF, Harris PJ, Fletcher PJ, Kelly DT (1985) Left ventricular response to exercise in coronary artery disease: relation to myocardial ischaemia and effects of nifedipine Eur Heart J 6: 1025–1031

Sherman LG, Liang CS (1983) Nifedipine in chronic stable angina: a double-blind placebo-controlled crossover trial Am J Cardiol 51: 706–711

Silber S (1986) Nitrat-Therapie Hohe oder niedrige Plasmaspiegel? Retardierte oder unretardierte Präparate? Konsequenzen des fehlenden 24–Stundenschutzes Herz und Gefäße 6: 372–380

Silke B, Verma SP, Frais MA, Reynolds G, Jackson N, Taylor SH (1985) Haemodynamic analysis of the effects of nicardipine and metoprolol alone and in combination in coronary artery disease Eur Heart J 6: 930–938

Specchia G, De Servi S, Falcone C, Angoli L, Gavazzi A, Bramucci E, Mussini A, Ferrario M, Salerno J, Montemartini C (1983) Effects of nifedipine on coronary hemodynamic findings during exercise in patients with stable exertional angina Circulation 68: 1035–1043

Stone PH, Antman EM, Muller JE, Braunwald E (1980) Calcium channel blocking agents in the treatment of cardiovascular disorders Part II: hemodynamic effects and clinical application Ann Intern Med 93: 886–904

Strauss WE, McIntyre KM, Parisi AF, Shapiro W (1982) Safety and efficacy of diltiazem hydrochloride for the treatment of stable angina pectoris: report of a cooperative clinical trial Am J Cardiol 49: 560–566

Strauss WE, Parisi AF (1985) Superiority of combined diltiazem and propranolol therapy for angina pectoris Circulation 71: 951–957

Subramanian VB (1982a) Calcium channel blockers in chronic stable angina Herz 7: 211–220

Subramanian VB, Bowles MJ, Davies AB, Raftery EB (1982b) Combined therapy with verapamil and propranolol in chronic stable angina Am J Cardiol 49: 125–132

Subramanian VB, Bowles MJ, Khurmi NS, Davies AB, O'Hara MJ, Raftery EB (1983) Calcium antagonist withdrawal syndrome: objective demonstration with frequency-modulated ambulatory ST-segment monitoring Br Med J 286: 520–521

Subramanian VB (1984) Comparative evaluation of seven calcium ion antagonists with placebo and propranolol in patients with chronic stable angina In: Althaus U, Burckhardt D, Vogt E (eds) Calcium-Antagonismus – Internat. Symposium on Calcium-Antagonism Universamed Frankfurt/Main pp 162–180

Subramanian VB (1986) Combined therapy with calcium-channel and beta blockers: facts, fiction, and practical aspects Cardiovascular Reviews & Reports 7: 259–274

Tan ATH, Sadick N, Kelly DT, Harris PJ, Freedman SB, Bautovich G (1982) Verapamil in stable effort angina: effects on left ventricular function evaluated with exercise randionuclide ventriculography Am J Cardiol 49: 425–430

Theisen F, Jahrmärker H (1983) Wirkung von Gallopamil (D 600) auf das Belastungs-EKG bei koronarer Herzerkrankung In: Kaltenbach M, Hopf R (eds) Gallopamil Springer Berlin pp 120–126

Tolins M, Weir EK, Chesler E, Pierpont GL (1984) „Maximal" drug therapy is not necessarily optimal in chronic angina pectoris J Am Coll Cardiol 3: 1051–1057

Tweddel AC, Beattie JM, Lawrie TDV, Hutton I (1980) Effects of nifedipine on physical performance in patients with angina on β-blockers In: Puech P, Krebs R (eds) 4th International Adalat Symposium Excerpta Medica Amsterdam pp 143–146

Waters DD, McCans JL, Crean PA (1985) Serial exercise testing in patients with effort angina: variable tolerance, fixed threshold J Am Coll Cardiol 6: 1011–1015

Weiner DA, McCabe CH, Cutler SS, Ryan TJ, Klein MD (1984) The efficacy and safety of high-dose verapamil and diltiazem in the long-term treatment of stable exertional angina Clin Cardiol 7:648–653

White HD, Polak JF, Wynne J, Holman BL, Antman EM, Nesto RW (1985) Addition of nifedipine to maximal nitrate and betaadrenoreceptor blocker therapy in coronary artery disease Am J Cardiol 55: 1303

WHO (1980) Cooperative trial on primary prevention of ischaemic heart disease using clofibrate to lower serum cholesterol: mortality follow-up: report of the Committee of Principal Investigators Lancet 2: 379–385

Wilcox RG, Hampton JR, Banks DC, Birkhead JS, Brooksby IAB, Burns-Cox CJ, Hayes MJ, Joy MD, Malcolm AD, Mather HG, Rowley JM (1986) Trial of early nifedipine in acute myocardial infarction: the Trent study Br Med J 293: 1204–1208

Winniford MD, Hillis LD (1982) Combined therapy with betaadrenergic blockers and calcium-antagonists in patients with ischemic heart disease Herz 7: 243–249

Winniford MD, Huxley RL, Hillis LD (1983) Randomized, double-blind comparison of propranolol alone and a propranolol-verapamil combination in patients with severe angina of effort J Am Coll Cardiol 1: 492–498

Yusuf S, Peto R, Lewis J, Collins R, Sleight P (1985) Beta blockade during and after myocardial infarction: an overview of the randomized trials Progr Cardiovasc Dis 27: 335–371

4.1.3 Unstable Angina Pectoris

Within the pathological spectrum of coronary heart disease, unstable angina pectoris lies somewhere between stable (effort) angina pectoris and acute myocardial infarction. By definition (Cohn and Braunwald 1984a), there is either a crescendo angina pectoris (in preexisting stable angina pectoris an increase in intensity, duration or frequency of anginal attacks), angina pectoris at rest (or following minimal stress), or newly manifested (within the previous 4 weeks) angina pectoris – all in all, then, an extremely heterogeneous category of disease. The chest pains accompanying unstable angina pectoris resemble those in stable angina pectoris, but they can be more intense and of longer duration (up to 30 min). In most cases, ECG changes can be observed during an attack (ST segment elevation, ST depression, T-wave changes); by definition, however, the typical ECG changes of acute myocardial infarction are absent, and cardiac enzymes remain normal (Cohn and Braunwald 1984a; Hugenholtz 1986). Unstable angina pectoris progresses in 10%–25% of cases (Lichtlen 1984; Lubsen et al. 1986) into an acute myocardial infarction ("preinfarction syndrome" according to Lichtlen).

In pathophysiological terms, coronary insufficiency arises in unstable angina pectoris largely as a result of reduced oxygen supply. The various mechanisms by which this occurs, discussed in the literature, include the following: progressive stenosis due to coronary artery atherosclerosis; narrowing of the lumen as a result of intimal oedema or haemorrhage into an atherosclerotic plaque; increased vasomotor tone in the region of eccentrically positioned coronary artery stenoses; intermittently occurring thrombus or platelet aggregation, especially in endothelial rupture of atherosclerotic plaques; coronary spasm; changes in blood viscosity (reviews by Lichtlen 1984; Opie 1984; Hugenholtz 1986; Fitzgerald et al. 1986; Fuster and Chesebro 1986). Although vasospasm and changes in vascular tone were formerly regarded as the main cause of unstable angina pectoris, the explanation currently favoured centres on plaque rupture and thrombus formation (Fuster and Chesebro 1986; Roskamm 1987, personal communication). Thus, the coronary morphology in unstable angina pectoris resembles the changes that characterize myocardial infarction and sudden cardiac death. Myocardial oxygen demand remains initially unchanged, but, owing to pain and activation of the sympathetic nervous system, heart rate and blood pressure usually increase, whereby myocardial oxygen demand is further increased and coronary insufficiency further reinforced.

Naturally, there is a continuum between stable and unstable angina pectoris; typical transitional forms are angina pectoris at varying levels of effort ("variable-threshold angina") and "walk-through angina pectoris" (angina pectoris which occurs on very slight effort, but which then disappears on more intensive effort). So-called vasospastic angina pectoris (Prinzmetal's angina pectoris) is today regarded as a syndrome sui generis and as such is discussed separately (Sect. 4.1.3.1).

The treatment of unstable angina pectoris (Table 15) embraces general measures (admission to hospital, possibly monitoring in an intensive care unit, bed rest, sedation, analgesics, oxygen) and the administration of platelet-aggregation inhibitors, anticoagulating agents and antianginal (or antiischaemic) medication (monotherapy or combination therapy); in individual cases invasive therapeutic measures such as emergency coronary arteriography with PTCA or intracoronary thrombolysis, emergency bypass operation or intraaortal balloon pump are necessary.

On the basis of convincing results in vasospastic angina pectoris (Sect. 4.1.3.1), calcium antagonists are repeatedly cited as agents of first choice in unstable angina pectoris (Meyer 1983; Lichtlen 1984; Fach and Becker 1985). This evaluation presupposes, however, that an unstable angina pectoris is initiated mainly by coronary artery spasm or an increase in vasomotor tone of the coronary vessels, and that β-receptor blockers primarily exercise a negative effect, as observed in individual cases of vasospastic angina pectoris (Yasue et al. 1976; Robertson et al. 1982; Kern et al. 1983). Neither supposition can be sustained by currently available data, however, and the role of calcium antagonists must therefore be considered in relative terms. A review of the literature for the past 10 years reveals that calcium antagonists have been used only in isolated cases as primary therapy (monotherapy) in unstable angina pectoris (Parodi et al. 1979: 480 mg verapamil; Previtale et al. 1980: 100–120 mg nifedipine; Mehta et al. 1983: 240–640 mg verapamil; Mauritson 1983: 480 mg verapamil; Repine 1983: various doses of diltiazem or verapamil; Theroux et al. 1985: 160 mg diltiazem; Rafflenbeul et al. 1985: 15–30 mg nifedipine i.v./24 h; Gottlieb et al. 1986: 80 mg nifedipine.

Table 15. Therapeutic measures in unstable angina pectoris

General	Admission to hospital, bed rest, sedation, analgesics, oxygen
Pharmacological	Anticoagulant agents (heparin) or Antiplatelet agents (acetylsalicylic acid) Nitrates (possibly parenteral) β-Receptor blockers and calcium antagonists[a] Thrombolysis (intravenously)
Invasive	Coronary arteriography with PTCA or intracoronary thrombolysis Bypass surgery

[a] The pros and cons of monotherapy with calcium antagonists are discussed in the text

Calcium antagonists reduce the frequency of angina pectoris attacks and thereby also reduce nitrate consumption, but in most cases they have been found to exert no influence on total mortality and/or the development of acute myocardial infarction or sudden cardiac death. In most of the completed studies calcium antagonists were introduced when the patient could not be stabilized by maximum doses of β-receptor blockers and nitrates. Since nitrates were mostly given in high doses spread over 24 h, the development of nitrate tolerance in individual cases cannot be excluded. In the context of combination therapy with calcium antagonists, nifedipine was mainly used (Moses et al. 1980; Gerstenblith et al. 1982; Sellers et al. 1983; Stone et al. 1983; Blaustein et al. 1983; Hagemeijer and van Mechelen 1983; Hugenholtz et al. 1984; Muller et al. 1984) in doses of 40–120 mg daily, in most cases subdivided into three or four smaller doses (in some cases – Stone et al. 1983 – daily doses of nifedipine up to 200 mg). In one study (Andre-Fouet et al. 1983) β-receptor blockers were combined with diltiazem (in most cases 240–360 mg daily). With a combination of nitrates, β-receptor blockers and a calcium antagonist (usually nifedipine), a high percentage of patients could be adequately stabilized within a few hours (i.e. freedom from angina pectoris, no development of acute myocardial infarction). It remains an open question to what extent longer-term prognosis is improved by the supplementary administration of calcium antagonists (Blaustein et al. 1983). If patients cannot be stabilized by combination therapy, acute myocardial infarction frequently develops, or prompt coronary angioplasty or bypass surgery is required (Hugenholtz et al. 1984; de Feyter et al. 1985).

The procedure practised in our hospital for some years is supported by two recent studies (Muller et al. 1984; Lubsen et al. 1986), which indicate – the latter study with statistical significance – that patients previously treated with a β-receptor blocker should be given supplementary nifedipine. On the other hand, patients who have *not* previously received such medication should be given a β-receptor blocker as primary therapy. This procedure can free patients of pain in a shorter time (Muller et al. 1984; Fig. 20) and prevent recurrent ischaemia as well as acute myocardial infarction (Lubsen et al. 1986). Monotherapy with nifedipine can in individual cases, by reducing blood pressure and increasing heart rate, further increase coronary insufficiency. In the Dutch HINT study (Holland Interuniversity Nifedipine/Metoprolol Trial, Lubsen et al. 1986) this treatment was even found inferior to placebo; a limiting factor to be borne in mind, however, is the "rigid" dosage of nifedipine used (6 × 10 mg daily), which conceivably contributed to the negative outcome of this study.

The details of our regimen are as follows: the patient with unstable angina pectoris is given, besides parenteral nitrates (to reduce preload) and a platelet-aggregation inhibitor (acetylsalicylic acid), a β-receptor blocker at an early stage. We favour the use of β_1-selective (cardioselective) β-receptor blockers because of the more favourable spectrum of side effects and to avoid the possibility of peripheral and coronary vasoconstriction through a combination of the blocking actions on β_1 and β_2 receptors (so-called unmasking of the α effect). If the patient's systolic blood pressure is above 100 mgHg, a calcium antagonist is also given, preferably (as stated in Sect. 4.1.2) nifedipine; after a test dose of 5–10 mg, a daily dose of 40–80 mg is usually given. If unstable angina pectoris develops in

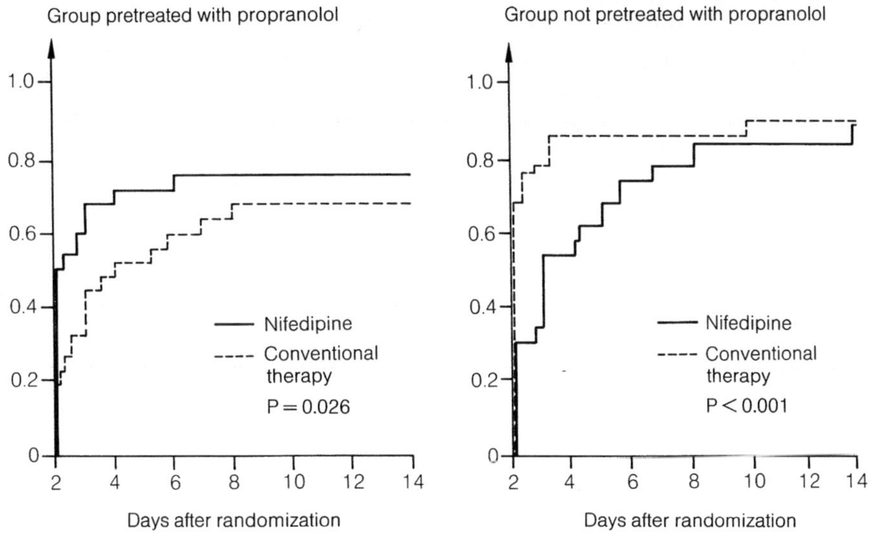

Fig. 20. Probability of freedom from pain for at least 48 h in patients previously treated with propranolol (nifedipine more effective than conventional therapy) and in patients not previously treated with propranolol (conventional therapy more effective). Conventional therapy = propranolol and/or nitrates. Probability 1.0 = 100%. (After Muller et al. 1984)

a patient currently under treatment with a β-receptor blocker, we continue the treatment, but supplement it with nifedipine. This combination therapy is generally well tolerated, but calls for close surveillance in order to recognize promptly any excessive drop in heart rate (due to β-receptor blockade) or in blood pressure (due to nifedipine), and to be able to make appropriate dose adjustment without further delay. If a patient cannot be stabilized under the maximum tolerated doses of this combination therapy (even up to 120 mg nifedipine daily), then invasive diagnosis and treatment (coronary arteriography, PTCA, bypass surgery) should be considered immediately (Lydtin 1989).

All attempts at optimization or standardization of treatment of unstable angina pectoris must take into account that the groups of patients included in studies to date are of extremely heterogeneous composition (with respect to severity of angina pectoris, exclusion criteria, etc.) and that treatment is defined to a major degree by local factors (e.g. proximity and availability of invasive diagnosis and treatment procedures).

4.1.3.1 Prinzmetal's Angina Pectoris

This form of angina pectoris, (vasospastic angina pectoris, variant angina pectoris) described by Prinzmetal et al. in 1959, is, by definition, present either when angina pectoris at rest is accompanied by an elevation of the ST segment or when coronary arteriography shows spontaneous or ergonovine-induced coronary spasm with anginal symptoms. The pain occurs mainly suddenly at night and in the early morning hours during a state of complete quiescence. Occasionally, a combination of vasospastic and stable angina pectoris is observed. Frequently, attacks are accompanied by arrhythmias (AV block, extrasystoles, ventricular tachycardia, ventricular fibrillation), which explains cases of sudden cardiac death. Development of acute myocardial infarction is also possible.

Data given in the literature concerning the incidence of Prinzmetal's angina pectoris show wide variation, ranging from 3% to around 20% in patients seeking hospital help for their anginal symptoms. The pathophysiological correlate of Prinzmetal's angina pectoris is the coronary spasm, which, as a rule, leads to reversible transmural ischaemia. In about two-thirds of cases coronary spasm develops in altered (atherosclerotic) coronary arteries (stenoses exceeding 75%), and in one-third of cases arteriographically normal coronary arteries are seen; the latter group frequently includes younger women and has, in general, a better prognosis (Selzer et al. 1976). There are numerous triggering factors in coronary spasm: cold, nicotine (90% of patients are smokers), hyperventilation, local release of thromboxane A_2, local deficiency of prostacyclin or EDRF (endothelium-derived relaxing factor), and experimentally or pharmacologically administered ergonovine, metacholine, histamine, serotonin, Tris-buffers, or non β_1-selective β-receptor blockers (Robertson et al. 1982; Kern et al. 1983). The occurrence of symptoms mainly at night or in the early hours is explained (Yasue 1984) by daytime fluctuations of coronary vasomotor tone (autonomic innervation, predominance of vagotonia). Alternatively, a changed ventricular geometry can be considered, with increased volume due to nocturnal redistribution of the blood (Meyer 1983). In about 25% of patients, there is a general tendency to vasospasm, and patients may also suffer from migraine or Raynaud's syndrome (Miller et al. 1981). For further details, see Hillis and Braunwald (1978), Luchie et al. (1978), Hugenholtz et al. (1984), Colin and Braunwald (1984b), Eichstädt (1984), Zähringer (1985).

In the treatment of Prinzmetal's angina pectoris, calcium antagonists are, beside nitrates, the agents of first choice. However, since in individual cases high doses of both substances may be indicated, calcium antagonists are to be preferred in the case of monotherapy as they do not induce tolerance. The effectiveness of calcium antagonists has been affirmed in numerous studies (reviews by Stone et al. 1980; Eichstädt 1984; Hugenholtz 1985; Fach and Becker 1985). In about 85% of patients the frequency and intensity of angina pectoris attacks markedly recede, or attacks cease completely, and nitroglycerin consumption thereby decreases. Isolated cases have been reported in which there is a decline in malignant ventricular arrhythmias (Ellrodt et al. 1980). Severi et al. (1980) found an improved prognosis for patients with Prinzmetal's angina pectoris under long-term treatment with calcium antagonists. However, unequivocal evidence for a prophylactic effect for calcium antagonists against sudden cardiac death or acute myocardial infarction remains to be demonstrated. Advanced coronary artery stenosis (greater than 75%) in patients with vasospastic angina pectoris can be recognized with fair certainty by exercise testing under continuing treatment with calcium antagonists (sensitivity and specificity greater than 90%; Araki et al. 1986). The daily doses stated in the literature with respect to individual calcium antagonists are: for verapamil 120–560 mg (Parodi et al. 1980; Maseri et al. 1980; Johnson et al. 1981; Hillis and Winniford 1980); for diltiazem 120–300 mg (Pepine et al. 1981; Feldmann et al. 1982; Schroeder et al. 1983); for nifedipine

40–160 mg (Mabuchi 1975; Muller and Gunther 1978; Delahaye et al. 1979; Antmann et al. 1980; Hill et al. 1982; Hillis and Winniford 1984); for gallopamil 50–200 mg. Among the various calcium antagonists there appears to be no significant difference (Hillis and Winniford 1984); Hill et al. (1982) even found an identical response in patients to calcium antagonists and nitrates.

If monotherapy with a calcium antagonist fails to relieve symptoms, a combination treatment with nitrates is recommended (Winniford et al. 1984), whereby an additive effect is to be expected on account of the different modes of action on vascular smooth muscle. Examples of effective daily doses are 360–480 mg verapamil, 180–360 mg diltiazem, or 60–80 mg nifedipine, together with 40–120 mg isosorbide dinitrate (Winniford et al. 1984). In confirmed Prinzmetal's angina pectoris, β-receptor blockers are to be avoided as monotherapy (Julian 1983); noncardioselective β-receptor blockers (propranolol, for example) can even trigger angina pectoris attacks, since by blockade of β_2-receptor-mediated coronary vasodilation, vasoconstriction of the coronary vessels via α-receptors predominates. This negative effect of β-receptor blockers can be avoided by the concomitant administration of a calcium antagonist (Tilmant 1983; Kugiyama 1986), or perhaps by giving a cardioselective β-receptor blocker (Kupper 1984). Combination therapy is the primary choice in all mixed forms between Prinzmetal's angina pectoris and stable angina pectoris.

There is no consensus in the literature concerning treatment with α-receptor blockers such as phentolamine and prazosin (positive: Tsivone et al. 1983; negative: Winniford et al. 1983). A bypass operation should be carried out only in advanced fixed coronary artery stenosis and, as far as at all practicable, should be combined with plexectomy (details in Bertrand 1980). Postoperatively, continued treatment with calcium antagonists or with nitrates is necessary in most cases, since there is a retained tendency to vasospasm.

Despite regular ECG recording during angina pectoris attacks, we have only rarely seen cases of Prinzmetal's angina pectoris among our patients. This form of angina pectoris, however, could possibly be diagnosed more frequently in the future through use of the Holter ECG (with computer-controlled ST segment evaluation).

4.1.3.2 Syndrome X

Syndrome X is characterized by angina pectoris, exercise-dependent ST segment depressions, arteriographically normal coronary arteries, and the absence of coronary spasm and cardiac hypertrophy. It is held to be caused by inappropriate vasoconstriction of small coronary resistance vessels, i.e. a pathologically raised arteriolar tone. Calcium antagonists improve symptoms and myocardial ischaemia, since they raise coronary blood flow during stress and prevent a further increase in coronary vasomotor tone (e.g. in hyperventilation). Nitrates, which mainly distend large epicardial arteries, are said to be less effective in this syndrome (Pozzati et al. 1987, personal communication).

References

Andre-Fouet X, Gayet C, Wilner C, Viallet M, Thizy JF, Apoil E, Pont M (1983) Comparison of short-term efficacy of diltiazem and propranolol in unstable angina. A randomized trial in 30 patients with spontaneous chest pain Eur Heart J 4: 691–698

Antmann W, Muller J, Goldberg S, MacAlpin R, Rubenfire M, Tababsnik B, Liang C, Heupler F, Achuff St, Reichek N, Geltman E, Kerin NZ, Neff RK, Braunwald E (1980) Nifedipine therapy for coronary-artery spasm N Engl J Med 302: 1269–1273

Araki H, Hayata N, Matsuguchi T, Takeshita A, Nakamura M (1986) Diagnosis of important fixed coronary stenosis in patients with variant angina by exercise tests after treatment with calcium antagonists Br Heart J 56: 138–145

Bertrand ME, Lablanche JM, Rousseau MF, Warembourg Jr HH, Stankowtak C, Soots G (1980) Surgical treatment of variant angina: Use of plexectomy with aortocoronary bypass Circulation 61: 877–882

Blaustein AS, Heller GV, Kolman BS (1983) Adjunctive nifedipine therapy in high risk, medically refractory, unstable angina pectoris Am J Cardiol 52: 950–954

Braunwald E (1982) Mechanism of action of calcium channel blocking agents N Engl J Med 307: 1618–1627

Cohn PF, Braunwald E, (1984a) Unstable angina In: Braunwald E (ed) Heart disease – a textbook of cardiovascular medicine Saunders Philadelphia pp 1355–1360

Cohn PF, Braunwald E (1984b) Variant angina pectoris (Prinzmetal's angina) In: Braunwald E (ed) Heart disease – a textbook of cardiovascular medicine Saunders Philadelphia pp 1360–1362

De Feyter PJ, Serruys PW, van den Brand M, Balakumaran K, Mochtar B, Soward AL, Arnold AER, Hugenholtz PG (1985) Emergency coronary angioplasty in refractory unstable angina N Engl J Med 313: 342–346

Delahaye JP, Touboul P, Cassagnes J, Gaspard P, Milon H (1979) The long-term treatment of Prinzmetal's angina with nifedipine In: Puech P, Krebs R (eds) 4th International Adalat Symposium Excerpta Medica Amsterdam pp 87–93

Ellrodt G, Chew CY, Singh BN (1980) Therapeutic implications of slow-channel blockade in cardiocirculatory disorders Circulation 62: 669–679

Eichstädt H (1984) Calciumantagonisten In: Roskamm H (ed) Handbuch der inneren Medizin IX/3 Koronarkrankheiten Springer Berlin pp 1069–1113

Fach WA, Becker HJ (1985) Kalzium-Antagonisten in der Therapie der stabilen Angina pectoris Herz und Gefäße 5: 347–358

Feldman RL, Pepine CJ, Whittle J, Conti CR (1982) Short and long-term responses to diltiazem in patients with variant angina Am J Cardiol 49: 554–559

Fitzgerald DJ, Roy L, Catella F, Fitzgerald GA (1986) Platelet activation in unstable coronary disease N Engl J Med 315: 983–989

Fuster V, Chesebro JH (1986) Mechanisms of unstable angina N Engl J Med 315: 1023–1025

Gerstenblith G, Ouyang P, Achuff SC, Bulkley BH, Becker LC, Mellits ED, Baughman KL, Weiss JL, Flaherty JT, Kallman CH Llewellyn M, Weisfeldt ML (1982) Nifedipine in unstable angina N Engl J Med 306: 885–889

Gottlieb SO, Weisfeldt ML, Ouyang P, Achuff SC, Baughman KL, Traill TA, Brinker JA, Shapiro EP, Chandra NC, Mellits ED, Townsend ST, Gerstenblith G (1986) Effect of the addition of propranolol to therapy with nifedipine for unstable angina pectoris: a randomized, double-blind, placebo-controlled trial Circulation 73: 331–337

Hagemeijer F, van Mechelen R (1983) Effects of nifedipine in unstable angina persisting despite conventional medical therapy In: Kaltenbach M, Neufeld HN (eds) 5th International Adalat Symposium Excerpta Medica Amsterdam pp 255–261

Hill JA, Feldman RL, Pepine CJ, Conti CR (1982) Randomized double-blind comparison of nifedipine and isosorbide dinitrate in patients with coronary arterial spasm Am J Cardiol 49: 431–438

Hillis LD, Braunwald E (1978) Coronary-artery spasm N Engl J Med 299: 695–702

Hillis LD, Winniford MD (1984) Verapamil and nifedipine in the therapy of variant angina pectoris In:Althaus U, Burckhardt D, Vogt E (eds) Calcium-Antagonismus-Internat. Symposium on Calcium-Antagonism Universimed Frankfurt/Main pp 147–151

Hugenholtz PG, Michels R, Serruys PW, Simoons ML (1984) Behandlung der instabilen Angina pectoris mit Calcium Antagonisten In: Althaus U, Burckhardt D, Vogt E (eds) Calcium-Antagonismus – Internat. Symposium on Calcium-Antagonism Universimed Frankfurt/Main pp 138–186

Hugenholtz PG (1985) Calcium antagonists In: Abshagen U (ed) Clinical pharmacology of antianginal drugs Springer Berlin pp 459–538

Hugenholtz PG (1986) Unstable angina revisited once more Eur Heart J 7: 1010–1013

Johnson SM, Mauritson DR, Willerson JT, Hillis LD (1981) A controlled trial of verapamil for Prinzmetal's variant angina N Engl J Med 304: 862–866

Julian DG (1983) Is the use of beta-blockade contraindicated in the patient with coronary spasm? Circulation 57 (Suppl 1): 92–96

Kern MJ, Ganz P, Horowitz JD, Gaspar J, Barry WH, Lorell BH, Grossman W, Mudge Jr GH (1983) Potentiation of coronary vasoconstriction by beta-adrenergic blockade in patients with coronary artery disease Circulation 67: 1178–1185

Kugiyama K, Yasue H, Horio Y, Morikami Y, Fujii H, Koga Y, Kojima A, Takahashi M (1986) Effects of propranolol and nifedipine on exercise-induced attack in patients with variant angina: assessment by exercise thallium-201 myocardial scintigraphy with quantitative rotational tomography Circulation 74: 374–380

Kupper W (1984) Betarezeptorenblocker und Koronarspasmus In: Lohmann FW (ed) Die klinische Bedeutung der β-1-Selektivität Verlag für angewandte Wissenschaften pp 99–112

Lichtlen P (1984) Klinik und medikamentöse Therapie der koronaren Herzkrankheit Deutscher Ärzte-Verlag Köln

Lubsen J, Tijssen JGP, Kerkkamp HJJ (1986) Early treatment of unstable angina in the coronary care unit: a randomised, double blind, placebo controlled comparison of recurrent ischaemia in patients treated with nifedipine or metoprolol or both (Report of the Holland Interuniversity Nifedipine/Metoprolol Trial – HINT – Research group) Br Heart J 56: 400–413

Luchi RJ, Chahine RA, Raizner AE (1979) Coronary artery spasm Ann Intern Med 91: 441–449

Lydtin H (1989) Konservative Therapie der stabilen und instabilen Angina pectoris In: Jahrmärker H, Theisen K, Holzgartner H (eds) Das kardiale Risiko AMV München pp 75–86

Mabuchi G, Kishida H, Suzuki K (1975) Clinical effect of nifedipine on variant form of angina pectoris In: Hashimoto K, Kimura E, Kobayashi T (eds) 1st International Nifedipine Adalat Symposium University of Tokyo Press Tokyo pp 177–184

Maseri A, Severi S, Parodi O, Distante A, Biagini A (1980) Application of calcium antagonists in patients with vasospastic angina pectoris In: Fleckenstein A, Roskamm H (eds) Calcium-Antagonismus Springer Berlin pp 243–245

Mauritson DR, Johnson SM, Winniford MD, Cary JR, Willerson JT, Hillis LD (1983) Verapamil for unstable angina at rest: a short-term randomized double-blind study Am Heart J 106: 652–658

Mehta J, Conti CP, Pepine CJ (1983) Long-term oral verapamil therapy in patients with rest angina Am Heart J 106: 1133–1137

Meyer J (1983) Der Einsatz von Calciumantagonisten in der Therapie der Koronaren Herzkrankheit Internist 24: 408–414

Miller D, Waters DD, Warnica W, Szlachcic J, Kreeft J, Théroux P (1981) Is variant angina the coronary manifestation of a generalized vasospastic disorder? N Engl J Med 304: 763–766

Moses J, Feldman MS, Helfart RH (1980) Efficacy of nifedipine in the intermediate syndrome refractory to propranolol and nitrate therapy Am J Cardiol 45: 390 (abstract)

Muller JE, Gunther SJ (1978) Nifedipine therapy for Prinzmetal's angina Circulation 57: 137–139

Muller JE, Turi ZG, Pearle DL, Schneider JF, Serfas DH, Morrison J, Stone PH, Rude RE, Rosner B, Sobel BE, Tate C, Scheiner E, Roberts R, Hennekens CH, Braunwald E (1984) Nifedipine and conventional therapy for unstable angina pectoris: a randomized, double-blind comparison Circulation 69: 728–739

Opie LH (1984) Angina at rest In: Opie LH (ed) The Heart Grune & Stratton London pp 346–349

Parodi O, Maseri A, Simonetti I (1979) Management of unstable angina at rest by verapamil, a double-blind crossover study in the coronary care unit Br Heart J 41: 167–174

Parodi O, Simonetti I, L'Abbate A (1980) Treatment of unstable angina: verapamil Hosp Physician 12: 12–21

Pepine CJ, Feldman RL, Whittle J, Curry RC, Conti CR (1981) Effects of the calcium antagonist diltiazem in patients with variant angina Am Heart J 101: 719–725

Pepine CJ, Feldman RL, Hill JA, Conti CR, Mehta J, Hill C, Scott E (1983) Clinical outcome after treatment of rest angina with calcium blockers: comparative experience during the initial year of therapy with diltiazem, nifedipine and verapamil Am Heart J 106: 1341–1347

Previtali M, Salerno JA, Tavazzi L, Ray M, Medici A, Chimienti M, Specchia G, Bobba P (1980) Treatment of angina at rest with nifedipine: a short term controlled study Am J Cardiol 45: 825–830

Prinzmetal M, Kennamer R, Merlis R, Wada T, Bor N (1959) Angina pectoris. 1. A variant form of angina pectoris Am J Med 27: 375–388

Rafflenbeul W, Bossaller C, Lichtlen PR (1985) Intravenöse Infusion von Nifedipin bei Patienten mit instabiler Angina pectoris In: Meyer J, Erbel R (eds) Intravenöse und intrakoronare Anwendung von Adalat Springer Heidelberg pp 125–131

Robertson RM, Wood AJJ, Vaughn WK, Robertson D (1982) Exacerbation of vasotonic angina pectoris by propranolol Circulation 65: 281–285

Schroeder JS, Lamb IH, Bristow MR, Ginsburg R, Hung J, McAuley BJ (1983) Prevention of cardiovascular events in variant angina by long-term diltiazem therapy J Am Coll Cardiol 1: 1507–1511

Sellers TD, Gibson RS, Taylor GJ, Beller GA, Martin RP, McGuire LB, Carabello BA, Gascho JA, Ayers CR, DiMarco JP, Beckwith JR, Burwell LR, Craddock GA, Crampton RS (1983) Relation of therapeutic response to nifedipine to coronary anatomy and motion of ST segment during unstable angina pectoris Am J Med 75: 57–64

Selzer A, Langston M, Ruggeroli C, Cohn K (1976) Clinical syndrome of variant angina with normal coronary arteriogram N Engl J Med 295: 1343–1347

Severi S, Davies GJ, Maseri A, Marzullo P, L'Abbate A (1980) Long-term prognosis of variant angina with medical treatment Am J Cardiol 46: 226–232

Stone PH, Antmann EM, Muller JE, Braunwald E (1980) Calcium channel blocking agents in the treatment of cardiovascular disorders. Part II: Hemodynamic effects and clinical application Ann Intern Med 93: 886–904

Stone PH, Muller JE, Turi ZG, Geltman E, Jaffe AS, Braunwald E (1983) Efficacy of nifedipine therapy in patients with refractory angina pectoris: significance of the presence of coronary vasospasm Am Heart J 106: 644–652

Théroux P, Tacymans Y, Marissette JM, Bosch X, Pelletier GB, Waters DD (1985) A randomized study comparing propranolol and diltiazem in the treatment of unstable angina J Am Coll Cardiol 5: 717–722

Tilmant PY, Lablanche JM, Thieuleux FA, Dupuis BA, Bertrand ME (1983) Detrimental effect of propranolol in patients with coronary arterial spasm countered by combination with diltiazem Am J Cardiol 52: 230–233

Tzivoni D, Keren A, BenHorin J, Gottlieb S, Atlas D, Stern S (1983) Prazosin therapy for refractory variant angina Am Heart J 105: 262–266

Winniford MD, Filipchu KN, Hillis LD (1983) Alpha-adrenergic blockade for variant angina: A long-term, double-blind, randomized trial Circulation 67: 1185–1188

Winniford MD, Gabliani G, Johnson SM, Mauritson DR, Fulton KL, Hillis LD (1984) Concomitant calcium antagonist plus isosorbide dinitrate therapy for markedly active variant angina Am Heart J 108: 1269–1273

Yasue H, Touyama M, Kato H, Tamaka S, Akiyama F (1976) Prinzmetal's variant form of angina as a manifestation of alpha-adrenergic receptor-mediated coronary artery spasm: Documentation by coronary arteriography Am Heart J 91: 148–155

Yasue H (1984) Coronary artery spasm and calcium ions In: Opie LH (ed) Calcium antagonists and cardiovascular disease Raven Press New York pp 117–128

Zähringer J (1985) Variant-Angina: Pathophysiologie und Therapie mit Calcium-Antagonisten Kolloquium 3: Nr 28

4.1.4 Myocardial Infarction (MI)

Myocardial infarction is the necrosis of a region of heart muscle due to hypoxia, blood flow to this muscular region being interrupted or critically restricted. In contrast to angina pectoris, coronary insufficiency in myocardial infarction is irreversible; after a critical period of ischaemia, myocardial necrosis ensues (according to Hugenholtz et al. 1986, this period lasts between 4 and 30 min, depending on remaining coronary blood flow, heart rate, afterload and collateral vessels).

Irreversible coronary ischaemia is, according to contemporary models, the result of a thrombotic occlusion of atherosclerotically altered coronary arteries (Laffel and Braunwald 1984; Epstein and Palmeri 1984). In a small percentage of cases (according to Braunwald, less than 5%) coronary artery spasm plays a role; such spasm can occur not only in atherosclerotically changed, but also in normal coronary arteries (as determined by coronary arteriography; Sect. 4.1.3.1) and secondarily can contribute to the development of a coronary artery thrombus (Maseri et al. 1982). The occurrence of acute myocardial infarction following discontinuation of a calcium antagonist (Kozeny et al. 1986) and in the context of withdrawal of nitrates, for instance nitroglycerin (Lange et al. 1977), is explained by coronary spasm.

Anatomically, two distinct forms of myocardial infarction are distinguished: transmural and nontransmural (subendocardial). Since classical ECG criteria are not always compatible with pathological–anatomical findings, it has now become popular to use the terms "Q-wave" and "non-Q-wave" infarction, according to whether classical Q-waves or only ST segment changes are discernible in the ECG. A Q-wave infarction denotes complete occlusion of a coronary artery without adequate collateral flow; a non-Q-wave infarction reflects partial occlusion of a coronary artery, or complete occlusion, but with good collateral flow (Kennedy 1986).

4.1.4.1 Primary Prevention of MI

Definitive proof of significant primary prevention of acute myocardial infarction has yet to be demonstrated for any group of drugs. No relevant studies have been published for calcium antagonists, neither for healthy subjects nor for patients with arterial hypertension or other risk factors. This issue will possibly be further clarified in the future through more intensive use of calcium antagonists in the treatment of hypertension (Sect. 4.2).

4.1.4.2 Calcium Antagonists in the Acute Phase of MI

In the laboratory (in isolated myocardial preparations and in animal tests) calcium antagonists reduce the extent of infarction. In such studies, however, calcium antagonists are given before, during or shortly after (up to 15 min) the ischaemic event (in most cases, coronary occlusion; Klein et al. 1984; Endo et al. 1986; Hugenholtz et al. 1986). The cardioprotective action of calcium antagonists thus manifests itself only during an ischaemic episode, not in the reperfusion phase. The following protective mechanisms of action are considered: calcium antagonists dilate collateral vessels, resolve vasospasm and thereby improve oxygen supply; they lower afterload and myocardial oxygen demand; they prevent intracellular calcium overload in the context of ischaemia and reperfusion, whereby mitochondrial function and ATP synthesis are sustained (Henry et al. 1980; Nayler and Ferrari 1980; Stone et al. 1980; Braunwald 1982; Eichstädt 1984). In addition to the limitation of the extent of myocardial infarction, calcium antagonists are also claimed to possess antiarrhythmic properties, preventing ventricular reentry arrhythmias during the early phase of infarction by suppressing slow-response action potentials. This type of action potential is triggered by a calcium influx through the so-called slow calcium channel and can consequently be inhibited by calcium antagonists (Naumann d'Alnoncourt et al. 1982; Peter et al. 1982; De Leiris et al. 1984). Furthermore, ventricular arrhythmias, induced by coronary artery spasm in the early phase of infarction, are also said to improve under calcium antagonists.

As experience has repeatedly shown, experimental findings and animal trials cannot be directly transferred to human conditions. A patient with acute myocardial infarction is invariably admitted to hospital only after a more or less considerable time lapse, so that therapy with calcium antagonists can usually be initiated only when myocardial necrosis is already in an advanced state.

It is also all too easily forgotten that the prognosis in acute myocardial infarction is determined by a range of factors which cannot be influenced therapeutically (Table 16). The only feasible therapeutic goal is to attempt to limit the extent of the infarction and thus to improve early and late prognosis in myocardial infarction. This is achieved by improving myocardial oxygen supply and/or by limiting myocardial oxygen consumption.

Investigations in humans have provided conflicting data: there are, on one hand, studies available which indicate that calcium antagonists have favourable haemodynamic actions in acute myocardial infarction (including limitation of the extent of infarction); on the other hand, five large-scale, randomized, double-blind studies with verapamil and nifedipine have been unable to demonstrate any significantly positive effect for calcium antagonists in acute myocardial infarc-

Table 16. Factors determining prognosis in acute myocardial infarction (AMI)

Not influenceable	(Possibly) Influenceable
Age	Extent of AMI
Sex	
Hypertension	
Previous MIs	
Congestive heart failure	
Diabetes mellitus	
Pulmonary disease	
etc.	
Localization of infarction	

tion. The haemodynamic effects of calcium antagonists in uncomplicated acute myocardial infarction have been thoroughly investigated: verapamil, diltiazem and nifedipine lower arterial blood pressure and peripheral vascular resistance; the heart rate increases with nifedipine, decreases with diltiazem and remains unchanged with verapamil.

The rate–pressure product (heart rate times blood pressure) decreased during therapy with all three substances: with diltiazem −20%; verapamil −11%; nifedipine (not significantly) −7%); signs of congestive heart failure were not observed in any patients (Singh et al. 1983). Just et al. (1986) reported a decrease in peripheral resistance and in central venous pressure with nifedipine; heart rate increased minimally. Bussmann (1984) has described a decrease in both blood pressure and peripheral vascular resistance for verapamil and nifedipine (the effect being stronger for nifedipine than for verapamil). Pulmonary arterial pressure and left ventricular filling pressure did not change; cardiac output increased with nifedipine, but decreased with verapamil. Myocardial decompensation did not occur with either of the two substances. Eichler et al. (1985) observed with tiapamil (a verapamil derivative) a decrease in heart rate and in peripheral vascular resistance, with an increase in stroke volume and cardiac output. Congestive heart failure did not arise in any of the 30 patients studied (exclusion criteria: AV block grade II and grade III, systolic blood pressure below 90 mmHg).

Bussmann et al. (1983) reported a reduction of CK and CK-MB (creatine phosphokinase MB isoenzyme) activity during 2 days of parenteral verapamil therapy (5–10 mg/h) in the early infarction phase, from which they inferred a reduction in infarction size; as a reference base they used investigations by Bleifeld et al. (1977), which showed a good correlation between CK activity and pathological-anatomically measured extent of infarction. Even earlier, Bleifeld (1975) and Wolf et al. (1977) had observed with parenteral verapamil in acute myocardial infarction reduced S-T segment elevation and also a smaller CK release. On the other hand, Just et al. (1986) found, with 3 days nifedipine therapy, no significant reduction in CK or CK-MB activity in comparison with placebo. In all four studies, calcium antagonists were given for only 2–3 days, and consequently they make no statement about total mortality during the patients' hospital stay. Gottlieb et al. (1985) reported, with 120 mg nifedipine over 10 days, a smaller "early infarct expansion", as determined by echocardiography, in comparison with placebo.

In contrast to these partly nonrandomized haemodynamic investigations with positive results are the actual intervention studies of calcium antagonists in acute myocardial infarction:

In the Danish verapamil study (1984), patients were given, immediately on admission to hospital, 0.1 mg verapamil per kg body weight intravenously, and thereafter 360 mg oral verapamil daily for 6 months. Only patients with manifest myocardial infarction were included in the study; mortality and reinfarction rate showed no significant differences among the treated and nontreated groups. In the Norwegian nifedipine study (Sirnes et al. 1984), patients received, on average 5.5 h after the onset of pain, the initial dose of nifedipine, thereafter 40 mg/day was given for 6 weeks. Mortality over the 6-week period and infarction size, assessed by CK-MB activity (Fig. 21), were identical in both the nifedipine and the placebo group. Muller et al. (1984) administered, on average 4.6 h after the onset of pain, nifedipine or placebo to patients in whom myocardial infarction was suspected. A daily dose of 120 mg was given over 14 days. The nifedipine group differed from the placebo group neither in the number of patients developing myocardial infarction nor with respect to the enzymatically determined infarction size or to mortality at the end of 6 months: early mortality after 2 weeks showed a tendency to be higher in the nifedipine-treated group. Two further recent studies with a daily dose of 4×10 mg nifedipine for 4 weeks (Wilcox et al. 1986; Branagen et al. 1986) also showed no difference from placebo with respect to expansion of infarction (enzymatically determined) and 4-week mortality. Even in a subgroup, which was treated within 4 h of onset of pain, no advantage for nifedipine could be shown.

On the basis of these results, the use of calcium antagonists in the early phase of acute myocardial infarction cannot be generally recommended, for in no study

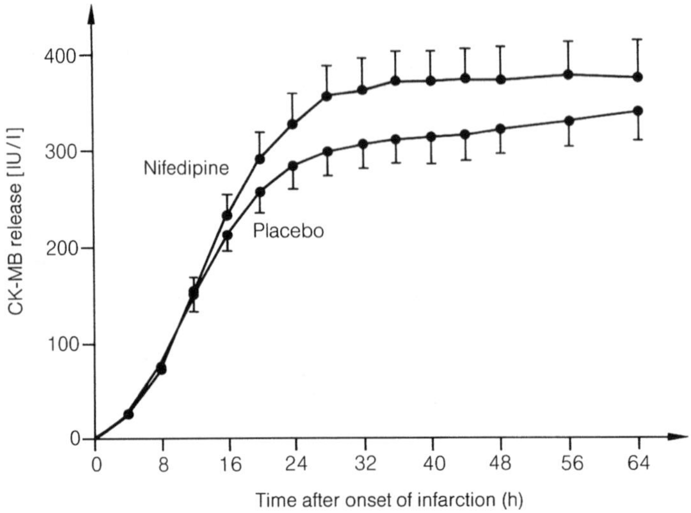

Fig. 21. CK-MB release in acute myocardial infarction under nifedipine and placebo. Differences not significant (average values are indicated, with standard deviation). (After Sirnes et al. 1984)

so far has it been possible to show that this treatment significantly influences mortality and reinfarction rate. Possible reasons for the lack of efficacy could be delayed initiation of calcium antagonist therapy, i.e. the period between the onset of symptoms and commencement of treatment is too long, as well as failure to influence the raised sympathetic tone in the early phase of myocardial infarction (Opie 1984). Nifedipine in particular can also lower coronary perfusion pressure (via a decrease in blood pressure) and increase oxygen demand (via increased heart rate; Ferlinz 1986). The absence of sympathetic nervous system blockade distinguishes calcium antagonists from β-receptor blockers, which are successfully employed in the early phase of myocardial infarction (review by Lohmöller and Lydtin 1984; Evans 1986). Future studies may clarify to what extent calcium antagonists can positively influence the course of infarction by early administration (1–2 h after onset of pain) and/or by careful parenteral administration (Cyran et al. 1987; nifedipine i.v.). Parallels to thrombolysis therapy are conceivable; here, too, positive influences on infarction size, left ventricular function and mortality are to be expected mainly after early administration (GISSI[4] 1986, 1987).

In a subgroup of patients with nontransmural myocardial infarction (criterion for inclusion: non-Q-wave infarction) it was found possible, by administering 360 mg diltiazem daily for 14 days, to reduce the reinfarction rate and the incidence of refractory postinfarction angina pectoris during that period, in comparison with placebo (Gibson et al. 1986). Mortality was the same in both groups; after 1 year, however, there was a reduction in mortality and the reinfarction rate

[4] GISSI: Gruppo italiano per lo studio della streptochinasi nell'infarto miocardio (see References, p. 100)

(Boden et al. 1988). Differences in coronary anatomy possibly play a role in transmural and nontransmural myocardial infarction; nontransmural infarctions in early coronary arteriography more rarely show complete occlusion of the infarct-related vessel, or even when occlusion is total, usually well-developed collaterals (DeWood et al. 1986). Geary et al. had already postulated in 1982 on the basis of animal trials that the efficacy of calcium antagonists in myocardial infarction was dependent on the availability of collaterals and/or only partial occlusion of coronary arteries. To what extent the positive effect of diltiazem in nontransmural myocardial infarction is substance-specific or a consequence of specific coronary anatomy in this type of infarction remains to be clarified.

Practical Considerations

Although, on the basis of the results discussed in the preceding pages, a general recommendation for the use of calcium antagonists in acute myocardial infarction cannot be made, there are various special indications in which calcium antagonists may be given in the early phase of infarction. In such situations we favour the combined administration of a calcium antagonist (usually nifedipine) and a β-receptor blocker, in order to avoid a deleterious rise in heart rate. In addition, we take care to ensure that systolic blood pressure remains above 100 mmHg, so that no critical reduction in coronary perfusion pressure occurs. The decision to treat with calcium antagonists is independent of whether the patient is receiving thrombolysis therapy (intravenous streptokinase, 1.5×10^6 IU in 1 h) or not. Immediately after a patient is admitted to the intensive care unit of our hospital, a calcium antagonist is administered in parallel with a nitrate and a β-receptor blocker – on the one hand, as supplementary antianginal therapy, and, on the other, in order to reduce afterload – with the aim of bringing blood pressure down to about 110–120 mmHg. This is especially advisable in patients with initially high blood pressure (cf. Gordon et al. 1986); raised blood pressure can be the consequence of raised sympathetic tone in the acute phase of infarction, or the expression of preexisting hypertensive disease. In most cases, in parallel therapy with a β-receptor blocker, an initial dose of 5–10 mg sublingually and thereafter a daily dose of 40–60 mg nifedipine is sufficient to achieve appropriate adjustment of blood pressure (alternatively, nifedipine parenterally: initial dose 0.5 mg i.v., thereafter 1 mg/h; according to Cyran et al. 1987). If there are contraindications to the administration of β-receptor blockers, we use instead a combination of a nitrate and a calcium antagonist. Since an excessive heart rate at rest is to be avoided, diltiazem or verapamil are therefore considered as an alternative to nifedipine.

A further indication for calcium antagonists is postinfarction angina pectoris. These attacks which occur at rest are often due to coronary spasm or raised vasomotor tone. The efficacy of calcium antagonists in this situation is well attested (Braunwald 1982; Nakamura and Koiwaya 1983; Wilcox et al. 1986). Postinfarction angina pectoris can, of course, also be a first indicator of reocclusion of the coronary artery following thrombolysis, or of a second myocardial infarction. In general, invasive diagnostic procedures should immediately be considered in patients with postinfarction angina pectoris.

If in acute myocardial infarction left ventricular heart failure develops, and if there is marked pulmonary congestion, nitrates and diuretics are prescribed; in the case of a low cardiac output, afterload reduction with calcium antagonists (usually nifedipine) may be cautiously attempted, under close haemodynamic surveillance (Swan-Ganz catheter). This procedure is mainly of value in patients with normal or raised blood pressure, and a lowering of the blood pressure below 100 mmHg is to be avoided (Gordon et al. 1986). All in all, in our experience calcium antagonists are well tolerated in the acute phase of myocardial infarction, and only rarely have we observed development of congestive heart failure among our patients. An increase in heart rate with nifedipine may be avoided by parallel administration of a β-receptor blocker, and an excessive fall in blood pressure by appropriately cautious dosage. The individualized application of calcium antagonists in acute myocardial infarction must be assessed against other forms of therapy currently available. Today there is strong evidence in favour of systemic (i.e. of intravenous) thrombolysis within the first 3–4 h (possibly up to 24 h) accompanied by acetylsalicylic acid (GISSI 1986, 1987, ISIS-2 [5] 1988) and of early administration of cardioselective β-receptor blockers (ISIS-1 [5] 1986; Yusuf et al. 1988), with due attention to contraindications; definitive assessment of combination therapy with systemic thrombolysis and β-receptor blockade is not yet available.

4.1.4.3 Long-term Therapy with Calcium Antagonists Following Acute MI (Secondary Prevention)

In two of the five studies with calcium antagonists in acute myocardial infarction, treatment was continued, respectively, for 6 weeks (Sirnes et al. 1984; nifedipine) and for 6 months (Danish verapamil study). In neither study could a significant decline in mortality and reinfarction rate against placebo be demonstrated. In the Danish verapamil study a reduced mortality was observed only in patients over 65 years of age (15%, as against 22% with placebo), and in the entire group there was a tendency, albeit not significant (8%), towards a reduced mortality (Fig. 22). Currently, two postinfarction intervention studies are in progress, one in Denmark (MCU-II [6]) and one in Italy (CRIS [7]), each involving more than 2000 patients. Both studies are comparing the effect of 360 mg verapamil daily with a placebo group. The results of an intervention study with diltiazem in comparison with placebo have recently been published (MDPIT [8]: 2466 patients; commencement of treatment 3–15 days following acute myocardial infarction, with follow-up supervision for 1–4 years; 240 mg diltiazem daily; parallel therapy with β-receptor blockers permitted). They show no reduction in total mortality, although there was a reduction in the number of "cardiac events" (death from cardiac causes or nonfatal reinfarction) in patients without pulmonary congestion (the Multicenter Diltiazem Postinfarction Trial Research Group 1988). For the subgroup of non-Q-wave infarctions preliminary data have been published show-

---

[5] ISIS: (First) International Study of Infarction Survival (see References, p. 101)
[6] MCU-II: Multicenter Investigation II (verapamil)
[7] CRIS: Calcium Antagonists Reinfarction Italian Study (verapamil)
[8] MDPIT: Multicenter Diltiazem Postinfarction Trial

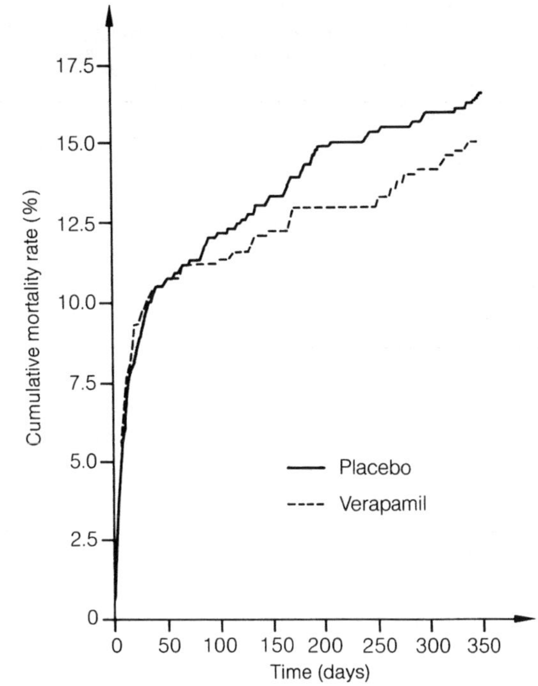

Fig. 22. Danish verapamil study (1984). Comparison of cumulative mortality rate over 1 year in patients with acute myocardial infarction, with verapamil or placebo. (After the Danish Study Group 1984)

ing a reduction in mortality and reinfarction rate after 1 year of treatment with diltiazem (Boden et al. 1988). The results of the SPRINT-1 study[9] in Israel (3×10 mg nifedipine, initiated 7–21 days following myocardial infarction and continued for 1 year) have shown no difference in respect to mortality and reinfarction rate between the nifedipine group and the placebo group (the Israeli Sprint Study Group 1988). An important constraining factor in the SPRINT-1 study is that a large proportion of the patients (in both nifedipine and placebo groups) were given parallel treatment with β-receptor blockers, which have meanwhile been shown to have a proven positive effect in secondary prevention following acute myocardial infarction (review by Yusuf et al. 1985). A follow-up study with 3×20 mg nifedipine daily is currently in progress (SPRINT-2-study).

In Summary. The studies so far available reveal no conclusive effect for calcium antagonists on either mortality or reinfarction rates in long-term treatment following acute myocardial infarction.

References

Bleifeld W (1974) Physiologische Grundlagen der medikamentösen Therapie der Koronarinsuffizienz Inn Med 1: 100

Bleifeld W , Mathey D, Hanrath P, Buss H, Effert S (1977) Infarct size estimated from serial serum creatine phosphokinase in relation to left ventricular hemodynamics Circulation 55: 303–311

---

[9] SPRINT-1, SPRINT-2: Secondary Prevention Reinfarction Nifedipine Trial

Boden WE, Krone RJ, Kleiger RE, Miller JP, Hager WD, Moss AJ, MDPIT Research Group (1988) Diltiazem reduces long-term cardiac event rate after non-Q-wave infarction: multicenter diltiazem post-infarction trial (MDPIT) Circulation 78 (Suppl II): 96

Branagan JP, Walsh K, Kelly P, Collins WC, McCafferty D, Walsh MJ (1986) Effect of early treatment with nifedipine in suspected acute myocardial infarction Eur Heart J 7: 859–865

Braunwald E (1982) Mechanism of action of calcium-channel blocking agents N Engl J Med 307: 1618–1627

Bussmann WD, Seher W, Grüngras M (1983) Reduktion der CK- und CKMB-Infarktgröße durch Verapamil Dtsch Med Wschr 108: 1047–1053

Bussmann WD (1984) Akute und chronische Herzinsuffizienz Springer Berlin pp 104–115

Cyran J, Kuthan P, Remien P, Fischer J, Henslemann L (1987) Kalzium-Antagonisten bei akutem Myokardinfarkt? Nifedipindauerinfusion bei Präinfarktangina und akutem Myokardinfarkt Internist 28: 100–107

DeLeiris J, Richard V, Pestre S (1984) Calcium antagonists and eperimental myocardial ischemia and infarction In: Opie LH (ed) Calcium Antagonists and cardiovascular disease Raven Press New York pp 105–115

DeWood MA, Stifter WF, Simpson CS, Spores J, Eugster GS, Judge TP, Hinnen ML (1986) Coronary arteriographic findings soon after non-Q-wave myocardial infarction N Engl J Med 315: 417–423

Eichler HG, Mabin TA, Commerford PJ, LLoyd EA, Beck W, Opie LH (1985) Tiapamil, a new calcium antagonist: hemodynamic effects in patients with acute myocardial infarction Circulation 71: 779–786

Eichstädt H (1984) Calcium-Antagonisten In: Roskamm H (ed) Handbuch der inneren Medizin Vol 9/3 Koronarerkrankungen Springer Berlin pp 1069–1113

Endo T, Nejima J, Fujita S, Kiuchi K, Iida N, Kikuchi K, Hayakawa H, Okumura H (1986) Comparative effects of nicardipine, a new calcium antagonist, on size of myocardial infarction after coronary artery occlusion in dogs Circulation 74: 420–430

Epstein SE, Palmeri ST (1984) Mechanisms contributing to precipitation of unstable angina and acute myocardial infarction: implications regarding therapy Am J Cardiol 54: 1245–1250

Evans AE (1986) Secondary prevention after myocardial infarction Lancet 2: 150–151

Ferlinz J (1986) Nifedipine in myocardial ischemia, systemic hypertension and other cardiovascular disorders Ann Intern Med 105: 714–729

Geary GG, Smith GT, Suehiro GT, McNamara JJ (1982) Failure of nifedipine therapy to reduce myocardial infarct size in the baboon Am J Cardiol 49: 331–338

Gibson RS, Boden WE, Theroux P, Strauss HD, Pratt CM, Gheorghiade M, Capone RJ, Crawford MH, Schlant RC, Kleiger RE, Young PM, Schechtman K, Perryman MB, Roberts R, the Diltiazem Reinfarction Study Group (1986) Diltiazem and reinfarction in patients with non-Q-wave myocardial infarction N Engl J Med 315: 423–429

GISSI (Gruppo italiano per lo studio della streptochinasi nell'infarto miocardico) (1986) Effectiveness of intravenous thrombolytic treatment in acute myocardial infarction Lancet 1: 397–402

GISSI (Gruppo italiano per lo studio della streptochinasi nell'infarto miocardico) (1987) Long-term effects of intravenous thrombolysis in acute myocardial infarction: final report of the GISSI study. Lancet 2: 871–874

Gordon GD, Mabin TA, Lloyd EA, Eichler HG, Commerford PJ, Opie LH (1986) Nifedipine in acute myocardial infarction In: Lichtlen PR (ed) 6th International Adalat symposium Excerpta Medica Amsterdam pp 271–277

Gottlieb SO, Gerstenblith G, Shapiro EP, Chandra N, Healy B, Weisfeldt ML, Weiss JL (1985) Nifedipine reduces early infarct expansion: results of a double blind,randomized trial Circulation 72 (Suppl 3): III 274

Henry PD, Shuchleib R, Clark RE, Perez JE (1980) Protection of ischemic myocardium by nifedipine In: Puech P, Krebs R (eds) 4th International Adalat symposium Excerpta Medica Amsterdam pp 189–196

Hugenholtz PG, Serruys PW, Fleckenstein A, Nayler W (1986) Why Ca^{2+} antagonists will be most useful before or during early myocardial ischaemia and not after infarction has been established Eur Heart J 7: 270–278

ISIS-1 (First International Study of Infarct Survival) (1986) Randomised trial of intravenous atenolol among 16 027 Cases of suspected acute myocardial infarction: ISIS-1 Lancet 2: 57–66

ISIS-2 Collaborative Study Group (1988) Randomized trial of intravenous streptokinase, oral aspirin, both, or neither among 17 187 cases of suspected acute myocardial infarction: ISIS-2 (Second international study of infarction survival) Lancet 2: 349–360

Just H, Heidelbach I, Jäckle B, Wollschläger H (1986) Nifedipin bei akutem Myokardinfarkt: Ist ein myokardprotektiver Effekt in einer doppeltblinden, placebokontrollierten Untersuchung nachweisbar? Intensivmed 23: 159–164

Kennedy JW (1986) Non-Q-wave myocardial infarction N Engl J Med 315: 451–453

Klein HH, Schubothe M, Nebendahl K, Kreuzer H (1984) The effect of two different diltiazem treatments on infarct size in ischemic, reperfused porcine hearts Circulation 69: 1001–1005

Kozeny GA, Ragona BP, Bansal VK, Hurley RM, Dixon DW, Vertuno LL, Hano JE (1986) Myocardial infarction with normal results of coronary angiography following diltiazem withdrawal Am J Med 80: 1184–1186

Laffel GL, Braunwald E (1984) Thrombolytic therapy: a new strategy for the treatment of acute myocardial infarction N Engl J Med 311: 710–717 and 770–776

Lange RL, Reid MS, Tresch DD, Keelan MH, Berhard VM, Coolidge G (1977) Non-atheromatous ischemic heart disease following withdrawal from chronic industrial nitroglycerin exposure Circulation 56: 366–378

Lohmöller G, Lydtin H (1984) Betarezeptorenblocker In: Roskamm H (ed) Handbuch der inneren Medizin, Vol 9/3 Koronarerkrankungen Springer Berlin pp 1007–1067

Maseri A, L'Abbate A, Baroldi G, Chierchia S, Marzilli M, Ballestra AM, Severi S, Parodi O, Biagini A, Distante A, Pesola A (1978) Coronary vasospasm as a possible cause of myocardial infarction-A conclusion derived from the study of „preinfarction" angina N Engl J Med 299: 1271–1277

Muller JE, Morrison J, Stone PH, Rude RE, Rosner B, Roberts R, Pearle DL, Turi ZG, Schneider JF, Serfas DH, Tate C, Scheiner E, Sobel BE, Hennekens CH, Braunwald E (1984) Nifedipine therapy for patients with threatened and acute myocardial infarction: a randomized, double-blind, placebo-controlled comparison Circulation 69: 740–747

Nakamura M, Koiwaya Y (1983) Effect of diltiazem on recurrent spontaneous angina after acute myocardial infarction Circ Res 52 (Suppl 1): 158–162

Naumann d'Alnoncourt C, Zierhut W, Lüderitz B (1982) Wirkung von Diltiazem auf abnorme Reizbildung in Purkinje-Fasern bei Ischämie In: Bender F, Greeff K (eds) Calciumantagonisten zur Behandlung der Angina pectoris, Hypertonie und Arrhythmie. Excerpta Medica Amsterdam pp 270–277

Nayler WG, Ferrari R (1980) The use of nifedipine in the protection of the ischemic heart. In: Puech P, Krebs R (eds) 4th International Adalat Symposium Excerpta Medica Amsterdam pp 197–205

Opie LH (1984) Myocardial infarction In: Opie LH (ed) The Heart Grune & Stratton London pp 351–371

Peter T, Fujimoto T, Hamamoto H, McCullen A, Mandel WJ (1982) Electrophysiologic effects of diltiazem in canine myocardium with special reference to conduction delay during ischemia Am J Cardiol 49: 602–605

Singh BW, Josephson MA, Nademanee K (1984) Elektrophysiologische und hämodynamische Wirkungen von Calcium-Antagonisten: klinischer Überblick In: Althaus U, Burckhardt D, Vogt E (eds) Calcium-Antagonismus Internat. Symposium Calcium-Antagonism Universimed Frankfurt/Main pp 57–85

Sirnes PA, Overskeid K, Pedersen TR, Bathen J, Drivenes A, Froland GS, Kjekshus JK, Landmark K, Rokseth R, Sirnes KE, Sundoy A, Torjussen BR, Westlund KM, Wik BA (1984) Evolution of infarct size during the early use of nifedipine in patients with acute myocardial infarction: the norwegian nifedipine multicenter trial Circulation 70: 638–644

Stone PH, Antman EM, Muller JE, Braunwald E (1980) Calcium channel blocking agents in the treatment of cardiovascular disorders Part 2: Hemodynamic effects and clinical applications Ann Intern Med 93: 886–904

The Danish Study Group on Verapamil in Myocardial Infarction (1984) Verapamil in acute myocardial infarction Eur Heart J 5: 516–528

The Israeli Sprint Study Group (1988) Secondary prevention reinfarction israeli nifedipine trial (SPRINT) A randomized trial intervention trial of nifedipine in patients with acute myocardial infarction Eur Heart J 9: 354–364
The Multicenter Diltiazem Postinfarction Trial Research Group (1988) The effect of diltiazem on mortality and reinfarction after myocardial infarction N Engl J Med 319: 385–392
Wilcox RG, Hampton JR, Banks DC, Birkhead JS, Brooksby IAB, Burns-Cox CJ, Hayes MJ, Joy MD, Malcolm AD, Mather HG, Rowley JM (1986) Trial of early nifedipine in acute myocardial infarction: the Trent study Br Med J 293: 1204–1208
Wolf R, Habel F, Witt E, Nötges A, Everling F, Hochrein H (1977) Wirkung von Verapamil auf die Hämodynamik und Größe des akuten Myokardinfarkts Herz 2: 110
Yusuf S, Reto R, Lewis J, Collins R, Sleight P (1985) Beta blockade during and after myocardial infarction: an overview of the randomised trials Progr Cardiovasc Dis 27: 335–371
Yusuf S, Wittes J, Friedman L (1988) Overview of results of randomized clinical trials in heart disease I. Treatments following myocardial infarction JAMA 260: 2088–2093

4.1.5 Arrhythmias in Coronary Heart Disease

The range of arrhythmias which may be observed in the various presenting forms of coronary heart disease is comparable to that seen in other organic heart diseases. The spectrum extends from single supraventricular or ventricular extrasystoles via various forms of exit or conduction block to ventricular flutter and fibrillation and sudden cardiac death. More severe ventricular arrhythmias (polymorphous extrasystoles, ventricular tachycardia, ventricular flutter/ventricular fibrillation) occur more frequently in impaired ventricular function (as is seen following large anterior wall infarctions, in ventricular aneurysm and in angina pectoris and two- or three-vessel coronary disease). The nature and intensity of functional disturbance of the ventricular myocardium (also of hypertrophy) determine to a large extent the prognosis in any given ventricular arrhythmia.

The treatment of arrhythmias in coronary heart disease corresponds in many cases to that of arrhythmias of different pathogenesis. The use of calcium antagonists in the treatment of arrhythmias is therefore presented separately in Sect. 4.3. In the presence of coronary heart disease, however, antiarrhythmic agents which have supplementary antianginal action will be preferentially selected: namely, calcium antagonists of the verapamil type, or diltiazem or β-receptor blockers. An antianginal effect is also being discussed for amiodarone, a class III antiarrhythmic agent (Kulbertus et al. 1982). The substances named here have, in addition to their direct antiarrhythmic properties, an indirect antiarrhythmic action, since they can reduce myocardial ischaemia. This indirect antiarrhythmic effect is of particular relevance in Prinzmetal's angina pectoris, where there are often arrhythmias during an anginal attack, which resolve under appropriate treatment with calcium antagonists or with nitrates (Sect. 4.1.3.1).

It has been surmised from experimental findings that arrhythmias are triggered in the early phase of a myocardial infarction by calcium-dependent mechanisms; an antiarrhythmic action of calcium antagonists (both in the early ischaemic phase and in the reperfusion phase) is under discussion, although as yet there has been no unambiguous demonstration of such an action in humans (Haverkampf et al. 1987; Opie 1987, personal communication). As stated in Sect. 4.1.4, neither in the early phase of infarction nor in secondary prevention following myocardial infarction do calcium antagonists show any positive effect in the context of the statistical end-points of "sudden cardiac death" and "total mortality".

References

Haverkamp W, Hindricks G, Kunert M, Thale J, Behrenbeck Th, Gülker H (1987) Wirkungen von Gallopamil und Nifedipin auf ventrikuläre Arrhythmien, Kammerflimmern, epikardiale Leitungsverzögerungen und den Zeitverlauf der ventrikulären Flimmerschwelle während Myokardischämie und Reperfusion In: Bender F, Meesmann W (eds) Therapie mit Gallopamil Steinkopff Darmstadt pp 31–49

Kulbertus HE, Demoulin CI, Bertholet M (1982) Amiodarone: Darstellung der antianginösen Eigenschaften und Analyse der Nebenwirkungen Herz 7: 317–324

4.1.6 Silent Myocardial Ischaemia – Therapy with Calcium Antagonists

"Silent myocardial ischaemia" is today, as a symptomless manifestation of coronary heart disease, juxtaposed to the symptomatic forms, especially angina pectoris. The significance of silent ischaemia lies first in the fact that about half of all patients with coronary heart disease experience as their first clinical manifestation either acute myocardial infarction or sudden cardiac death, and, secondly, in the fact that silent ischaemias may have a prognostic significance for patients following myocardial infarction or those with angina pectoris.

While it has long been possible, by means of an exercise ECG or radionuclide scanning (thallium scintigraphy) to detect signs of ischaemia, even in symptomless patients under stress, only the technical possibilities offered by the Holter ECG (with computer-assisted evaluation) now permit diagnosis of silent ischaemias under normal day-to-day conditions (review by Arnim 1986). To obtain an adequate degree of sensitivity and specificity by means of Holter ECG supervision, strict criteria must be adhered to with respect to the type of apparatus employed, the care in its assembly and the evaluation of the results. These criteria are at the present time certainly not met, when the widespread use of this technology in outpatients is considered.

According to Cohn (1984, 1985), three groups of patients with silent ischaemia can be distinguished:

a) Absolutely symptomless patients (according to preliminary estimates, 2–4% of middle-aged men belong to this group)
b) Patients who are symptomless following myocardial infarction (this group is thought to comprise 20% of symptom-free patients who have suffered an infarction)
c) Patients who present with a mixture of anginal symptoms and silent myocardial ischaemia (this is thought to apply to 80% of patients with angina pectoris; the major proportion of myocardial ischaemias are thought to be "silent", that is, to take a symptomless course).

From a pathophysiological standpoint, a reduced or variable pain threshold (level of pain receptors, of pain conduction and of central pain processing), or myocardial ischaemia below a critical angina pectoris threshold, is thought to underlie silent myocardial ischaemia (review by Cohn 1983; von Arnim 1987).

The prognostic significance of silent myocardial ischaemia and the appropriate procedure differ from one group to another (a, b, c):

a) For the group of patients without any symptoms there are as yet no unequivocal data on the natural course of the disease. Particularly in this group, in which treatment can be assessed only by objective end-points (myocardial infarction, mortality rate) and not by subjective symptoms, especially rigorous requirements must be imposed with respect to sensitivity and specificity of the diagnostic methods employed. In our opinion, at the present time these requirements are, in general, not met either for exercise testing or for Holter ECG (see also Berman et al. 1987). A diagnosis of silent myocardial ischaemia can be made only if the results of several different methods (exercise testing, Holter ECG, radionuclide scanning) are reproducibly positive. If, for example, an exercise ECG alone is used as a means of detecting coronary heart disease among symptomless 50-year-old men, only 17% of positive test results are due to coronary heart disease (assuming, according to coronary arteriographic findings, an incidence of 5% for the condition in this group of patients, Berman et al. 1987). Treatment initiated on the basis of a positive exercise ECG would, therefore, be inappropriate in many cases.

Selection of therapy must always be based on consideration of the total risk profile, including the family history and concomitant symptoms (e.g. arrhythmias). For such antiischaemic therapy both β-receptor blockers and calcium antagonists are appropriate. As with patients presenting with angina pectoris, the response to a given treatment is not predictable, and the therapeutic course must be monitored by exercise testing or Holter ECG (von Armin and Erath 1986). In patients with protracted symptomless ischaemic episodes, patients with a high risk profile, and especially in those who respond poorly to a drug regimen, coronary arteriography should not be delayed. There are repeated reports in the literature (Cohn 1983) of symptomless patients with left main stenosis or three-vessel coronary artery disease (with a correspondingly poor prognosis).

b) If periods of silent myocardial ischaemia are detected in patients following acute myocardial infarction – in most cases during exercise testing prior to discharge from hospital – coronary arteriography should be undertaken. On the basis of the catheterization result, the decision can then be made whether drug therapy, an operative procedure or PTCA is appropriate. If signs of ischaemia appear during drug therapy, the regimen should be either changed or intensified. A combination of a cardioselective β-receptor blocker and a calcium antagonist has frequently been the treatment of choice in such patients in our clinic (details of combination therapy in Sect. 4.1.2.3). A decision in favour of bypass surgery must take into account both the available large-scale comparative studies (VA Study [10], CASS [11]; Varnauskas et al. 1988) and also the facts relating to the individual patient (review by Califf et al. 1986). All in all, silent myocardial ischaemias, as

[10] VA Study: Veterans Administration Study
[11] CASS: Coronary Artery Surgery Study

well as the occurrence of postinfarction angina pectoris, negatively affect the prognosis for patients following myocardial infarction (Gottlieb et al. 1986a).

c) In the group of patients with angina pectoris and silent myocardial ischaemia, antianginal or antiischaemic therapy should aim not only to reduce the frequency of anginal attacks, but also to prevent episodes of silent myocardial ischaemia (during exercise testing or Holter ECG). The treatment plan does not, therefore, differ in principle from that for stable angina pectoris (Sect. 4.1.2). The investigations carried out by Deanfield et al. (1983) demonstrated that silent myocardial ischaemia frequently occurs, even without an increase in heart rate; that is, it is possibly evoked via a change of coronary vasomotor tone (due, perhaps, to coronary spasm). This would favour monotherapy or combination therapy with calcium antagonists or nitrates. To what extent a treatment which eliminates (or significantly reduces) both symptomatic and symptomless myocardial ischaemia also improves the prognosis for a patient remains at present an open question.

In patients with unstable angina pectoris, silent myocardial ischaemia is said to occur frequently in spite of intensive medication and indicates unfavourable outcome (Gottlieb et al. 1986b). Such patients should undergo coronary arteriography without delay, in order to prevent acute myocardial infarction or sudden cardiac death either by means of bypass surgery, PTCA or by changing and intensifying drug therapy (possibly also including thrombolysis).

Particularly in patients with unstable angina pectoris, the decision process will in future be influenced by the incidence of silent myocardial ischaemia, whereby the strict requirements for Holter ECG monitoring must be met.

In Summary. Calcium antagonists are also employed in the treatment of silent myocardial ischaemia. In principle, the treatment plan does not differ radically from that for stable and unstable angina pectoris. Frequent ischaemic episodes, unaccompanied by a change in heart rate, favour treatment with calcium antagonists, alongside nitrates.

References

v. Arnim T (1986) ST-Segment-Analyse im Langzeit-EKG Dtsch Med Wschr 110: 1047–1051

v. Arnim T, Erath A (1986) Medikamentöse Therapie bei transienten – überwiegemd stummen – Ischämie-Episoden. Cross-over-Vergleichsstudie doppelblind zwischen Nifedipin und IS-5-Mononitrat Z Kardiol 75 (Suppl 4): 25 (p 59)

v. Arnim T (1987) Stumme Myokardischämie Dtsch Ärztebl 84: B 332– B 335

Berman DS, Rozanski A, Knoebel SB (1987) The detection of silent ischemia: cautions and precautions Circulation 75: 101–106

Califf RM, Pryor DB, Greenfield JC (1986) Beyond randomized clinical trials: applying clinical experience in the treatment of patients with coronary artery disease Circulation 74: 1191–1194

Cohn PF (1983) Silent myocardial ischemia: to treat or not to treat Hosp Pract 18: 125–129 and 133–135

Cohn PF, Brown EJ, Cohn JK (1984) Detection and management of coronary artery disease in the asymptomatic population Am Heart J 108: 1064–1067

Cohn PF (1985) Silent myocardial ischemia as a manifestation of asymptomatic coronary artery disease: What is appropriate therapy? Am J Cardiol 56: 28 D- 34 D

Deanfield JE, Selwyn AP, Chierchia S, Maseri A, Ribeiro P, Krikler S, Morgan M (1983) Myocardial ischaemia during daily life in patients with stable angina: its relation to symptoms and heart rate changes Lancet 2: 753–758
Gottlieb SH, Gerstenblith G, Achuff SC, Mellits ED, Gottlieb SO (1986 a) Ischemic ST segment changes by ambulatory Holter predict one year mortality in high risk post-infarct patients Circulation 74 (Suppl II): 58
Gottlieb SO, Weisfeldt ML, Ouyang P, Mellits ED, Gerstenblith G (1986 b) Silent ischemia as a marker for early unfavorable outcomes in patients with unstable angina N Engl J Med 314: 1214–1219
Varnauskas E, and the European Coronary Surgery Study Group (1988) Twelve-year follow-up of survival in the randomized European coronary surgery study. N Engl J Med 319: 332–337

4.2 Hypertension

4.2.1 Fundamentals of Antihypertensive Therapy with Calcium Antagonists

4.2.1.1 Definition

Arterial hypertension is not a clearly defined pathogenetic entity, but rather a syndrome without uniform aetiology and pathogenesis. Regardless of whether it is primary (essential) or secondary, the hypertension syndrome promotes the development of atherosclerosis (involving the brain, eye, heart and kidneys, with corresponding organ damage) and reduces life expectancy. Primary (essential) hypertension accounts for 94%–95% of all cases of hypertension; only about 5% can be classified as secondary, due to renal disease (2.4%–4%), renal artery stenoses (0.2%–1.0%), coarctation of the aorta (0.1%–0.2%), phaeochromocytoma (0.1%–0.2%), hyperaldosteronism (about 0.1%), Cushing's syndrome (0.1%–0.2%) or associated with the use of oral contraceptives (0.2%–0.8%) (Kaplan 1984). Secondary forms of hypertension should always be considered, but above all in the following situations: in younger age groups; in newly manifested hypertension; in critical fluctuations in blood pressure; in extremely marked electrolyte disturbances (hypokalaemia); when there is a family history of renal disease; in overtly refractory hypertension; and/or when hypertensive organ damage is inappropriate for current blood pressure levels.

It is extremely difficult to designate absolutely normal blood pressure levels, above which a diagnosis of hypertension is required: the World Health Organization cites as upper limits 140–160 mmHg for systolic blood pressure and 90–95 mmHg for diastolic pressure (WHO 1959). According to life insurance statistics, however, cardiovascular mortality rises significantly when in repeated measurements blood pressure levels exceed the following suggested norms:

140/90 for men under 45 years
140/95 for men over 45 years
150/95 for women of all age groups

Although these thresholds lie below or within the WHO data, they are perhaps still too high, since in the Framingham Study, for instance, individuals with occasional blood pressure levels between 140/90 and 160/95 (so-called borderline hypertensives) were shown to have a doubling of the standard cardiovascular risk.

4.2.1.2 Hypertension as a Risk Factor or Risk Indicator

In virtually every industrialized country hypertension is the commonest cause of death and invalidity among adults. Hypertension is, besides cigarette smoking and hypercholesterolaemia, the most important risk factor for coronary heart disease, and it is the leading risk factor in cerebrovascular and renovascular disease. The commonest cause of death among hypertensive individuals is myocardial infarction, such deaths occurring 2–3 times more frequently than those due to stroke (HDFP 1982a, b; MRC 1985; Cruickshank et al. 1987). Although many studies have shown that a lowering of blood pressure reduces the numbers of fatal and nonfatal strokes, there are to date only scant data available (Wikstrand et al. 1988) which reliably attest that the incidence of myocardial infarction and of cardiac death following coronary heart disease is also reduced by antihypertensive treatment.

The indication for treatment of severe hypertension is unequivocal: if diastolic pressure is greater than 105 mmHg, drug therapy significantly reduces mortality and morbidity (numbers of fatal and nonfatal strokes, VA 1967, 1970). It is more difficult to demonstrate a positive effect for antihypertensive therapy in mild forms of hypertension: the term "mild hypertension" embraces both borderline and stable hypertension, with diastolic blood pressure being frequently between 95 and 105 mmHg, independent of the systolic pressure, which in most cases is also raised. This category comprises 75%–85% of hypertensive individuals, and in the adult population aged between 30 and 70 years about 23% of men and 12% of women are affected. Although life expectancy for this large group of people (more than 20 million in the United States) with mild hypertension is diminished as a result of cardiovascular complications, the benefits of antihypertensive medical treatment cannot be clearly documented. This can be explained partly by the design of available studies, perhaps by the medications used sofar in these investigations, and also by the fact that hypertension is diagnosed by a raised blood pressure and not by its underlying or antecedent mechanisms. Furthermore, the level and extent of hypertension preceding intervention or diagnosis may already have produced irreversible organ damage, which cannot be influenced by treatment. On the other hand, however, no treatment is free from unwanted side effects. Measured by the most stringent criterion for successful preventive therapy, i.e. a reduction in total mortality, the well-attested benefits of medical therapy in mild hypertension do not justify indiscriminate treatment of all affected individuals (Anlauf et al. 1986; see also Table 17).

In our opinion, at the present time the following situations favour initiation of medical therapy in the diastolic pressure range 90–105 mmHg: failure of nonmedicinal, general measures (i.e. weight reduction, low-salt diet, less than 40 g alcohol per day, nicotine abstinence); diastolic pressure in the range 95–105 mmHg in repeated blood pressure measurements; concomitant high systolic pressure; evidence of organ complications (kidney, eye, heart, central nervous system); presence of other risk factors (e.g. hypercholesterolaemia, diabetes mellitus); younger age group (under 50 years) and male sex; a family history of hypertensive complications; and the patient's subjective tolerance of medication and compliance.

Table 17. Cerebrovascular (cerebr.) and coronary (coron.) or cardiac (card.) morbidity and mortality compounded (after Anlauf et al. 1986). Evaluation: "intention-to-treat" method for ANBPS (1980, 1982, 1984), HDFP (1979 a, b, 1982 a, b) and MRC (1975, 1985), "on randomized-treatment" method for Oslo (Helgeland 1980) and EWPHE (Amery et al. 1985); Editoral Lancet 1985). Calculation of confidence intervals uniformly according to Shipley (do not necessarily coincide with published data)

	Study	*n*		P = Events/1000 patients years		ΔP $P_r - P_k$	$\Delta P\%$	95%-confidence interval		Significance in the study
		Treatment	Control	Treatment	Control					
ANBPS	cerebr.	17	31	2.4	4.5	−2.1	−47	−70	−4	$P<0.025$
	coron.	98	109	14.0	15.9	−1.9	−12	−33	+16	n. s.
HDFP	cerebr.	59	86	3.1	4.5	−1.4	−31	−51	−4	n. d.
	coron.	–	–	–	–	–	–	–	–	n. d.
Oslo	cerebr.	0	7	0	3.4	−3.4	−100	–	–	$P<0.02$
	coron.	20	13	9	6.2	+2.8	+45	−28	+192	n. s.
MRC	cerebr.	60	109	1.4	2.6	−1.2	−46	−61	−26	$P<0.01$
	coron.	222	234	5.2	5.5	−0.3	−5.5	−21	+14	n. s.
EWPHE	cerebr.	16	22	12	17	−5	−29	−63	+34	n. d.
	card.	24	46	17	36	−19	−53	−71	−23	n. d.

n. s. not significant; n. d. no data.

4.2.1.3 Stepped Care Versus Individualized Therapy

Daily medical practice calls for a clear and simple classification of the many therapeutic alternatives in arterial hypertension. The 1986 recommendations of the German League Against Hypertension (Deutsche Liga zur Bekämpfung des hohen Blutdrucks; DL 1986a, b) advise that the treatment of the hypertensive individual without additional diseases or risk factors be based on the stepped care approach (Fig. 23). Newer antihypertensive agents such as calcium antagonists are here cited as alternatives in initial monotherapy to diuretics and β-receptor blockers. This is certainly justifiable when diuretics or β-receptor blockers do not produce a satisfactory reduction of blood pressure, or if they produce intolerable side effects. Indeed, diabetes mellitus, asthma, or intermittent claudication favour the use of calcium antagonists as first-line medication (Vetter 1986). Further concomitant diseases and supplementary criteria that can modify the stepped care approach are summarized in Table 18.

The 1986 recommendations of the German League Against Hypertension are clearly distinct from the stepped care plan of hypertensive therapy published in 1984 in the United States (Fig. 23a). Figure 23b shows the latest, almost identical recommendations of both groups published in 1988 (DL 1988; The 1988 report of the Joint National Committee on Detection, Evaluation and Treatment of High Blood Pressure 1988). In two editorials (Saunders 1986; Dustan 1987) doubts have been voiced in the United States about the stepped care plan recommended in 1984. To summarize, Dustan called for preference to be given in the treatment of mild hypertension to drugs with a suppressive effect on sympathetic activity: reserpine, methyldopa, but also mixed α, β-receptor blockers (e.g. labetalol). Conversely, she also stressed that the heterogeneity of mild hypertension should be borne in mind. Hypertension in black people should be treated initially with diuretics. In severe hypertension, vasodilators should be the drugs of choice, and here a plea is made mainly for calcium antagonists of the dihydropyridine type, possibly in combination with a β-receptor blocker, or for minoxidil in combination with a diuretic. Saunders recommended above all individualized treatment for the hypertensive patient (profiled care) with appropriate attention being given not only to age, ethnic background, sex etc., but also to cost and probable patient compliance.

Besides the provisional therapeutic recommendations derived mainly from pathophysiological considerations, today attention is increasingly given both to the tolerance of drugs, and to the more or less high acceptance of the treatment by the patient (compliance) (Chobanian 1986; Croog et al. 1986; Williams 1987). ACE inhibitors have here, together with calcium antagonists, brought a decisive advance: in many cases they have made superfluous the once necessary simultaneous administration of three or four antihypertensive agents in pursuance of a stepped care plan. Calcium antagonists and ACE inhibitors are subjectively well tolerated. This is an extremely important consideration in a disease which may often run a symptomless course over a long time, or at least exhibit only very few symptoms.

Recent studies have provided evidence for the possibility and/or necessity of antihypertensive treatment in the elderly (Amery et al. 1985; Editorial Lancet 1985; see also Table 19). The EWPHE study[12] demonstrated that cardiovascular mortality can be reduced by diuretic therapy, possibly in combination with methyldopa. Total mortality among patients over 60 years of age was, however,

---

[12] EWPHE: European Working Party on High Blood Pressure in the Elderly

Stepped Care 1984

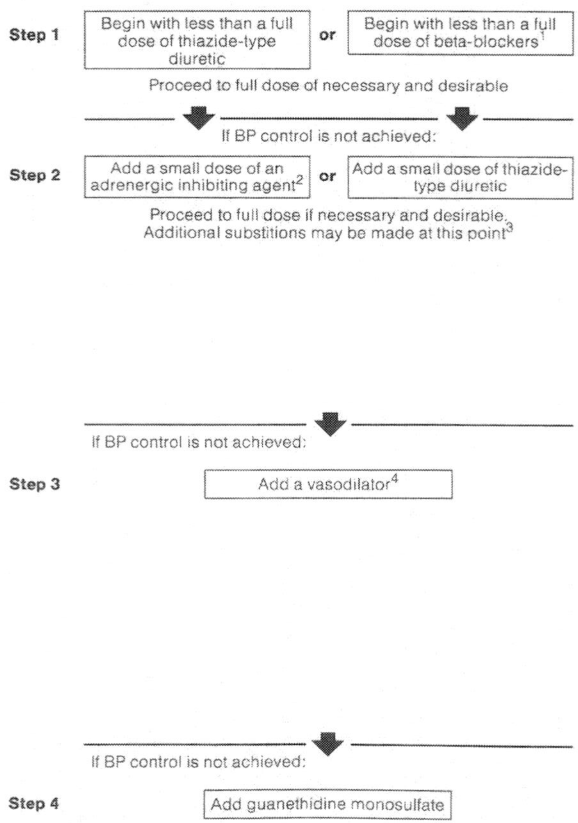

Stepped Care Plan for Drug Treatment of Hypertension 1986

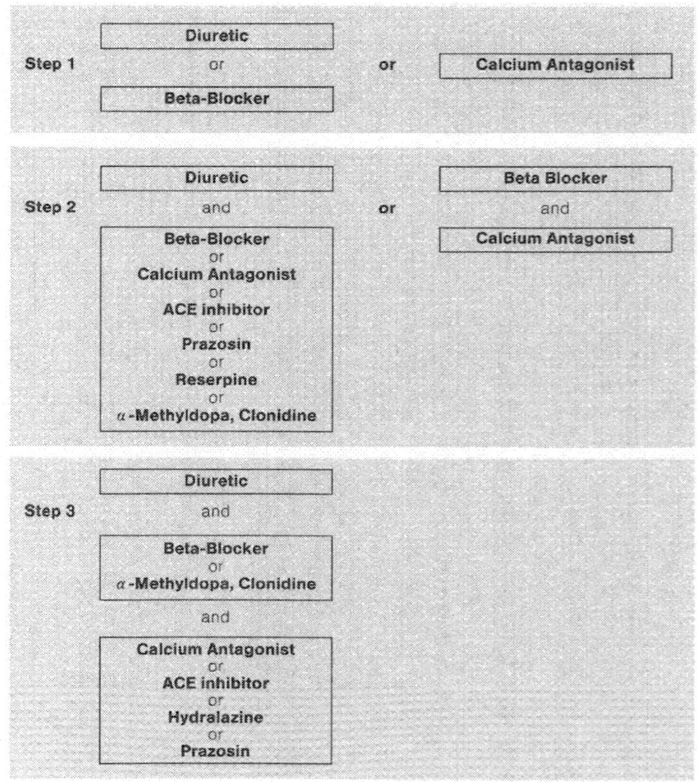

[1] Beta-blockers include: atenolol, metoprolol tartrate, nadalol, oxprenolol hydrochloride, pindolol, propranolol hydrochloride, and timolol maleate.
[2] These include: centrally acting adrenergic inhibitors (clonidine hydrochloride, guanabenz acetate, and methyldopa hydrochloride); peripherally acting adrenergic inhibitors (guanadrel sulfate and reserpine); an alpha-1 adrenergic blocker (prazosin hydrochloride); and the beta-blocking agents listed above.
[3] An angiotensin-converting enzyme inhibitor may be substituted at Step 2, Step 3 or Step 4 if side effects limit use of other agents or if other agents are ineffective. Slow channel calcium-entry blockers have not been approved for therapy in hypertension but may be acceptable as Step 2 or Step 3 drugs.
[4] Hydralazine, or minoxidil for resistant cases.

Fig. 23a. Comparison of *Stepped Care 1984* (The 1984 Report of the Joint National Committee on the Detection, Evaluation, and Treatment of High Blood Pressure) and *Stepped Care Plan for Drug Treatment of Hypertension* (German League Against Hypertension 1986; DL 1986 a)

Table 18. Concomitant diseases and supplementary criteria modifying the stepped care plan for antihypertensive therapy (German League Against Hypertension 1986 a; DL 1986)

Coronary heart disease	β-Receptor blockers, calcium antagonists
Congestive heart failure	Preference for diuretics, perhaps combined with postsynaptic α-receptor blockers, or ACE inhibitors
Diabetes mellitus	Restraint towards diuretics and β-receptor blockers
Gout	Restraint towards diuretics
Renal failure	If serum creatine above 2.0 mg/dl, administration of loop diuretics. Potassium-sparing diuretics contraindicated. Reduce dose of atenolol, nadolol and sotalol, and of ACE inhibitors.
Younger hypertensive individuals	Preference for β-receptor blockers
Older hypertensive individuals (>65 years)	Preference for diuretics or calcium antagonists

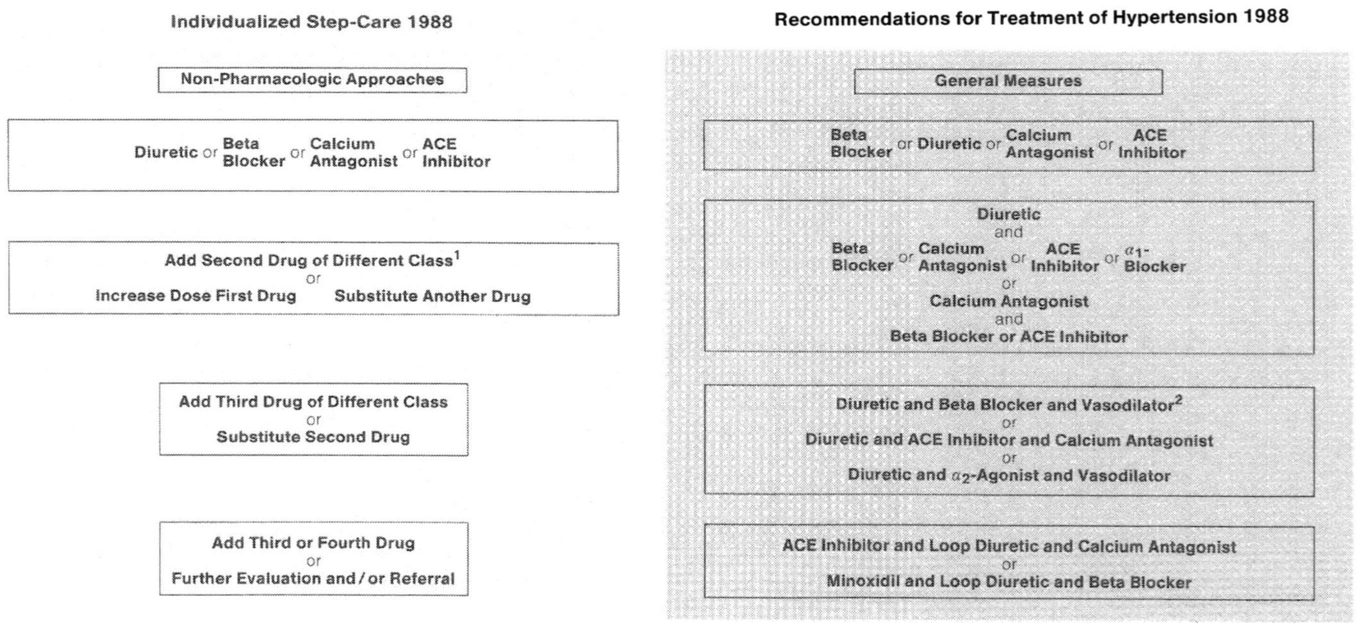

[1] Drugs such as diuretics, β-blockers, calcium antagonists, ACE-inhibitors, α-blockers, centrally acting α_2-agonists, rauwolfia serpentina, and vasodilators.

[2] Vasodilators: calcium antagonists, ACE-inhibitors, postsynaptic α_1-blockers, hydralazine.

Fig. 23 b. Comparison of *Individualized Step Care 1988* (The 1988 Report of the Joint National Committee on the Detection, Evaluation, and Treatment of High Blood Pressure) and *Recommendations for Treatment of Hypertension 1988* (German League Against Hypertension 1988; DL 1988)

Table 19. Reduction of cardiovascular events in treated hypertensive patients over 65 years of age. All studies without calcium antagonists (After Moser 1986)

EWPHE	−36%
VA study	−59%
ANBPS	−39%

not significantly influenced by the treatment. This study has by no means settled the controversy concerning the need for treatment of the elderly; it has nevertheless produced some strong arguments in favour of a cautious, individually tailored initial treatment, with a critical control for possible side effects. Especially in the older patient calcium antagonists appear to be well suited as antihypertensive therapy (Bühler et al. 1982a).

In all discussions concerning the preferential choice of certain drugs in both monotherapy and combined therapy for mild and severe forms of hypertension, it must always be borne in mind that in the large-scale intervention studies there are definite indicators of the annulment or diminution of therapeutic benefit by drug-related iatrogenic effects (Oliver 1986). All promising new therapeutic principles must therefore be subjected to efficacy studies (in respect of long-term effects on mortality and morbidity); for all antihypertensive agents used in the treatment of mild hypertension, the probability of fatal or irreversible side effects must be demonstrated not to exceed a rate of 1:1000 patient years. Seen even from this standpoint, the recommendations of the German League Against Hypertension in 1986 represent a carefully weighed pragmatic compromise, which, like the 1988 recommendations, of course still requires large-scale scientific confirmation. In this context, it must also be mentioned that the high expectations placed on β-receptor blockers in primary prevention in respect of a larger reduction in the mortality rate compared with diuretics have not been fulfilled either in the MRC[13] or the IPPPSH[14] or the HAPPHY (Wilhelmson et al. 1987) studies[15]. Only the MAPHY study[16], completed in the summer of 1987 (Wikstrand et al. 1988) showed a significantly lower mortality in patients treated with the β-blocker metoprolol (total mortality: −48%; cardiovascular mortality: −58%) than under diuretics (with comparable lowering of blood pressure in both groups).

4.2.1.4 Discovery of the Antihypertensive Action of Calcium Antagonists

During the 1960s various groups of workers observed that the coronary vasodilator and antiarrhythmic agent iproveratril (verapamil) produced a reduction in

[13] MRC: Medical Research Council Working Party
[14] IPPPSH: International Prospective Primary Prevention Study in Hypertension
[15] HAPPHY: Heart Attack Primary Prevention in Hypertension
[16] MAPHY: Metoprolol Atherosclerosis Prevention in Hypertensives

blood pressure as a side effect (Heidland et al. 1962). More specific investigations into the acute hypotensive action of verapamil were subsequently published by Brittinger et al. (1969, 1970), by Bender (1970), and by Belz (1970). These authors held the consensus view that rapid intravenous injection of 5 mg verapamil in various forms of hypertension, including renal hypertensive crisis, markedly lowers blood pressure, with a simultaneous reduction in central venous pressure. Signs of beginning or deteriorating heart failure due to the negative inotropic action of the drug were not observed.

Brittinger et al. recognized as early as 1970 that the hypotensive effect of verapamil was stronger with a high than with a low initial blood pressure, and that by infusion of verapamil the blood pressure levels attained with an initial intravenous dose could be maintained until the end of the hypertensive crisis. However, the primary development of calcium antagonists as selective coronary vasodilators and the antiarrhythmic action associated with both verapamil and diltiazem prevented acceptance and transfer of these findings to general antihypertensive therapy. Observations by Japanese authors at the beginning of the 1970s concerning an antihypertensive effect of verapamil and also of nifedipine in chronic hypertension were largely ignored since they did not readily fit the concept of an antiischaemic action for calcium antagonists. There was concern that, by lowering blood pressure, coronary perfusion pressure would be critically reduced and that the profile of coronary effects of these substances would be distorted.

In 1972, Grün and Fleckenstein presented an experimental study demonstrating the calcium dependence of the tone of systemic (arterial) resistance vessels and the acute relaxation of these vessels by verapamil and nifedipine. They drew attention to the fact that "calcium antagonists are vasodilators which lend themselves to the treatment of hypertension in a unique manner, in that they block the last pharmacologically influenceable link in the chain of vasoconstricting reactions in hypertension". Not until the close of the 1970s and start of the 1980s did general interest, outside Japan (Aoki et al. 1976) and physiologico-pharmacological laboratories, awake to the use of calcium antagonists in antihypertensive therapy. Initial investigations in this area concerned both the overall haemodynamic profile of individual calcium antagonists (with emphasis on the peripheral vasodilatory effect of the dihydropyridine group in particular, including their action on the cerebral vessels, without manifestation of a "steal" effect; Lydtin et al. 1975; Schmitz et al. 1975), and the demonstration of the compatibility of combining calcium antagonists with β-receptor blockers in healthy individuals and in patients with coronary heart disease (Lydtin et al. 1975; Ekelund and Orö 1976). The use of calcium antagonists as antihypertensive agents was advanced in a major way by Bühler and his co-workers in Basle (Bühler et al. 1982a, b), and during the last few years there has been a growing international awareness and acceptance of the antihypertensive effectiveness of calcium antagonists (Frohlich 1985; Gifford 1986; Opie and Jennings 1986; Ray et al. 1986).

The relatively late inclusion of calcium antagonists among the antihypertensive drugs explains their virtual omission from the large-scale intervention studies. Numerous investigations document their efficacy in reducing blood pressure in both monotherapy and combination therapy as well as in specific indications.

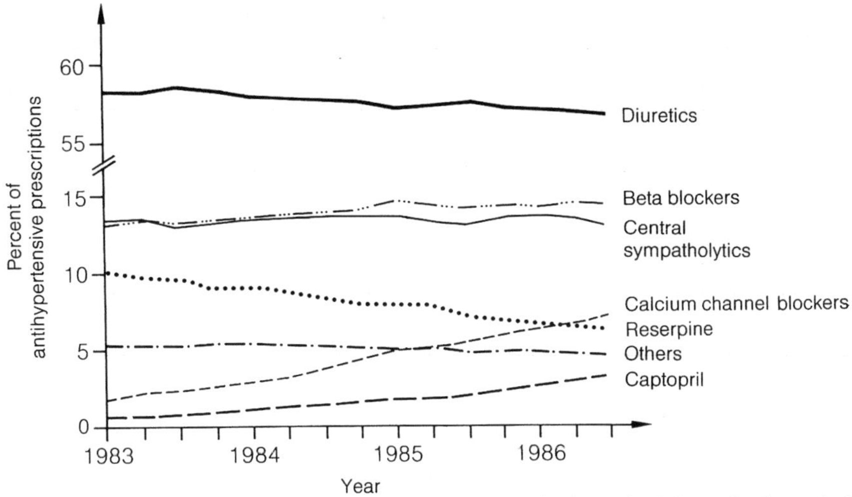

Fig. 24. Trends in the prescription of antihypertensive agents in the United States in the period between January 1983 and July 1986. The figure shows the relative percentages for each class of substances. (After Ray et al. 1986)

Furthermore, they are well tolerated and have a high degree of patient acceptance. However, cost–benefit considerations, particularly in cases of mild hypertension, call for proof of their effectiveness in respect to mortality since their increasing use will have considerable economic consequences. For instance, there was a 42% increase in the cost of antihypertensive medication in the German Federal Republic between 1982 and 1984, although the annual numbers of prescriptions remained virtually unchanged; similar figures apply to the United States (Ray et al. 1986; Moser 1986). Figure 24 shows the increasing use of calcium antagonists in comparison with other antihypertensive agents over the last 3 years in a region of the United States, which is probably typical of the country as a whole. In German-speaking countries calcium antagonists have probably captured an even greater share of the relevant market.

References

Amery A, Birkenhäger W, Brixko P, Bulpitt C, Clement D, Deruyttere M, De Schaepdryver A, Dollery C, Fagard R, Forette F, Forte J, Hamdy R, Henry JF, Joossens JV, Leonetti G, Lund-Johansen P, O'Malley K, Petrie J, Strasser T, Tuomilehto J, Williams B (1985) Mortality and morbidity results from the European working party on high blood pressure in the elderly Lancet 1: 1349–1354

ANBPS (1980) Management Committee: The Australian Therapeutic Trial in Mild Hypertension Lancet 1: 1261–1267

ANBPS (1982) Management Committee of the Australian Therapeutic Trial in Mild Hypertension: untreated mild hypertension Lancet 1: 185–191

ANBPS (1984) Management Committee of the Australian National Blood Pressure Study: prognostic factors in the treatment of mild hypertension Circulation 69: 668–676

Anlauf M, Weber F, Bock KD (1986) Milde Hypertonie: Risiko und Chance – Ergebnisse kontrollierter Studien Münch Med Wschr 128: 81–86

Aoki K, Yoshida T, Kato S, Tazumi K, Sato I, Takikawa K, Hotta K (1976) Hypotensive action and increased plasma renin activity by Ca^{2+}-antagonist (nifedipine) in hypertensive patients Jap Heart J 17: 479–484

Belz GG (1970) Die Wirkung des a-Isopropyl-a(N-methyl-N-homoveratril)-g-aminopropyl)-3,4,5–trimethoxyphenyl-acetonitril auf den Blutdruckdes Menschen bei oraler Medikation im akuten Versuch Arzneimittelforschung 20: 799–802

Bender F (1970) Die Behandlung der tachykarden Arrhythmien und der arteriellen Hypertonie mit Verapamil Arzneimittelforschung 20: 1301–1316

Brittinger DW, Strauch M, Huber W, Koch WD, von Henning GE, Wittenmeier KW, Twittendorf WD (1969) Iproveratril als Antihypertonikum bei krisenhafter renaler Hypertonie Dtsch Med Wschr 94: 945–948

Brittinger WD, Schwarzbeck A, Wittenmeier KW, Twittenhoff WD, Stegaru B, Huber W, Ewald RW, von Henning GE, Fabricius M, Strauch M (1970) Klinisch-experimentelle Untersuchungen über die blutdrucksenkende Wirkung von Verapamil Dtsch Med Wschr 95: 1871–1877

Bühler FR, Hulthen UL, Bolli P (1982a) Calcium channel inhibitors for identification of mechanisms and treatment of hypertension J Cardiovasc Pharmacol 4 (Suppl 3): 267–280

Bühler FR, Hulthen UL, Kiowski W, Müller FB, Bolli P (1982b) The place of the calcium antagonist verapamil in antihypertensive therapy J Cardiovasc Pharmacol 4 (Suppl 3): 350–357

Chobanian AV (1986) Antihypertensive therapy in evolution N Engl J Med 314: 1701–1702

Croog SH , Levine S, Testa MA, Brown B, Bulpitt CJ, Jenkins CD, Klerman GL, Williams GH (1986) The effects of antihypertensive therapy on quality of life N Engl J Med 314: 1657–1664

Cruickshank JM, Thorp JM, Zacharias FJ (1987) Benefits and potential harm of lowering high blood pressure Lancet 1: 581–584

DL (1986a) Deutsche Liga zur Bekämpfung des hohen Blutdruckes e.V. Empfehlungen zur Hochdruckbehandlung in der Praxis und zur Behandlung hypertensiver Notfälle 7. Auflage Heidelberg

DL (1986b) Deutsche Liga zur Bekämpfung des hohen Blutdruckes e.V. Kochsalz und Bluthochdruck 2. Auflage Heidelberg

DL (1988) Deutsche Liga zur Bekämpfung des hohen Blutdruckes e.V. Empfehlungen zur Hochdruckbehandlung in der Praxis und zur Behandlung hypertensiver Notfälle 8. Auflage Heidelberg

Dustan HP (1987) Rational therapies for hypertension-is step 1 of stepped care archaic? Circulation 75: 96–100

Editorial (1985) Treatment of hypertension in the over-60s Lancet 1: 1369–1370

Ekelund LG, Orö L (1976) Antianginal efficiency of Adalat with and without a beta-blocker. A subacute study with exercise tests In: Jatene AD, Lichtlen PR (eds) 3rd International Adalat Symposium Excerpta Medica Amsterdam pp 218–225

Frohlich ED (1985) Antihypertensive therapy: newer concepts and agents Cardiology 72: 349–365

Gifford Jr R W (1986) Essential hypertension: cost effective evaluation and treatment Am J Med 81 (Suppl 6C): 33–38

Grün G, Fleckenstein A (1972) Die elektromechanische Entkopplung der glatten Gefäß-Muskulatur als Grundprinzip der Coronardilatation durch 4-(2′-Nitrophenyl)-2,6-dimethyl-1,4-dihydropyridin-3,5-dicarbonsäure-dimethylester (Bay 1040, Nifedipin) Arzneimittelforschung 22: 334–344

HDFP (1979a) Hypertension detection and follow-up program cooperative group:five year findings of the hypertension detection and follow-up program. I. Reduction in mortality of persons with high blood pressure, including mild hypertension JAMA 242: 2562–2571

HDFP (1979b) Hypertension detection and follow-up program cooperative group: five year findings of the hypertension detection and follow-up program. II. Mortality by race, sex and age JAMA 242: 2572–2577

HDFP (1982a) Hypertension detection and follow-up program cooperative group: five year findings of the hypertension detection and follow-up program. III. Reduction in stroke incidence among persons with high blood pressure JAMA 247: 628–633

HDFP (1982b) Hypertension detection and follow-up program cooperative group: the effect of treatment on mortality in „mild" hypertension N Engl J Med 307: 976–980
Heidland A, Klütsch K, Öbek A (1962) Myogenbedingte Vasodilatation bei Nierenischämie Münch Med Wschr 104: 1636–1637
Helgeland A (1980) Treatment of mild hypertension: a five-year controlled drug trial – the Oslo study Am J Med 69: 725–732
IPPPSH (1985) Cardiovascular risk and risk factors in a randomized trial of treatment based on the beta blocker oxprenolol: the international prospective primary prevention study in hypertension. The IPPPSH collaborative group J Hypertens 3: 379–392
Kaplan NM (1984) Systemic hypertension: mechanisms and diagnosis – therapy In: Braunwald E (ed) Heart disease – a textbook of cardiovascular medicine. 2nd edition WB Saunders Philadelphia pp 849–927
Lydtin H, Lohmöller G, Lohmöller R, Walter I (1975) Hemodynamic studies on nifedipine in man In: Hashimoto K, Kimura E, Kobayashi T (eds) 1st International Nifedipine „Adalat" Symposium University of Tokyo press Tokyo pp 97–106
Moser M (1986) Treating hypertension – a review of clinical trials Am J Med 81 (Suppl 6C): 25–32
MRC (1977) Medical Research Council Working Party: randomised controlled trial of treatment for mild hypertension: design and pilot trial Br Med J1: 1437–1440
MRC (1985) Medical Research Council Working Party: Medical Research Council Trial of treatment for mild hypertension: principal results Br Med J 291: 97–104
Oliver MF (1986) Prevention of coronary heart disease – propaganda, problems and prospects Circulation 73: 1–9
Opie LH, Jennings AA (1986) Role of calcium channel blockade in chronic hypertension following successful management of hypertensive emergency Am J Med 81 (Suppl 6A): 35–42
Ray WA , Schaffner W , Oates JA (1986) Therapeutic choice in the treatment of hypertension initial treatment of newly diagnosed hypertension and secular trends in the prescribing of antihypertensive medications for Medicaid patients Am J Med 81 (Suppl 6C): 9–19
Saunders E (1986) Stepped care and profiled care in the treatment of hypertension: considerations for black americans Am J Med 81 (Suppl 6C): 39–44
Schmitz H, Schierl W, Beck O, Lydtin H (1975) Über die Wirkung von Nifedipin (Adalat) auf regionale Hirndurchblutung und Unterschenkeldurchblutung In: Schlegel B (ed) Verhandlungen der Deutschen Gesellschaft für Innere Medizin Bergmann München pp 888–890
The 1984 report of the Joint National Committee on Detection, Evaluation and Treatment of High Blood Pressure (1984) Arch Int Med 144: 1045–1057
The 1988 report of the Joint National Committee on Detection, Evaluation and Treatment of High Blood Pressure (1988) Arch Int Med 148: 1023–1038
VA (1967) Veterans Administration Cooperative Study Group on Antihypertensive Agents: effects of treatment on morbidity in hypertension. I. Results in patients with diastolic blood pressures averaging 115 through 129 mmHg JAMA 202: 1028–1034
VA (1970) Veterans Administration Cooperative Study Group on Antihypertensive Agents. II. Results in patients with diastolic blood pressure averaging 90 through 114 mmHg JAMA 213: 1143–1152
Vetter H (1986) Fortschritte in der Therapie der Hypertonie? Münch Med Wschr 128: 823
Wikstrand J, Warnold I, Olsson G, Tuomilehto J, Elmfeldt D, Berglund G (1988) Primary prevention with metoprolol in patients with hypertension – mortality results from the MAPHY study JAMA 259: 1976–1982
Wilhelmsen L, Berglund G, Elmfeldt D, Fitzsimons T, Holzgreve H, Hosie J, Hörnkvist PE, Pennert K, Tuomilehto J, Wedel H (1987) Beta-blockers versus diuretics in hypertensive men: main results from the HAPPHY trial J Hypertension 5: 561–572
Williams GH (1987) Quality of life and its impact on hypertensive patients Am J Med 82: 98–105

4.2.2 Mechanism of Antihypertensive Action of Calcium Antagonists

In borderline and established hypertension, cardiac output and blood viscosity are, as a rule, normal; elevated blood pressure is the consequence of an increased peripheral vascular resistance. The basic abnormality remains an inappropriately high peripheral resistance, even though in the early stages of essential hypertension cardiac output can also be increased. Vascular resistance depends directly on the tone of the arteriolar smooth muscle. The contractile process in the vascular muscle cell is activated by an increase in free cytoplasmic calcium concentration. Calcium antagonists inhibit transmembrane entry of calcium ions and thus lower the free intracellular calcium ion concentration, and thereby the tone of the resistance vessels. The inhibitory effect of calcium antagonists on the vascular smooth muscle cell involves: basal tone ("myogenic tone"); α-adrenergic activation by adrenaline (epinephrine) and noradrenaline (norepinephrine); activation by nonadrenergic mediators (e.g. serotonin, angiotensin II); and stimulation of the system by electrical and mechanical impulses, or by other factors such as metabolites of arachidonic acid. The elucidation of this basic action was the result of extensive studies by Fleckenstein and his co-workers (for details see Sects. 1.1.3, 1.2.3, 3.1.5, 3.1.6; Vanhoutte 1985; Fleckenstein et al. 1986). Thus, calcium antagonists lower excessive peripheral vascular resistance by targeted intervention at the critical effector of vasoconstriction. As a consequence of vasodilation with a concomitant reduction in blood pressure there is, via the baroreceptor reflex, vagal withdrawal and activation of the sympathetic nervous system. These effects diminish the primary antihypertensive action (predominantly on the systolic blood pressure), above all among young hypertensive individuals (Bolli et al. 1986), and overshadow and mask any direct effects of these substances on the heart. This applies to healthy individuals, patients with coronary heart disease, hypertensive test animals and hypertensive patients – to the dihydropyridine group, verapamil and diltiazem (Lydtin and Lohmöller 1972; Lydtin et al. 1975; Bolt and Saxena 1984; Rosatti et al. 1984; Muir et al. 1985; Nordlander 1985; Wathen et al. 1985; Cody 1987).

An analysis of the antihypertensive action mechanism of calcium antagonists must today take into account the topical and controversial discussion of the role of calcium metabolism in the pathophysiology and pathogenesis of essential hypertension (Cohn 1983; Editoral Lancet 1983; Bühler et al. 1984a, b; McCarron and Morris 1985; McCarron 1985; Resnick 1985, 1986; Editoral Lancet 1986; Ferlinz 1986; Kaplan and Meese 1986). Such an analysis has to start from the paradoxical constellation of two sets of findings (a and b):

a) An excess of calcium can raise blood pressure: population surveys have shown a positive correlation between blood pressure and serum calcium levels (Keteloot and Geboers 1982). Excessively high calcium levels have been demonstrated in the platelets (cytosol) of hypertensive individuals (Erne et al. 1984); plasma from patients with essential hypertension contains a substance which increases the calcium concentration in the platelet cytosol of normal individuals (Lindner et al. 1987). The chronic hypercalcaemia in hyperparathyroidism and vitamin D intoxication is associated with a raised incidence of hypertension. In spontaneously hypertensive test animals and also in the elderly, the calcium content of the arterial walls increases (Fleckenstein et al. 1986). Hypercalcaemia is statistically associated with raised mortality from cardiovascular diseases (Palmer et al. 1987).

b) A reduced supply of calcium can also raise blood pressure: independent researchers have reported findings that an increase in dietary calcium supply can lower blood pressure in young normotensive adults, in elderly hypertensive individuals with osteoporosis, and in a proportion of patients with essential hypertension (McCarron and Morris 1985; Grobbee and Hofman 1986). A negative correlation between dietary calcium intake and blood pressure has even been observed (McCarron et al. 1984, quoted by Resnick 1985). The "calcium deficiency hypothesis" derived from these findings was finally subjected to criticism by Kaplan and Meese (1986), who conclude that the available data are by no means sufficient to risk premature general recommendations for dietary calcium therapy in hypertension.

The following points must also be considered:

c) Intracellular calcium concentration is regulated by a sodium/calcium exchange mechanism: an increase in the intracellular sodium content in hypertension hampers calcium outflow, thereby raising the intracellular calcium concentration, which ultimately controls the contractile process in vascular smooth muscle cells (Blaustein 1977). This defect in cellular sodium elimination is found especially in older hypertensive individuals with low initial renin levels. Physiological vasodilators can, by stimulating Na^+, K^+-ATPase, activate the sodium/calcium exchange mechanism, thereby reducing the concentration of calcium dissolved in the cytosol. The hypertensive effect of table salt appears to go hand-in-hand with a lowering of serum calcium ions by shifting calcium into the cells; vascular resistance is thereby raised (Resnik 1986).

d) The antihypertensive effect of calcium antagonists is positively correlated with the initial blood pressure level, the individual's age and the pretreatment plasma adrenaline (epinephrine) level, but is negatively correlated with initial levels of renin and of angiotensin II (MacGregor et al. 1983; Bühler et al. 1984a, b; Bolli et al. 1986). Diuretics show a similar antihypertensive profile, while β-receptor blockers are more effective in younger hypertensive individuals, who also exhibit raised renin levels more frequently (Bühler et al. 1984a, b).

e) Calcium antagonists can increase sodium and water elimination (Kazda et al. 1984a, b); the renin–angiotensin–aldosterone system is stimulated by calcium antagonists to a much lesser degree than by other, nonspecific vasodilators; furthermore, calcium antagonists appear to inhibit release of aldosterone by angiotensin II (Hollenberg 1986; Sluiter et al. 1986).

f) The antihypertensive effect of calcium antagonists can be diminished by a low-salt diet (Kingswood and Thompson 1986; Morgan et al. 1986).

g) Reduction of blood pressure by calcium antagonists (as by β-receptor blockers and ACE inhibitors) leads, in contrast to antihypertensive therapy with, for instance, diuretics, hydralazine and minoxidil, to a regression of an established hypertrophy of the heart muscle (Tarazi 1983; Strauer et al. 1984, 1986; Zähringer 1985; Zähringer et al. 1986; Wetzchewald et al. 1989). This holds not only for the dihydropyridine group, but also for verapamil (Muiesan et al. 1986).

As to (a) and (b): it is difficult to imagine that small changes in extracellular calcium concentration directly influence intracellular availability of calcium ions, since there is a 10000:1 gradient between extracellular calcium and the concentration in the cytosol. Far more likely is suppression of the parathormone by dietary calcium supply. Parathormone could have a "trailer" function, taking calcium into the cell, and its suppression could also reduce blood pressure by reducing available intracellular calcium (Palmieri et al. 1986).

Hypertensive individuals with low renin levels (i.e. the elderly, black people and those with hyperaldosteronism) also manifest a certain "calcium deficiency": they have, on average, lower extracellular calcium concentrations and higher parathormone and calcitrol levels than normal or hypertensive individuals with high renin levels. On the other hand, patients with high renin exhibit the characteristics of an "excess of calcium": relatively high levels of ionized calcium and calcitonin (Resnick 1986).

The findings cited in (a) and (b) may not be so contradictory: they could be explained by an inhomogeneity in the pathogenesis of the hypertension in the subjects examined. Only a proportion of hypertensive individuals respond to an increase in dietary calcium supply with a reduction in blood pressure: those with low renin, low ionized calcium, high parathormone and low calcitonin. It is hardly surprising that the hypertensive target group which will respond to a raised dietary calcium supply cannot yet be clearly defined: even though the effect of salt restriction in lowering blood pressure has been known for nearly 50 years, it has not been possible to define the target group with any degree of precision (Laragh and Pecker 1983).

As to (d): calcium antagonists are, like diuretics, especially effective in hypertensive black people. This can be explained by the high proportion of low-renin hypertensive individuals among this group.

As to (e) and (f): although calcium antagonists are vasodilators, they do not produce an increase in body weight (as, for instance, does minoxidil). Sodium retention with corresponding weight gain is a side effect of many antihypertensive agents, and this is one of the reasons why diuretics are often concomitantly prescribed. It is thus not immediately surprising if in the absence of sodium retention the effect of salt restriction is also very small. On the other hand, sodium restriction and calcium antagonists could both use the same antihypertensive "final pathway". According to the Blaustein–DeWardener hypothesis, an excess of sodium leads to a rise in intracellular calcium concentration and thereby to vasoconstriction. Sodium restriction reverses this effect and thereby deprives calcium antagonists of their site of action (Morgan et al. 1986). Conversely, one may also interpret the results of Morgan et al. (1986) in the sense that salt intake reinforces the antihypertensive effect of calcium antagonists, since salt supply reduces renin activity and the concentration of calcium ions in the plasma. Thus, low renin activity and low calcium concentration in the plasma, which occurs particularly with low renin levels, become indicators of a good response to antihypertensive therapy with calcium antagonists.

In Summary. The preceding discussion of the mechanism of antihypertensive action of calcium antagonists in the context of calcium metabolism allows the following practical inferences to be drawn regarding the use of calcium antagonists as antihypertensive agents:

1. Calcium antagonists inhibit vasconstriction in a targeted manner and thereby reduce elevated vascular resistance in hypertension.
2. Their effect is positively correlated with the initial blood pressure level and the age of the patient.
3. Their effect is negatively correlated with initial renin and angiotensin II levels.
4. Dietary salt restriction can annul the antihypertensive action of calcium antagonists. Calcium antagonists do not usually produce salt and water retention.
5. In younger patients receiving antihypertensive monotherapy with calcium antagonists, reactive sympathetic nervous activation weakens the antihypertensive effect; in this situation an antihypertensive combination therapy (or monotherapy) with β-receptor blockers can be considered. In older patients consideration should first be given to a combination with a diuretic.

6. Regression of cardiac hypertrophy under calcium antagonists provides a strong physiological argument for their inclusion in the antihypertensive treatment plan, irrespective of the patient's age.
7. The antihypertensive effect of an increased dietary calcium intake in a still imprecisely defined subgroup of hypertensive individuals is only apparently in contradiction to the antihypertensive effectiveness of calcium antagonists.

References

Blaustein MP (1977) Sodium ions, calcium ions, blood pressure regulation and hypertension: a reassessment and a hypothesis Am J Physiol 233: C 165–173

Bolli P, Bühler FR, Kiowski W, Müller FB (1986) Betarezeptorenblocker, Kalziumantagonisten und Diuretika in der Hochdrucktherapie In: Distler A (ed) Ca^{++}-Antagonisten und Beta-Rezeptorenblocker in der Behandlung von Hypertonie und koronarer Herzkrankheit Schattauer Stuttgart 41–52

Bolt GR, Saxena PR (1984) Acute systemic and regional hemodynamic effects of felodipine, a new calcium antagonist, in conscious renal hypertensive rabbits J Cardiovasc Pharmacol 6: 707–712

Bühler FR, Bolli P, Hulthén UL (1984a) Calcium influx-dependent vasoconstrictor mechanisms in essential hypertension In: Opie LH (ed) Calcium antagonists and cardiovascular disease Raven Press New York pp 313–321

Bühler FR, Erne P, Müller FB, Hulthen UL, Kiowski W, Block LH, Bolli P (1984b) Calcium-antagonists in hypertension – a new therapeutic concept In: Althaus U, Burckhardt D, Vogt E (eds) Calcium-Antagonismus -Internat. Symposium on Calcium-Antagonism Universimed Frankfurt pp 116 – 123

Cody RJ (1987) The hemodynamics of calcium-channel antagonists in hypertension: vascular and myocardial responses Circulation 75 (Suppl 1): 175–179

Cohn JN (1983) Calcium, vascular smooth muscle, and calcium entry blockers in hypertension Ann Intern Med 98 (Part 2): 806–809

Editorial (1983) Calcium antagonists and blood-pressure Lancet 2: 22–24

Editorial (1986) Hypertension: Is there a place for calcium Lancet 1: 359–361

Erne P, Bolli P, Bürgisser E, Bühler FR (1984) Correlation of platelet calcium with blood pressure: effect of antihypertensive therapy N Engl J Med 310: 1084–1088

Ferlinz J (1986) Nifedipine in myocardial ischemia, systemic hypertension, and other cardiovascular disorders Ann Intern Med 105: 714–729

Fleckenstein A, Frey M, Zorn J, Fleckenstein-Grün G (1986) Antihypertensive, anticalcinotische und antiarteriosklerotische Effekte von Calcium-Antagonisten In: Distler H (ed) Calcium-Antagonisten in der Hochdrucktherapie Schattauer Stuttgart New York pp 3–24 A

Grobbee DE, Hofman A (1986) Effect of calcium supplementation on diastolic blood pressure in young people with mild hypertension Lancet 2: 703–706

Hollenberg NK (1986) Strategies in antihypertensive therapy: implications of the kidney Am J Med 81 (Suppl 6A): 15–19

Kaplan NM, Meese RB (1986) The calcium deficiency hypothesis of hypertension: A critique Ann Intern Med 105: 947–955

Kazda S, Garthoff B, Luckhaus G (1984) Mode of antihypertensive action of nitrendipine J Cardiovasc Pharmacol 6: 956–962

Kesteloot H, Geboers J (1982) Calcium and blood pressure Lancet 1: 813–815

Kingswood JC, Thompson FD (1986) Sodium, blood pressure, and calcium antagonists Lancet 1: 1102

Laragh JH, Pecker MS (1983) Dietary sodium and essential hypertension: some myths, hopes and truths Ann Intern Med 98: 735–743

Lindner A, Kenny M, Meacham AJ (1987) Effects of a circulating factor in patients with essential hypertension on intracellular free calcium in normal platelets N Engl J Med 316: 509–513

Lydtin H, Lohmöller R (1972) Hämodynamische Wirkungen von Bay a 1040 vor und nach Atropingabe In: Schlegel B (ed) Verhandlungen der deutschen Gesellschaft für innere Medizin Bergmann München pp 1554–1559

Lydtin H, Lohmöller G, Lohmöller R, Walter I (1975) Hemodynamic studies on nifedipine in man In: Hashimoto K, Kimura E, Kobayashi T (eds) 1st International Nifedipine „Adalat" Symposium University of Tokyo Press Tokyo pp 97–106

McCarron DA (1985) Calcium in the pathogenesis and therapy of human hypertension Am J Med 78 (Suppl 2B): 27–29

McCarron DA, Morris CD (1985) Blood pressure response to oral calcium in persons with mild to moderate hypertension Ann Intern Med 103: 825–831

MacGregor GA, Markandu ND, Rotellar C, Smith SJ, Sagnella GA (1983) Contrasting effects of nifedipine in normotensive and hypertensive subjects: a functional abnormality of smooth muscle in essential hypertension? In: Kaltenbach M, Neufeld HN (eds) 5th International Adalat Symposium Excerpta Medica Amsterdam pp 111–117

Morgan T, Anderson A, Wilson D, Myers J, Murphy J, Nowson C (1986) Paradoxical effect of sodium restriction on blood pressure in people on slow-channel calcium blocking drugs Lancet 1: 793

Muiesan G, Agabiti-Rosei E, Romanelli G, Muiesan ML, Castellano M, Beschi M (1986) Adrenergic activity and left ventricular function during treatment of essential hypertension with calcium-antagonists Am J Cardiol 57:44D–49D

Muir AL, Wathen CG, Hannan WJ (1985) Effects of felodipine on resistance and capacitance vessels in patients with essential hypertension Drugs 29 (Suppl 2): 59–65

Nordlander M (1985) Haemodynamic effects of short and long term adminisitration of felodipine in spontaneously hypertensive rats Drugs 29 (suppl 2) 90–101

Palmer M, Adami HO, Bergström R, Jakobsson S, Akerström G, Ljunghall S (1987) Survival and renal function in untreated hypercalcaemia Lancet 1: 59–62

Palmieri GMA, Bertorini TE, Nutting DF (1986) Calcium and blood pressure Lancet 2: 1106

Resnick LM (1985) Calcium and hypertension: The emerging connection Ann Intern Med 103: 944–946

Resnick LM, Müller FB, Laragh JH (1986) Calcium-regulating hormones in essential hypertension Ann Intern Med 105: 649–654

Resnick LM (1986) Calcium metabolism, renin activity, and the antihypertensive effects of calcium channel blockade Am J Med 81 (Suppl 6A): 6–14

Rosatti F, Cella PL, Fici F, Rossi F, Chieppa S, Marmo E (1984) Diltiazem in the therapy of arterial hypertension: Polygraphic assessment of left ventricular performance Current Therapeutic Research 36: 701–711

Sluiter HE, Wetzels JFM, Huysmans FTM, Koene RAP (1986) The dihydropyridine calcium antagonist, felodipine, antagonizes the renal effects of exogenous angiotensin II in normotensive volunteers J Hypertension 4 (Suppl 6): 174–176

Strauer BE, Mahmoud MA, Bayer F, Bohn I, Motz U (1984) Reversal of left ventricular hypertrophy and improvement of cardiac function in man by nifedipine Eur Heart J 5 (Suppl F): 53–60

Strauer BE (1986) Der Stellenwert pharmakotherapeutischer Maßnahmen in der Rückbildung der Herzhypertrophie In: Distler A (ed) Ca^{++}-Antagonisten und Beta-Rezeptorenblocker in der Behandlung von Hypertonie und koronarer Herzkrankheit Schattauer Stuttgart pp 41–52

Tarazi R (1983) Regression of left ventricular hypertrophy by medical treatment:present status and possible implications Am J Med 75: 80–86

Vanhoutte PM (1985) Calcium-entry blockers, vascular smooth muscle and systemic hypertension Am J Cardiol 55: 17B-23B

Wathen CG, Hannan WJ, Turnbull LW, Muir AL (1985) Afterload reduction by nifedipine – the acute haemodynamic response to exercise in hypertensive subjects Eur Heart J 6: 358–364

Wetzchewald D, Klaus D, Garanin G, Hoffmann J, Lohr E (1989) Regression of left ventricular hypertrophy by felodipine or a combination of felodipine and metoprolol J Cardiovasc Pharmacol (in press)

Zähringer J (1985) Biochemische Veränderungen bei Herzmuskelhypertrophie – Regression Z Kardiol 74 (Suppl 1): 1–8

Zähringer J , Stangl E, Aschauer W, Motz W, Strauer BE (1986) Stellenwert der Calcium-Antagonisten hinsichtlich Herzmuskelhypertophie-Regression bei arterieller Hypertonie In: Distler A (ed) Calcium-Antagonisten in der Hochdrucktherapie Schattauer Stuttgart pp 357–364

4.2.3 Monotherapy with Calcium Antagonists in Hypertension

The available literature convincingly documents that all calcium antagonists can independently lower raised blood pressure. Numerous studies of various design and quality, above all with nifedipine and verapamil, and more recently with nitrendipine, felodipine, nisoldipine, diltiazem and gallopamil, demonstrate the antihypertensive effectiveness of this group of substances. Table 20 summarizes, without any claim to completeness, some recent studies involving various substances and doses. Several reviews in this area are also available: Stone et al. (1980), Guazzi et al. (1983), Rahn (1984), Ferlinz (1986), and Saltiel et al. (1988).

Practically all investigations have shown that the antihypertensive effect of calcium antagonists increases with increasing initial blood pressure levels. In normotensive subjects blood pressure is generally unaffected or lowered only very slightly. The calcium antagonists studied all have comparable antihypertensive effectiveness in monotherapy, independent of their chemical structure and specific pharmacological profile.

Although, following administration of dihydropyridines, the heart rate tends to rise during the initial phase of treatment, it returns to baseline values after a period of continued treatment. In the case of extremely high initial blood pressure levels, a decline in heart rate is also observed following the administration of nifedipine (perhaps explainable by unloading in congestive heart failure). No doubt, in older patients autonomic counter-regulation via the baroceptor reflex with activation of the sympathetic nervous system is not so pronounced (Abernethy et al. 1986). With the use of the verapamil group and diltiazem as antihypertensives the heart rate is more likely to be reduced, or to remain unchanged (Lewis et al. 1981).

Calcium antagonists lower blood pressure independent of body position (supine, sitting or standing) and of whether the individual is resting or exercising, without any tendency to orthostatic hypotension (Midtbo and Hals 1980). Cardiac output and stroke volume increase under calcium antagonists in both hypertensive and normotensive individuals as a consequence of the reduction in afterload (Amodeo et al. 1986). The parameters of cardiac contractility are positively influenced by afterload reduction and partly no doubt also by reflex sympathetic activation; even the negative inotropic action of the verapamil group is partially masked by their peripheral action.

All calcium antagonists appear in antihypertensive application to have in common the ability to produce an initial elevation of renal blood flow and renal sodium excretion (Amodeo et al. 1986). Although there have to date been no truly critically evaluated, sufficiently large-scale and sufficiently long comparative studies of calcium antagonists against other antihypertensive substances, calcium antagonists have nevertheless fared well in almost all the hitherto small-scale comparative investigations with respect to side effects (Leonetti et al. 1981). With

Table 20. Antihypertensive monotherapy with calcium antagonists (selection of representative studies since 1980). The quantities stated refer to total daily dose (usually divided into two or three separate doses)

Diltiazem		
Amodeo et al.	1986	240–360 mg
Brandt and Klein	1982	90–360 mg
Giesecke et al.	1982	90–270 mg
Pool et al.	1986	120–360 mg
Schulte et al.	1983	180–270 mg
Trimarco et al.	1984	240 mg
Yamuchi et al.	1986	180 mg
Felodipine		
Aberg et al.	1985	15–30 mg
Cooperative Study Gp.	1987	5–20 mg
Elmfeldt and Hedner	1985	15 mg
Harrington et al.	1987	5–20 mg
Krönig and Flygt	1987	10–20 mg
Leonetti et al.	1985	(37.5–150 mg[a]) 20 mg
Lorimer et al.	1985	7.5–22.5 mg
Gallopamil		
Franz et al.	1986	100 mg
Sucic	1983	100 mg
Nifedipine		
Aberg et al.	1985	30–60 mg
Klein et al.	1982	30–60 mg
Opie and Jee	1984	10–30 mg
Pacheco et al.	1986	30–60 mg
Rosendorff	1984	15–60 mg
Rosendorff and Goodman	1985	40 mg
Schulte et al.	1983	40–60 mg
Ueda and Murakami	1983	30–80 mg
Nisoldipine		
Deeg et al.	1987	5/10/20 mg
Rosendorff and Godman	1985	10 mg
Nitrendipine		
Black and Vlachakis	1984	20–40 mg
Ferrara et al.	1985	20 mg
Franz and Wiewel	1985	20 mg
Hansson et al.	1984	20 and 40mg
Johnson et al.	1986	20 mg
M'Buyamba-Kabangu et al.	1986	20–60 mg
Müller et al.	1986	20–80 mg
Rüddel et al.	1986	20–40 mg
Stoepel et al.	1986	10–40 mg
Vitacolonna et al.	1986	10–20 mg
Verapamil		
Belz and Spiesu	1985	480 mg
Bühler et al.	1982	240–720 mg
Corea et al.	1983	240 mg
De Leeuw and Birkenhäger	1984	120–480 mg
Doyle et al.	1981	360 mg
Edmonds et al.	1986	240 mg
Gould et al.	1982	360–480 mg
Hedbäck and Hermann	1984	360 mg
Leonetti et al.	1980	240–360 mg
Leonetti et al.	1981	360–480 mg
Lewis et al.	1981	240–360 mg
MacGregor	1984	240–480 mg
Mitbo and Hals	1980	480 mg

[a] Provisional tablet form (no longer used)

substances of the dihydropyridine group there is, especially during the first few days of treatment, a slightly increased incidence of headache, dizziness and flushing; with verapamil (and probably also with diltiazem) constipation seems to be slightly more frequent (Hedbäck and Hermann 1984).

With regard to their antihypertensive effectiveness, calcium antagonists in monotherapy are judged to be comparable to diuretics and β-receptor blockers (Doyle et al. 1981; Bühler et al. 1982, Holmes et al. 1988; Dunlay et al. 1988; Fig. 25). In a more focused comparative analysis of the antihypertensive effects of calcium antagonists, β-receptor blockers and diuretics in monotherapy with respect to age, calcium antagonists and diuretics in monotherapy seem to be more effective in older patients and hypertensive individuals with low renin activity than do β-receptor blockers, which, in turn, have a stronger antihypertensive

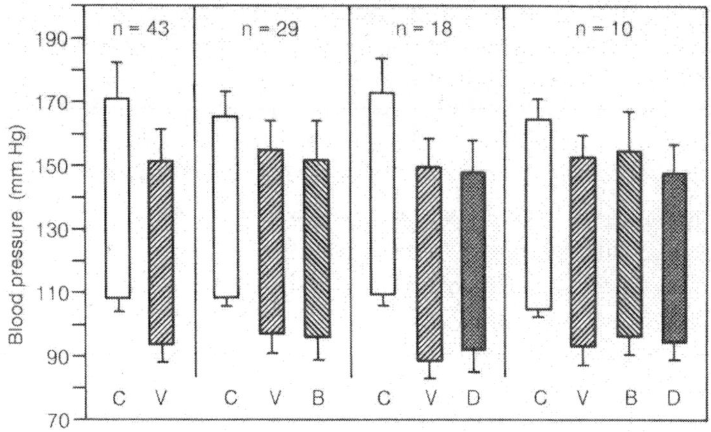

Fig. 25. Antihypertensive monotherapy. Comparison of the effect of verapamil (*V*), of a *β*-receptor blocker (*B*), and of a diuretic (*D*); *C* = control. (After Bühler et al. 1984)

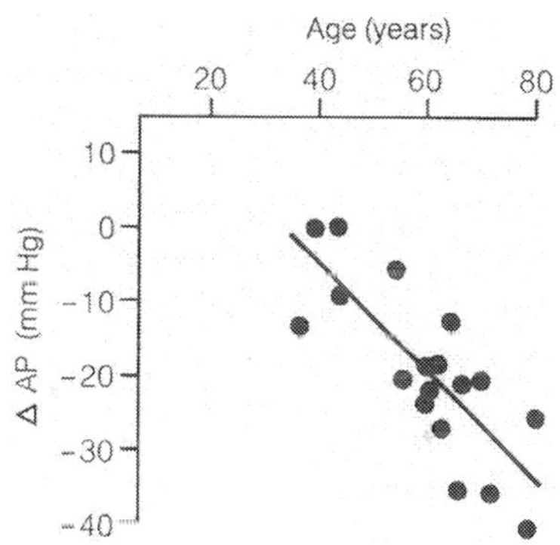

Fig. 26. Relationship between age and lowering of blood pressure by a calcium antagonist (verapamil). ΔAP change in mean arterial pressure. (After Bühler et al. 1984)

effectiveness in younger patients. *β*-Receptor blockers show a positive correlation between antihypertensive effect and pretreatment plasma renin levels; for calcium antagonists (and diuretics) the correlation is negative (Bühler et al. 1982, 1984). This favours the use of calcium antagonists as antihypertensive monotherapy in older patients, who frequently have low renin levels (Fig. 26).

From a pathophysiological standpoint, the ideal uniform lowering of blood pressure over a 24-h period presupposes a relatively constant plasma level of the antihypertensive agent. Slow-release formulations (Sect. 2.2.6) were therefore designed for calcium antagonists with short elimination half-life values (Sect. 2.2). These have also been used in antihypertensive therapy and the results achieved correspond in general with the effects of non-slow-release medication, but dosage intervals can be extended by the use of such sustained-release formulations, the total dose being given as one or two single doses. Calcium antagonists with long half-life values (e.g. nitrendipine and felodipine) offer the advantage of easier pharmacokinetic control; the total daily dose can be given as one or two single doses. In our own experience an equal antihypertensive effect can also be obtained with slow-release nifedipine. Overall, experience in medical practice (Table 20) confirms that the total daily dose required in monotherapy with verapamil and diltiazem varies to a greater extent than with substances of the dihydropyridine group (especially the newer dihydropyridines, nitrendipine and felodipine). With regard to patient compliance, the advantage of taking a medicine only once (or at most twice) a day is a distinct advantage.

In Summary. The following situations favour the use of calcium antagonists as initial antihypertensive monotherapy:

1. Age, the higher the better
2. Low plasma renin levels
3. Lipid metabolism disorders
4. Diabetes mellitus
5. Peripheral vascular disease, especially Raynaud's syndrome
6. When other antihypertensive medication is poorly tolerated.

The antihypertensive treatment plan for the patient with manifest coronary heart disease will invariably include a β-receptor blocker.

The choice of an individual calcium antagonist should be determined principally by the extent of general experience with a particular substance. Uncommon, but substance-specific side effects are often recognized only after protracted trials, although their consideration in the cost – benefit analysis of primary prophylaxis is essential. In the context of monotherapy of hypertension, this favours, after diuretics and β-receptor blockers, the use of established dihydropyridine derivatives (above all nifedipine), verapamil and diltiazem.

References

Aberg H, Lindsjö M, Mörlin B (1985) Comparative trial of felodipine and nifedipine in refractory hypertension Drugs 29 (Suppl 2): 117–123

Abernethy DR, Schwartz JB, Todd EL, Luchi R, Snow E (1986) Verapamil pharmacodynamics and disposition in young and elderly hypertensive patients – altered electrocardiographic and hypotensive responses Ann Intern Med 105: 329–336

Amodeo C, Kobrin I, Ventura HO, Messerli FH, Frohlich ED (1986) Immediate and short-term hemodynamic effects of diltiazem in patients with hypertension Circulation 73: 108–113

Belz GG, Spies G (1985) Kontrollierte Studie zur Behandlung der Hypertonie mit Verapamil in Retardform Z Kardiol 74: 453–459

Black HR, Vlachakis N (1984) Once and twice daily nitrendipine therapy in essential hypertension In: Scriabine A, Vavov S, Deck K (eds) Nitrendipine Urban & Schwarzenberg Baltimore pp 509–518

Brandt D, Klein W (1982) Dosiswirkungsbeziehung von Diltiazem bei Patienten mit essentieller Hypertonie In: Bender F, Greeff K (eds) Calciumantagonisten zur Behandlung der Angina pectoris, Hypertonie und Arrhythmie Excerpta Medica Amsterdam pp 202

Bühler FR, Hulthén UL, Kiowski W, Müller FB, Bolli P (1982) The place of the calcium antagonist verapamil in antihypertensive therapy J Cardiovasc Pharmacol 4: 350–357

Bühler FR, Erne P, Müller FB, Hulthen UL, Kiowski W, Block LH Bolli P (1984) Calcium-antagonists in hypertension – a new therapeutic concept In: Althaus U, Burckhardt D, Vogt E (eds) CalciumAntagonismus – Internat. Symposium on Calcium-Antagonism Universimed Frankfurt pp 116–123

Cooperative Study Group (1987) Felodipine, a new calcium antagonist, as monotherapy in mild or moderate hypertension Drugs 34 (Suppl 3): 139–148

Corea L, Bentivoglio M, Verdecchia P, Bianchini C (1983) Long term verapamil therapy in mild-to-moderate hypertension – randomized study of normal and slow release formulations Acta Ther 9: 263–274

Deeg P, Weiss KH, Schmitz H (1987) Nisoldipin in der Therapie der essentiellen Hypertonie Dtsch Med Wschr 112: 429–433

De Leeuw PW, Birkenhäger WH (1984) Effects of verapamil in hypertensive patients Acta Med Scand Suppl 681: 125–128

Doyle AE, Anavekar SN, Oliver LE (1981) A clinical trial of verapamil in the treatment of hypertension In: Zanchetti A, Krikler DM (eds) Calcium antagonism in cardiovascular therapy – experience with verapamil Excerpta Medica Amsterdam pp 252–258

Dunlay MC, Holmes GI, Hesney M, Irvin JD, Shapiro DA (1988) A comparative multicenter study of felodipine, hydrochlorothiazide and placebo in mild hypertensive patients Am J Hypertens 1: 23A (abstract)

Edmonds D, Würth JP, Knorr M, Hany S Baumgart P, Vetter H, Vetter W (1986) 24-Stunden-Blutdruckmessung unter Verapamil 240 mg. Effektive Blutdrucksenkung durch eine Einmaldosis? Klin Wschr 64 (Suppl 5): 211–212

Elmfeldt D, Hedner T (1985) Antihypertensive effects of felodipine compared with placebo Drugs 29 (Suppl 2): 109–116

Ferlinz J (1986) Nifedipine in myocardial ischemia, systemic hypertension, and other cardiovascular disorders Ann Intern Med 105: 714–729

Ferrara LA, Fasano ML, Soro S (1985) Age related antihypertensive effect of nitrendipine, a new calcium entry blocking agent Eur J Clin Pharmacol 28: 473–474

Franz IW, Wiewel D (1985) Antihypertensive Wirkung von Nitrendipin, Nifedipin und Acebutolol und deren Kombination auf den Ruhe- und Belastungsblutdruck bei Hochdruckkranken Z Kardiol 74: 111–116

Franz IW, Behr U, Ketelhut R, Wiewel D (1986) Antihypertensive Wirkung von Gallopamil und Metoprolol auf den Ruhe- und Belastungsblutdruck bei Hochdruckkranken Herz Kreislauf 18: 632–637

Giesecke HJ, Guckenbiehl W, Hagemann I (1982) Ergebnisse einer Multicenter-Studie mit Diltiazem bei Hypertonie In: Bender F, Greeff K (eds) Calciumantagonisten zur Behandlung der Angina pectoris, Hypertonie und Arrhythmie Excerpta Medica Amsterdam pp 220

Gould BA, Mann S, Kieso H, Bala Subramanian V, Raftery EB (1982) The 24-hour ambulatory blood pressure profile with verapamil Circulation 65: 22–27

Guazzi MD, Polese H, Fiorentini C, Bartorelli A, Moruzzi P (1983) Treatment of hypertension with calcium antagonists Review Hypertension 5 (Suppl 2): 1–85

Hansson L, Andrén L, Orö L, Ryman T (1984) The antihypertensive effects and pharmacokinetics of nitrendipine in patients with essential hypertension In: Scriabine A, Vanov S, Deck K (eds) Nitrendipine Urban & Schwarzenberg Baltimore pp 423–432

Harrington K, Fitzgerald P, O'Donnell P, Hill KW, O'Brien E, O'Malley K (1987) Short and long term treatment of essential hypertension with felodipine as monotherapy Drugs 34 (Suppl 3): 178–185

Hedbäck B, Hermann LS (1984) Antihypertensive effect of verapamil in patients with newly discovered mild to moderate hypertension Acta Med Scand Suppl 681: 129–135

Holmes GI, Slobodzian DK, Lipschutz KE, Irvin JD, Shapiro DA (1988) A comparative multicenter study of felodipine, atenolol and placebo in mildly to moderately hypertensive patients Am J Hypertens 1: 25A (abstract)

Johnson BF, Romero L, Marwaha R (1986) Hemodynamic and metabolic effects of the calcium channel blocking agent nitrendipine Clin Pharmacol Ther 39: 389–394

Klein W, Mlekusch E, Härringer M (1982) Kalziumantagonisten zur Behandlung der chronischen arteriellen Hypertonie. Ein Plazebo-kontrollierter, doppelblinder Vergleich von Diltiazem und Nifedipin Herz Kreislauf 5: 247–255

Krönig B, Flygt G (1987) Felodipine versus moduretic – a double-blind parallel-group multicenter study Drugs 34 (Suppl 3): 162–169

Leonetti G, Sala C, Bianchini C, Terzoli L, Zanchetti A (1980) Antihypertensive and renal effects of orally administered verapamil Eur J Clin Pharmacol 18: 375–382

Leonetti G, Pasotti C, Ferrari GP, Zanchetti A (1981) Double-blind comparison of the antihypertensive effects of verapamil and propranolol In: Zanchetti A, Krikler DM (eds) Calcium antagonism in cardiovascular therapy – experience with verapamil Excerpta Medica Amsterdam pp 260–267

Leonetti G, Fruscio M, Terzoli L, Rupoli L, Gradnik R, Sampieri L, Cuspidi C, Boselli L, Bolla G, Zanchetti A (1985) Antihypertensive and water and sodium balance effects of felodipine, a new vasodilating calcium antagonist, in hypertensive patients Drugs 29 (Suppl 2): 185–191

Lewis GRJ, Stewart DJ, Lewis BM, Bones PJ, Morley KD, Janus ED (1981) The antihypertensive effect of oral verapamil – acute and long-term administrations and its effects on the high density lipoprotein values in plasma In: Zanchetti A, Krikler DM (eds) Calcium antagonism in cardiovascular therapy – experience with verapamil Excerpta Medica Amsterdam pp 270–279

Lorimer AR, McAlpine HM, Rae AP, Simpson IA, Barbour MP, Forret EA, Hill TWK, Lawrie TDV (1985) Effects of felodipine on rest and exercise heart rate and blood pressure in hypertensive patients Drugs 29 (Suppl2): 154–164

Lydtin H, Lohmöller G, Lohmöller R, Walter I (1975) Hemodynamic studies on nifedipine in man In: Hashimoto K, Kimura E, Kobayashi T (eds) 1st international nifedipine „Adalat" symposium University of Tokyo Press Tokyo pp 97–106

MacGregor GA (1984) A randomized dose study of verapamil in patients with essential hypertension In: Althaus U, Burckhardt D, Vogt E (eds) CalciumAntagonismus – Internat. Symposium on Calcium-Antagonism Universimed Frankfurt/Main pp 125–126

M'Buyamba-Kabangu JR, Ditu M, Tshiani KA, Lepira B, Lijnen P, Fagard R, Amery A (1986) Nitrendipine and acebutolol as first line antihypertensive drugs in black patients In: Distler A (ed) Calcium-Antagonisten in der Hochdruck Therapie Schattauer Stuttgart pp 303–310

Midtbo K, Hals O (1980) Verapamil in the treatment of hypertension Current Therapeutic Research 27: 830–838

Muiesan G, Agabiti-Rosei E, Romanelli G, Muiesan ML, Castellano M, Beschi M (1986) Adrenergic activity and left ventricular function during treatment of essential hypertension with calcium antagonists Am J Cardiol 57: 44D–49D

Müller FB, Bolli P, Erne P, Bock LH, Kiowski W, Bühler FR (1986) Antihypertensive Therapie mit dem langwirkenden Calcium-Antagonisten Nitrendipin In: Distler A (ed) Calcium-Antagonisten in der Hochdrucktherapie Schattauer Stuttgart pp 133–140

Opie LH, Jee LD (1984) Nifedipine: Expanding indications in hypertension In: Opie LH (ed) Calcium antagonists and cardiovascular disease Raven Press New York pp 333–337

Pacheco JP, Fan F, Wright RA, Corkadel LK, Horwitz ID (1986) Monotherapy of mild hypertension with nifedipine Am J Med 81 (Suppl 6A): 20–24

Pool PE, Massie BM, Venkataraman K, Hirsch AT, Samant DR, Seagren SC, Gaw J, Salel AF, Tubau JF (1986) Diltiazem as monotherapy for systemic hypertension: A multicenter, randomized, placebo-controlled trial Am J Cardiol 57: 212–217

Rahn KH (1984) Calciumantagonisten in der Hochdrucktherapie Dtsch Med Wschr 109: 933–934

Rosendorff C (1984) Calcium channel blockers and hypertension In: Opiel LH (ed) Calcium antagonists and cardiovascular disease Raven Press New York pp 323–330

Rosendorff C, Goodman C (1985) Double-blind double-dummy crossover-study of the efficacy and safety of nisoldipine (Bayk 5552) versus nifedipine Curr Ther Res 37: 912–920

Rüddel H, Schmieder R, von Eiff AW (1986) Wirkung einer Langzeit-Monotherapie mit Nitrendipin auf hämodynamische Veränderungen unter psychischer Belastung In: Distler A (ed) Calcium-Antagonisten in der Hochdrucktherapie Schattauer Stuttgart pp 165–181

Saltiel E, Ellrodt AG, Monk JP, Langley MS (1988) Felodipine – a review of its pharmacodynamic and pharmacokinetic properties, and therapeutic use in hypertension Drugs 36: 387–428

Schulte KL, Meyer-Sabellek WA, Distler A, Gotzen R (1983) Antihypertensive Langzeitbehandlung mit Diltiazem und Nifedipin bei essentieller Hypertonie Therapiewoche 33: 6011–6014

Stoepel K, Deck K, Ingram C, Vanov S (1986) Multizentrische Doppelblind-Studien mit Nitrendipin aus den USA In: Distler A (ed) Calcium-Antagonisten in der Hochdrucktherapie Schattauer Stuttgart pp 121–129

Stone PH, Antman EM, Muller JE, Braunwald E (1980) Calcium channel blocking agents in the treatment of cardiovascular disorders. Part II: Hemodynamic effects and clinical applications Ann Intern Med 93: 886–904

Sucic M (1983) Die antihypertensive Wirksamkeit von Gallopamil Therapiewoche 33: 6016–6019

Trimarco B, DeLuca N, Ricciardelli B, Volpe M, Veniero A, Cuocolo A, Cicala M (1984) Diltiazem in the treatment of mild or moderate essential hypertension. Comparison with metoprolol in a crossover double-blind trial J Clin Pharmacol 24: 218–227

Ueda K, Murakami M (1983) A controlled, comparative multicenter study on the clinical efficacy of nifedipine in the treatment of severe hypertension In: Kaltenbach M, Neufeld HN (eds) 5th International Adalat Symposium Excerpta Medica Amsterdam pp 146–154

Vitacolonna E, Guagnano MT, Capani F, Sensi S (1986) Nitrendipine treatment of mild-to-moderate hypertension Curr Ther Res 39: 414–420

Yamuchi K, Furui H, Taniguchi N, Sotobata I (1986) Effects of diltiazem hydrochloride on cardiovascular response, platelet aggregation and coagulating activity during exercise testing in systemic hypertension Am J Cardiol 57: 609–612

4.2.4 Calcium Antagonists in Antihypertensive Combination Therapy (with Inclusion of Special Treatment Criteria)

In the past few years calcium antagonists have also been used in combination with other antihypertensive agents in the treatment of hypertension. The basic pros and cons which apply to all combination therapies must be considered.

The combination of drugs of differing chemical structures can be criticized on the grounds of an increased hazard of specific side effects. The combination of different action profiles can lead to undesirable pharmacodynamic interactions (with reduced target effect) and also to unwanted pharmacokinetic interactions (lowered absorption and/or increased metabolism of one of the substances involved). A detailed account concerning the combination of calcium antagonists with other compounds, focusing in particular on already known interactions, is to be found in Sects. 3.1, 4.1.2.3 and 5.8.

The following arguments can be used in favour of combination therapy in hypertension:

a) The side effects of various antihypertensive agents can, by selection of an appropriate combination of drugs, either partially or completely counteract one another. This applies in particular to sodium and water retention in antihypertensive therapy (hence the inclusion of diuretics in many antihypertensive combinations); it also applies to activation of the sympathetic nervous system by calcium antagonists, an activation which can be inhibited by the use of β-receptor blockers.

b) The occurrence of side effects is dose-dependent for various substances: by combining agents to produce the desired antihypertensive effect, therefore, the dose ranges which produce side effects may be avoided.

c) The antihypertensive effect of a substance is usually limited to a given dosage range. The dose–effect curve flattens out at higher drug concentration levels: combinations of effective single substances of differing pharmacological profiles can avoid uneconomic dosage.

d) A certain percentage of hypertensive patients fail to respond to a given antihypertensive medication: these patients are often missed in medical practice because the patient becomes accustomed to measurement of blood pressure and practice conditions, and this goes hand-in-hand with initiation of treatment: starting treatment with a combination therapy can therefore reduce the proportion of patients receiving an ineffective form of treatment.

Points (a) to (c) are the strongest arguments in favour of the use of calcium antagonists in combination therapy of hypertension, with the reservation that calcium antagonists have hitherto not been included in any large-scale intervention studies. For good reasons antihypertensive treatment follows, as a rule, a stepped care approach (Sect. 4.1.2.3).

4.2.4.1 Combination with β-Receptor Blockers

Taken from the standpoints of hypertensive pathophysiology and of pharmacodynamics, β-receptor blockers pair ideally with calcium antagonists in antihypertensive combination therapy, especially in younger patients with strong reactive sympathetic nervous system activation due to vasodilation. Numerous studies have demonstrated conclusively the compatibility and antihypertensive effectiveness of such combinations (Opie et al. 1982; Ekelund et al. 1982; Elmfeldt and Hedner 1983; Yagil et al. 1983; Orö and Ryman 1984; Collste et al. 1985; DeDivitiis et al. 1985; Höffler 1985; Bolli et al. 1986; Franz et al. 1986; Liebau et al. 1986; Rost et al. 1986). The authors conclude unanimously that combinations are more effective than the single substances, and that the incidence of side effects is no greater than in monotherapy (Opie et al. 1982; Yagil et al. 1983). Müller et al. (1986) successfully treated moderate to severe hypertension with a combination of nifedipine (50.0 ± 5.2 mg/day) and atenolol (89.0 ± 25.7 mg/day) in patients aged between 41 and 68 years. In our experience, too, patients over 65 are still good candidates for antihypertensive combination therapy consisting of a calcium antagonist of the dihydropyridine type and a β_1-selective receptor blocker. Even elderly patients can show considerable autonomic counter-regulation to a vasodilator. However, the initiation of such combination therapy in the elderly, who often exhibit multiple organ damage, calls for very close surveillance, which is often possible only on an inpatient basis. Headache and flushing under dihydropyridine derivatives are alleviated by combination with β-receptor blockers. Ankle oedema without weight gain, referable to dilation of precapillary resistance vessels by calcium antagonists, is not increased by administration of a β-receptor blocker. In the formulation of fixed antihypertensive combinations, the individual pharmacokinetic profiles must be taken into consideration. In long-term treatment one must take into account that the biological effects of β-receptor blockers do not reflect their plasma half-life values: the "functional" half-lives of the various β-receptor blockers are markedly longer than their plasma elimination half-lives.

4.2.4.2 Combination with Diuretics

Several groups of workers have combined calcium antagonists with diuretics for treating hypertension, in most cases combining a dihydropyridine with a thiazide diuretic (Hallin et al. 1983; Höffler 1985; Hedner et al. 1986; Müller et al. 1986; Stoepel et al. 1986). The investigations were unequivocal in their finding that the combination is well tolerated. In some of the studies the combination was found to provide more effective antihypertensive therapy than the single substances, although no clear-cut therapeutic recommendations (e.g. for varying categories of hypertension, precise dosage guidelines) can be drawn from their data. A distinct advantage of such combinations is that the effective dose of diuretics can be reduced and thereby also their specific and clearly dose-dependent side effects (especially loss of potassium). On the other hand diuretics continue, on account of the convincing intervention studies and by no means least on account of their relative cheapness, to be a basic treatment for hypertension (Sect. 4.2.1).

Triple-combination therapies (β-receptor blockers, diuretics and calcium antagonists) have also been used in the treatment of moderate and severe hypertension. Such combinations are relatively well tolerated, but the effects of the individual components can often not be quantified. In most studies the antihypertensive effectiveness of calcium antagonists was investigated in patients who had already received basic therapy; component studies were conducted only as exceptions (Laaser et al. 1975; Bayley et al. 1983; Murphy et al. 1983, 1984; Evans et al. 1984; Andersson et al. 1985; Höffler 1986; Wathen et al. 1986). However, all in all, the additional hypotensive effect of calcium antagonists of the dihydropyridine type and of verapamil in combination with diuretics and β-receptor blockers is sufficiently well documented. Calcium antagonists offer a genuine alternative to hydralazine (or dihydralazine), minoxidil and ACE inhibitors in combination therapy of severe hypertension, together with diuretics, β-receptor blockers or clonidine (Evans et al. 1984; Collste et al. 1985; Wathen et al. 1986).

4.2.4.3 Combination with Other Antihypertensives

Calcium antagonists of the dihydropyridine type have also been successfully used in combination with ACE inhibitors. Guazzi et al. (1984) combined nifedipine with captopril and observed a synergistic antihypertensive action. In investigations by Brouwer et al. (1985), a group of patients initially received a combination of a β-receptor blocker with a diuretic and a vasodilator, which produced an insufficient blood pressure reduction. Monotherapy was then initiated with either nitrendipine (40–120 mg/day) or verapamil (360–720 mg/day). Since calcium antagonists were unable to reduce diastolic pressure below 95 mmHg in any of the patients, a combination of nitrendipine in a dose of 50–100 mg/day and/or verapamil in a dose of 360–480 mg/day with captopril (50–100 mg/day) was introduced. This calcium antagonist–ACE inhibitor combination lowered systolic and diastolic pressure more effectively than a combination of the ACE inhibitor and a diuretic, which at a later date was also investigated in the same group of patients; this was also more effective than the previous triple combination. A combination of calcium antagonists with ACE inhibitors is thus, in the opinion of the investigators, a worthwhile alternative in the treatment of refractory hypertension. In these situations, calcium antagonists can replace thiazide diuretics. It remains, however, an open question, and in our view doubtful, whether this also applies to loop diuretics.

Besides the combinations discussed already, calcium antagonists have been used together with practically all other antihypertensive agents (including clonidine, methyldopa, prazosin), albeit in mostly small, poorly documented, and unrepeated series of investigations. In a study by Guazzi et al. (1980), a combination with methyldopa was tested: the authors reported that the combination was more effective than monotherapy with the individual substances. Confirmation of these findings, however, by means of a critically designed comparative study is still lacking. Calcium antagonists need to be used extremely cautiously in patients who have previously received treatment with other vasodilators, especially with the α_1-receptor blocker prazosin (Jee and Opie 1983); on account of their common mode of action, a combination of calcium antagonists with other vasodilators appears to be undesirable.

In Summary. Analysis of the available investigations regarding combination therapy with calcium antagonists in hypertension allows the following conclusions to be drawn: there are strong arguments for the combination of calcium antagonists with β-receptor blockers in younger hypertensive individuals (as step 2), and, in the case of refractory hypertension, also with diuretics (step 3); in older patients calcium antagonists are an effective alternative to diuretics, and in combination therapy the dosage of diuretics can thus be reduced. Furthermore, in severe hypertension calcium antagonists are an effective alternative to hydralazine, α_1-receptor blockers and minoxidil; a combination of calcium antagonists and ACE inhibitors can be effective in severe hypertension.

4.2.4.4 Special Treatment Criteria and Marginal Indications

The almost total absence of effect of calcium antagonists on lipid and/or carbohydrate metabolism (Lohmann 1986) establishes them as agents of choice in diabetes and in patients with disorders of lipid metabolism. Calcium antagonists lower blood pressure during exercise to a similar degree as β_1-receptor blockers (Franz 1986). In contrast, methyldopa, clonidine, diuretics, α-receptor blockers and reserpine produce no significant change in systolic pressure during exercise (Franz 1986). These findings favour the inclusion of calcium antagonists, alongside β-receptor blockers, in the treatment of hypertension in the physically active younger patient (without changing their special status in the treatment of the older hypertensive patient with low renin levels; Ben-Ishay et al. 1986; Bolli et al. 1986; Müller et al. 1986). Calcium antagonists (as diuretics) also appear to be especially effective in black people (who frequently exhibit low renin levels; Halperin et al. 1984); however, here as elsewhere, proof that they are more effective in respect of hard end-points is still lacking.

Coronary heart disease in the hypertensive patient also favours the inclusion of calcium antagonists together with β-receptor blockers in the treatment plan (Tammen et al. 1984). This obviously applies also to all other indications for calcium antagonists (Chap. 4) and/or if there are any contraindications to the use of other antihypertensive agents (e.g. β-receptor blockers in asthma). Anavekar et al. (1982) conducted a double-blind comparative trial of verapamil and the α,β-receptor blocker, labetalol, in hypertensive individuals with obstructive pulmonary disease, and have thereby shown that, with an equally good antihypertensive effect, labetalol, in contrast to verapamil, increases the bronchial obstruction (and also weakens the effect of the β_2-receptor agonist, salbutamol).

A possible indication for nifedipine is in the temporary increase in blood pressure following coronary surgery (and following other cardiac surgical procedures, Sect. 4.5.3): a single, unrepeated intranasal application of 10 mg nifedipine reduces a critically raised blood pressure without any unwanted hypotension (Iyer and Russel 1986).

Calcium antagonists also lower elevated blood pressure in primary hyperaldosteronism (Nadler et al. 1985). In one case nitrendipine also normalized the plasma hormone level, indicating an interference with aldosterone release (Stimpel et al. 1986). This interesting finding does not yet permit any therapeutic recommendations for treatment of primary (or secondary) hyperaldosteronism, but it could well provide another explanation for the absence of sodium and water retention during treatment with calcium antagonists.

Calcium antagonists have not to date been used in hypertension during early pregnancy. Animal trials showing embryonal damage with nifedipine speak strongly against its use in pregnancy; β_1-selective receptor blockers and methyldopa, and in severe hypertension hydralazine, are preferred (Girndt 1985). Whether these reservations are also valid for verapamil (or diltiazem) remains an open question. Verapamil has been used to inhibit premature labour and for cardioprotection during tocolysis (Sect. 4.10.3), without unwanted side effects to mother or child.

In the treatment of hypertension in the presence of obstructive lung disease (asthma, chronic obstructive pulmonary disease) calcium antagonists are the agents of choice; β-receptor blockers are contraindicated in this situation, and diuretics should be used only with reservation because of the danger of mucus impaction due to increased mucus viscosity. (For a discussion of the use of calcium antagonists in the regulation of blood pressure in anaesthesiology see Sect. 4.12.)

References

Anavekar SN, Barter C, Adam WR, Doyle AE (1982) A double-blind comparison of verapamil and labetalol in hypertensive patients with coexisting chronic obstructive airway disease J Cardiovasc Pharmacol 4: 374–377

Andersson OK, Granérus G, Hedner T (1985) Felodipine: a calcium-inhibiting vasodilator in refractory hypertension Drugs 29 (Suppl 2): 102–108

Bayley S, Dobbs RJ, Robinson BF (1982) Nifedipine in the treatment of hypertension:report of a double-blind controlled trial Br J Clin Pharmacol 14: 509–512

Ben-Ishay D, Leibel B, Stessman J (1986) Calcium channel blockers in the management of hypertension in the elderly Am J Med 81 (Suppl 6A): 30–34

Bolli P, Bühler FR, Kiowski W, Müller FB (1986) Betarezeptorenblocker, Kalziumantagonisten und Diuretika in der Hochdrucktherapie In:Distler A (ed) Ca^{++}-Antagonisten und Beta-Rezeptorenblocker in der Behandlung von Hypertonie und koronarer Herzkrankheit Schattauer Stuttgart pp 41–52

Brouwer RML, Bolli P, Erné P, Conen D, Kiowski W, Bühler FR (1985) Antihypertensive treatment using calcium antagonists in combination with captopril rather than diuretics J Cardiovasc Pharmacol 7 (Suppl 1): 88–91

Collste P, Danielsson M, Elmfeldt D, Feleke E, Gelin A, Hedner T, Rydén L (1985) Long term experience of felodipine in combination with β-blockade and diuretics in refractory hypertension Drugs 29 (Suppl 2): 124–130

DeDivitiis O, Petitto M, Di Somma S, Galderisi M, Villari B (1985) Nitrendipine and atenolol: Comparison and combination in the treatment of arterial hypertension Arzneimittelforschung 35: 727–729

Ekelund LG, Ekelund C, Rössner S (1982) Antihypertensive effects at rest and during exercise of a calcium blocker, nifedipine, alone and in combination with metoprolol Acta Med Scand 212: 71–75

Elmfeldt D, Hedner T (1983) Felodipine – a new vasodilator, in addition to β-receptor blockade in hypertension Eur J Clin Pharmacol 25: 571–575

Evans MG Jr, Olanoff LS, Hurwitz G, Cowart TD, Conradi EC (1984) Use of nifedipine as an adjunct to current antihypertensive therapy Arch Intern Med 144: 985–987

Franz IW (1985) Beeinflussung der Hypertonie unter Belastung durch antihypertensive Pharmakotherapie In: Distler A (ed) Ca^{++}-Antagonisten und Beta-Rezeptorenblocker in der Behandlung von Hypertonie und koronarer Herzkrankheit Schattauer Berlin pp 71–74

Girndt J (1985) Therapie der Schwangerschaftshypertonie Dtsch Med Wschr 110: 1579–1580

Guazzi MD, Fiorentini C, Olivari MT, Bartorelli A, Necchi G, Polesi A (1980) Short-and-long-term efficacy of a calcium-antagonistic agent (nifedipine) combined with methyldopa in the treatment of severe hypertension Circulation 61: 913–919

Guazzi MD, DeCesare N, Galli C, Salvioni A, Tramontana C, Tamborini G, Bartorelli A (1984) Calcium-channel blockade with nifedipine and angiotensin converting-enzyme inhibition with captopril in the therapy of patients with severe primary hypertension Circulation 70: 279–284

Hallin L, Andrén L, Hansson L (1983) Controlled trial of nifedipine and bendroflumethiazide in hypertension J Cardiovasc Pharmacol 5: 1083–1085

Halperin AK,Gross KM, Rogers JF,Cubeddu LX (1984) Verapamil and propranolol in essential hypertension Clin Pharmacol Ther 36:750–758

Hedner T, Samuelsson O, Sjörgen E, Elmfeldt T (1986) Treatment of essential hypertension with felodipine in combination with a diuretic Eur J Clin Pharmacol 30: 133–139

Höffler D (1985) Nitrendipin bei schwer einstellbarer Hypertonie Therapiewoche 35: 3638–3642
Iyer VS, Russel WJ (1986) Nifedipine for postoperative blood pressure control following coronary artery vein grafts J Roy Coll Surg (Edin) 68: 73–75
Jee LD, Opie LH (1983) Acute hypotensive response to nifedipine added to prazosin in treatment of hypertension Br Med J (Clin Res) 287: 1514
Laaser U, Meurer KA, Krüger H, Kaufmann W (1975) Clinical studies on the combined effect of various antihypertensive agents and Adalat In: Lochner W, Braasch W, Kroneberg G (eds) 2nd International Adalat Symposium Springer Berlin pp 287–288
Liebau H, Fassl H, Riepl B (1985) Behandlung der arteriellen Hypertonie mit einer Kombination von Atenolol 50 mg und Nifedipin retard 20 mg – Ergebnisse einer multizentrischen Praxisstudie In:Distler A (ed) Ca^{++}-Antagonisten und Beta-Rezeptorenblocker in der Behandlung von Hypertonie und koronarer Herzkrankheit Schattauer Berlin pp 110–111
Lohmann FW (1985) Metabolische Auswirkungen der antihypertensiven Therapie mit β-Rezeptorenblockern und Kalziumantagonisten In: Distler A (ed) Ca^{++}-Antagonisten und Beta-Rezeptorenblocker in der Behandlung von Hypertonie und koronarer herzkrankheit Schattauer Berlin pp 87–88
Müller FB, Bolli B, Lindner L, Kiowski W, Erné P, Bühler FR (1986) Calcium antagonists and the second drug for antihypertensive therapy Am J Med 81 (Suppl 6a): 25–29
Murphy MB, Scriven AJI, Dollery CT (1984) Efficacy of nifedipine as a step 3 antihypertensive drug Hypertension 5 (4pt 2): II 118–121
Murphy MB, Bulpitt CJ, Dollery CT (1984) Role of nifedipine in hospital outpatients Am J Med 77 (Suppl 2B): 16–21
Nadler JL, Hsueh W, Horton R (1985) Therapeutic effect of calcium channel blockade in primary aldosteronism J Clin Endocr 60: 896
Opie LH, Jee L, White D (1982) Antihypertensive effects of nifedipine combined with cardioselective beta-adrenergic receptor antagonism by atenolol Am Heart J 104: 606–612
Orö L, Ryman T (1984) A long-term study of the combined antihypertensive action of nitrendipine and metoprolol In: Scriabine A, Vanov S, Deck K (eds) Nitrendipine Urban & Schwarzenberg Baltimore pp 527–534
Rost R, Völker K, Reinke A, Tegenthoff H (1985) Der Einfluß von Nifedipin und Atenolol bzw. der fixen Kombination beider Substanzen auf physiologische Belastungsreaktionen sowie auf den Belastungsblutdruck bei Hypertonikern In: Distler A (ed) Ca^{++}-Antagonisten und Beta-Rezeptorenblocker in der Behandlung von Hypertonie und koronarer Herzkrankheit Schattauer Berlin pp 122–124
Stimpel M, Dralle H, von zur Mühlen A (1986) Therapie des primären Aldosteronismus Dtsch Med Wschr 111: 1487–1488
Tammen AT, Stoepel K, Vollmer H, Blümchen G (1984) Efficacy and safety of nitrendipine in patients with essential hypertension associated with coronary heart disease In: Scriabine A, Vanov S, Deck K (eds) Nitrendipine Urban & Schwarzenberg Baltimore pp 469–475
Ueda K, Murakami M (1983) A controlled, comparative multicenter study on the clinical efficacy of nifedipine in the treatment of severe hypertension In: Kaltenbach M, Neufeld HN (eds) 5th International Adalat Symposium Excerpta Medica Amsterdam pp 146–155
Yagil Y, Kobrin I, Stessman J, Ghanem J, Leibel B, Ben-Ishay D (1983) Effectiveness of combined nifedipine and propranolol treatment in hypertension Hypertension 4 (a pt 2): II 113–117
Wathen CG, MacLeod D, Tucker L, Muir AL (1986) Felodipine as a replacement for minoxidil in the treatment of severe hypertension Eur Heart J 7: 893–897

4.2.5 Calcium Antagonists in Hypertensive Emergencies

There is a hypertensive emergency when either malignant (permanent) hypertension or a hypertensive crisis evokes immediately threatening organ damage and/or symptoms in respect of the eye, the central nervous system, the heart or the kidneys (Table 21).

Table 21. Findings and symptoms in hypertensive emergency

Diastolic blood pressure generally above 140 mmHg	
Eye	Bleeding, exudates, papillary oedema, disturbances of vision
Central nervous system	Headache, confusion, somnolence, stupor, focal neurological deficits, seizures, coma, vertigo
Heart	Enlargement of heart, angina pectoris, congestive heart failure
Kidneys	Oliguria, proteinuria, haematuria, increase in serum creatinine
Gastrointestinal tract	Nausea and emesis
Blood	Microangiopathic, haemolytic anaemia (with erythrocyte fragmentation and intravascular coagulation)

Life-threatening consequences of hypertension include severe arterial epistaxis and dissecting aortic aneurysm. During the acute stage outside the hospital, and even during hospital admission, it is often difficult to distinguish between an acute hypertensive crisis and rapidly developing malignant hypertension. Causes of hypertensive crisis are summarized in Table 22, and syndromes relevant to differential diagnosis in Table 23.

In the development of vascular organ complications, the essential feature of a hypertensive crisis is not the absolute level of blood pressure, but the rate of increase in pressure. With an acute rise in blood pressure autoregulation (i.e. elevation of vascular resistance with rising blood pressure in order to maintain constant perfusion) no longer occurs in the arterioles, especially the cerebral arterioles (and those of the eye): there is thus excessive blood flow with exudation of fluid from the vascular bed. Simultanously, as a result of the abrupt rise in

Table 22. Causes of hypertensive crisis. (Grabensee 1985)

Abrupt discontinuation of antihypertensive treatment (particularly of clonidine)
Essential hypertension (oral contraceptives, climacteric)
Renal hypertension
Eclampsia
Phaeochromocytoma
Tyramine during therapy with MAD inhibitors
Primary malignant nephrosclerosis

Table 23. Differential diagnosis of hypertensive crisis. (After Kaplan 1984)

Acute left ventricular failure
Uraemia of all origins (with hypervolaemia)
Cerebrovascular diseases
Subarachnoid haemorrhage
Brain tumour
Head injury
Epilepsy
Connective tissue diseases (e.g. lupus erythematosus) with cerebral vasculitis
Encephalitis
Overdosage and withdrawal of narcotics, amphetamines etc.
Acute anxiety syndrome (with hyperventilation)

blood pressure, changes in the vascular wall develop which are collectively labelled as fibrinoid necrosis. Disturbed autoregulation and fibrinoid necrosis are direct sequelae of excessively elevated blood pressure; other mechanisms (such as a rise in noradrenaline or in angiotensin II) can contribute to the rise in blood pressure, but are not essential elements of the pathogenetic mechanism.

Hypertensive emergencies are rare. In medical practice they are probably too frequently diagnosed (either from blood pressure alone or from misinterpretation of the syndromes listed in Table 23). On the other hand, there are no doubt patients who have few or no symptoms in spite of an excessive elevation in blood pressure and incipient organ damage, in whom the dangers of the presenting situation and the urgency of therapeutic intervention are underestimated. The practical significance of retinal examination in the work-up of the hypertensive patient cannot be overemphasized. In a suspected hypertensive emergency immediate admission to hospital is indicated, and as a rule initiation of treatment both outside the hospital and immediately on admission. Specific immediate treatment is often not possible because of differential diagnostic difficulties in the acute situation. Drastic reduction of blood pressure is contraindicated in atherothrombotic or thromboembolic strokes, and certainly in some patients with congestive heart failure (imminent cardiogenic shock) and coronary insufficiency (triggering of acute myocardial infarction). Outside the hospital or intensive care unit substances with unpredictably powerful, not readily controllable hypotensive effects and which might paradoxically increase blood pressure should never be used. Even the intravenous administration of drugs in such situations can be difficult, and continuous infusion demands intensive surveillance. The onset of action may occur only after a matter of hours and be too late, and sedation may mask neurological complications. When the spectrum of antihypertensive agents suitable for the hypertensive emergency (Table 24) is considered from this standpoint, calcium antagonists of the dihydropyridine type have clear advantages in that they exhibit: no central nervous side effects (as with reserpine, clonidine, methyldopa); no increase in heamatocrit with consequent disturbances in microcirculation (as with the powerful loop diuretics); no paradoxical increase in blood pressure (as occasionally occurs with clonidine and with β-receptor blockers in phaeochromocytoma); no, or in relation to the peripheral effect negligible, direct effects on the heart (as with verapamil, diltiazem and β-receptor blockers).

Verapamil (intravenously) was the first calcium antagonist to be used successfully in hypertensive emergencies (Brittinger et al. 1969; Sect. 4.2.1.4). Intravenously administered diltiazem and gallopamil also lower critically raised blood pressure (Rosenthal 1982; Rameis 1983). The cardiac side effects of these substances prevented wider use of verapamil, gallopamil and diltiazem in such situations. Moreover, the reduction in blood pressure following intravenous administration of these substances lasts only for a few minutes, thus making repeated follow-up injections necessary. The real breakthrough in the treatment of the hypertensive emergency came with nifedipine, which is rapidly absorbed following oral administration (Lydtin et al. 1975); oral application is possible even in the unconscious patient.

Table 25 summarizes studies with nifedipine in the hypertensive emergency. The doses used lie between 10 and 20 mg (Fig. 27). Capsules containing nifedipine

Table 24. Drugs used in hypertensive emergencies (after Kaplan 1984; Stumpe 1984; Grabensee 1985; Tschollar and Belz 1985; Hauger-Klevene 1986)

Substance	Dose	Onset of effect	Duration of effect	Side effects
Clonidine	0.15 mg i. v.	5–10 min	2–5 h	Rarely, increase in blood pressure; sedation
Hydralazine	6.25–25 mg i. v.	5–10 min	3–6 h	Tachycardia
Diazoxide	75/150/300 mg i. v.	1–2 min	2–12 h	Tachycardia, hyperglycaemia
Sodium nitroprusside	10–500 µg/min (i. v. with perfusor)	Immediate	Duration of infusion	Tachycardia, cyanide or thiocyanate intoxication
Phentolamine[a]	5–10 mg i. v. or 0.2–0.8 mg/min (i. v. with perfusor)	30–60 s	15–30 min	Tachycardia, angina pectoris
Reserpine	0.5–5 mg i. v.	90–180 min	6–24 h	Sedation
α-Methyldopa	250–1000 mg i. v.	120–180 min	6–12 h	Sedation
Captopril	6.25–25 mg orally 25 mg sublingual	20–40 min 15 min	3–6 h	*Caution:* renal artery stenosis
Urapidil	25 mg i. v.	1–2 min	–	
Labetalol	50 mg i. v. (over 2 min)	3–6 min	6–12 h	
Furosemide	40–80 mg i. v.	15–30 min	–	
Ethacrynic acid	25–50 mg i. v.	15–30 min	–	
Nifedipine	10–20 mg sublingual	2–4 min	1–4 h	Tachycardia, headache
Verapamil	5 mg i. v.	2–4 min	10–15 min	AV block, negative inotropic effect

[a] In phaeochromocytoma.

Table 25. Nifedipine in the hypertensive emergency

Reference		Dose	Onset of effect
Bertel and Conen	1986	10/20/60 mg oral 10 mg sublingual	2–4 min
Brandt	1985	20 mg sublingual	2–4 min
Davidson et al.	1985	10–20 mg oral and sublingual	2–4 min
Ellrodt et al.	1985	10–20 mg sublingual	2–4 min
Hauger-Klevene	1986	10 mg sublingual	2–4 min
Ito et al.	1984	10 mg sublingual	2–4 min
Klein et al.	1980	10 mg sublingual	2–4 min
Opie and Jennings	1985/1986	10 mg sublingual	2–4 min
Wolff et al.	1985	20 mg oral	10 min

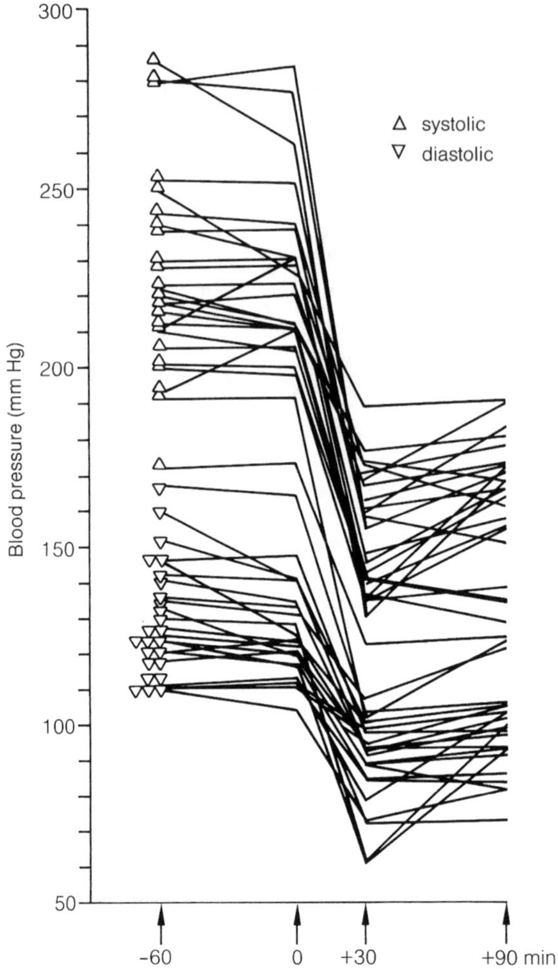

Fig. 27. Hypertensive emergency. Effect of 10–20 mg nifedipine (orally or sublingually) on systolic and diastolic blood pressure in 25 patients (maximum effect after 30 min, duration of effect approximately 90 min). (After Bertel and Conen 1986)

are either bitten and the contents retained in the mouth or, preferably, swallowed, or the capsule contents can be drawn into a hypodermic syringe and injected into the buccal cavity. An effect is usually elicited within 2–4 min, in most cases the only side effect, besides flushing, being an increase in heart rate. However, caution is necessary here, too, since in rare cases nifedipine (like all quick-acting antihypertensive medication) can trigger an unexpected or undesired sharp drop in blood pressure. This applies especially if the patient is known to or assumed to have received antihypertensive medication prior to the emergency (e.g. with preexisting α- and/or β-receptor blockade, volume reduction by diuretics, previous dilation by, for instance, minoxidil or hydralazine and also by previous treatment with ACE inhibitors). Furthermore, every lowering of blood pressure can, given a critical central nervous system blood supply (owing to vascular stenoses), trigger neurological deficits or visual disturbances (as a consequence of diminished retinal blood flow; Opie and Jennings 1986). We therefore recommend in unclear situations initiation of treatment with only 5 mg nifedipine sublingually; if there is no response, a supplementary dose of 10 (to 20) mg can be given after 5–10 min. In the acute situation, the blood pressure should usually not be lowered below 160 (180) mmHg in systole and 90 (95) mmHg in diastole.

If there is a very fast heart rate (over 130/min), we also cautiously administer dihydropyridines, despite the fact that in such cases the reduced afterload often produces a decrease rather than a further increase in heart rate. Nifedipine is now also available as an intravenous presentation (for controlled infusions) for hospital use. Other dihydropyridines (e.g. felodipine, nitrendipine) appear to have an effect similar to nifedipine (Sluiter et al. 1985), but their onset of effect seems to be delayed (with oral administration).

References

Bertel O, Conen LD (1986) Treatment of hypertensive emergencies with the calcium channel blocker nifedipine Am J Med 79 (suppl 4A): 31–35

Brandt D (1985) Akute hämodynamische Effekte von Nifedipin bei Patienten mit Hochdruckkrisen Herz/Kreislauf 1: 24–27

Brittinger DW, Strauch M, Huber W, Koch WD, von Henning GE, Wittenmeier KW, Twittenhoff WD (1969) Iproveratril als Antihypertonikum bei krisenhafter renaler Hypertonie Dtsch Med Wschr 94: 945–948

Davidson RC, Bursten SL, Keeley PA, Kenny MA, Stewart DK (1985) Oral nifedipine for the treatment of patients with severe hypertension Am J Med 79 (suppl 4A): 26–30

Ellrodt AG, Ault MJ, Riedinger MS, Murata GH (1985) Efficacy and safety of sublinqual nifedipine in hypertensive emergencies Am J Med 79 (suppl 4A): 19–25

Grabensee B (1985) Therapie der hypertensiven Krise Dtsch Med Wschr 110: 1738–1740

Hauger-Klevene JH (1986) Comparison of sublingual captopril and nifedipine Lancet 1: 219

Ito H, Arakawa M, Shibasaki T, Hirakawa S, Fulenda T, Yasue T Kondo Y, Watanabe I, Mori N, Fueki T, Oda M, Ota N, Kobayashi H (1984) Acute antihypertensive effect of nifedipine by sublingual route in cases with clinically severe systolic hypertension Arzneimittelforschung 34: 630–635

Kaplan NM (1984) Systemic hypertension: mechanisms and diagnosis – therapy In: Braunwald E (ed) Heart disease – a textbook of cardiovascular medicine – 2nd edition Saunders Philadelphia pp 849–927

Klein N, Beer N, Gallegos I, Cohen A, Sonnenblick E, Frishman W (1980) Efficacy of sublingual nifedipine in the acute management of systemic hypertension Circulation 62 (suppl 3): 36

Lydtin H, Lohmöller G, Lohmöller R, Schmitz H, Walter I (1975) Hemodynamic studies on nifedipine in man In: Hashimoto K, Kimura E, Kobayashi T (eds) 1st International nifedipine „Adalat" Symposium University of Tokyo Press Tokyo pp 97–106

Opie LH, Jennings AA (1985) Sublingual captopril versus nifedipine in hypertensive crises Lancet 2: 555

Opie LH , Jennings AA (1986) Role of calcium channel blockade in chronic hypertension following succesful management of hypertensive emergencies Am J Med 81 (suppl 6A): 35–42

Rameis H (1983) Unterschiede in der akuten blutdrucksenkenden Wirkung der Kalziumantagonisten Diltiazem, Verapamil,Nifedipin In: Kalziumantagonisten in der Therapie der arteriellen und pulmonalen Hypertonie. Magometschnigg D (ed) Uhlen Wien pp 45 – 55

Rosenthal J (1982) Die Behandlung der hypertensiven Krise mit Diltiazem In: Bender F, Greeff K (eds) Calciumantagonisten zur Behandlung der Angina pectoris, Hypertonie und Arrhythmie Excepta Medica Amsterdam pp 227–234

Sluiter HE, Huysmans FTM, Thien TA, Koene RAP (1985) Haemodynamic effects of intravenous felodipine in normotensive and hypertensive subjects Drugs 29 (Suppl 2): 144–153

Stumpe KO (1984) Klinik und Therapie des hypertensiven Notfalls Internist 25: 359–366

Tschollar W, Belz GG (1985) Sublingual captopril in hypertensive crisis Lancet 2: 34–35

Wolff E, Neubauer F, Schikora B, Albertus J (1985) Die Behandlung des hypertensiven Notfalls mit Nifedipin Zschr Klin Med 40: 725–727

4.3 Cardiac Arrhythmias

Cardiac arrhythmias are the result of disturbed impulse formation and/or conduction (Fig. 28). According to the heart rate, bradycardia and tachycardia can be distinguished and, according to the site of origin, supraventricular and ventricular forms. Heart rate can be either regular or irregular in both bradycardia and tachycardia; arrhythmias can occur as paroxysms or continuously. Many different forms of cardiac arrhythmia can be encountered, the spectrum ranging from occasional "harmless" extrasystoles to higher degree AV block and to ventricular tachycardia or ventricular fibrillation. Aetiological classification should be attempted in every presenting arrhythmia (e.g. coronary heart disease, cardiomyopathy, myocarditis, hyperthyroidism, electrolyte disturbances, Wolff–Parkinson–White (WPW) syndrome, arrhythmogenic medication) and whenever possible a causal treatment should be initiated (e.g. reduction of myocardial ischaemia, antiinflammatory therapy, reinstatement of electrolyte balance, treatment of hyperthyroidism). The main issue for the physician is in deciding whether or not therapy is needed; the principle aim must be to improve the patient's quality of life (i.e. to alleviate symptoms) and/or to prolong the patient's life, i.e. to prevent sudden cardiac death. The patient's subjective complaints (symptoms) and the type of underlying heart disease are of greater importance in deciding whether treatment be initiated than a single ECG finding.

The antiarrhythmic agents applied in the treatment of cardiac arrhythmias are nowadays grouped into four classes according to the scheme devised by Vaughan-Williams and Singh (Table 26). Here calcium antagonists occupy a class of their own. Verapamil, gallopamil and diltiazem exhibit an antiarrhythmic effect in vivo; the verapamil derivative tiapamil is still under clinical investigation. Perhexiline and bepridil (not yet commercially available) also show antiarrhythmic effects, but both substances are, because of their supplementary blocking effect on the fast sodium channel, not regarded as pure calcium antagonists (Sects. 1.2.4, 1.2.5, 2.1.4).

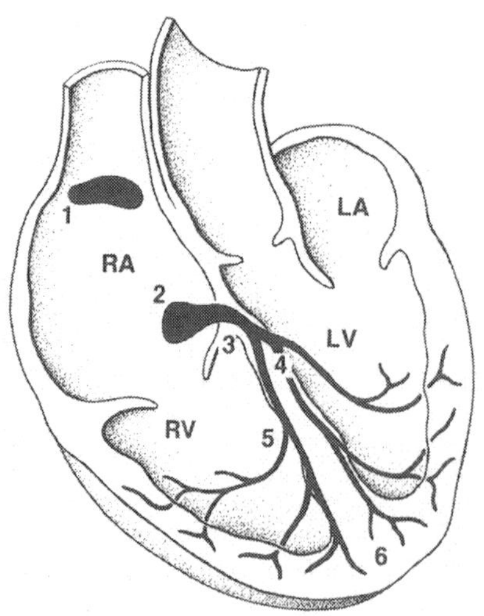

Fig. 28. Formation and conduction of impulses in the normal heart. 1 sinoatrial node, 2 AV node, 3 bundle of His, 4 left bundle branch, 5 right bundle branch, 6 Purkinje fibres

Table 26. Classification of antiarrhythmic agents according to Vaughan-Williams and Singh

Class I:	*Antagonists of fast sodium influx* ("Membrane stabilizing agents")	*Further examples*
	I a: Prolongation of action potential Prototype: quinidine	Procainamide Disopyramide Ajmaline Aprindine
	I b: Shortening of action potential Prototype: lidocaine	Mexiletine Tocainide Phenytoin
	I c: No essential effect on duration of action potential Prototype: propafenone	Flecainide
Class II:	*β-Blocking agents* Prototype: propranolol	Metoprolol Atenolol etc.
Class III:	*Agents increasing action potential duration* (Prolongation of action potential) Prototype: amiodarone	Bretylium Sotalol
Class IV:	*Calcium antagonists* Prototype: verapamil	Gallopamil Diltiazem

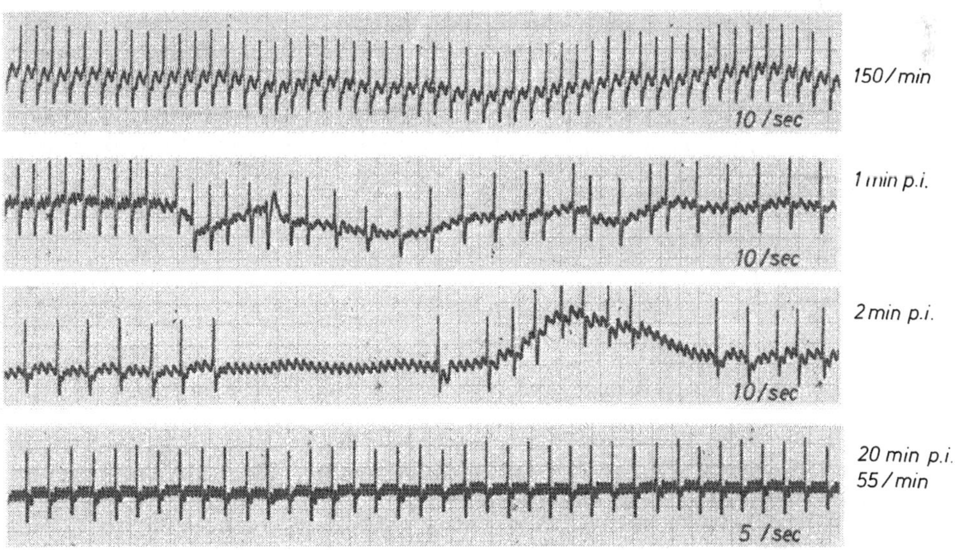

Fig. 29. First antiarrhythmic application of verapamil (Bender 1965). Inhibition of AV conduction in atrial flutter by 2.5 mg intravenously: *p. i.* postinjection. (Original recording made available by courtesy of Professor Bender, Münster)

Calcium antagonists with antiarrhythmic action principally block AV conduction, and their main field of indication is in supraventricular tachycardias. In bradycardias (higher degree AV block, sinus bradycardia, sick sinus syndrome) these calcium antagonists are contraindicated. Substances of the dihydropyridine group (prototype: nifedipine) exhibit no antiarrhythmic effect in vivo, partly due to their powerful peripheral vasodilation with reflex autonomic counter-regulation and partly due to vascular selectivity (Sect. 3.1.2). These substances can sometimes be used in patients with bradycardia. The antiarrhythmic effect of verapamil was described by Bender as long ago as 1965 (Fig. 29). Towards the close of the 1960s Bender (1968; see also Belz et al. 1970) initiated the use of the combination of verapamil and quinidine.

4.3.1 Mechanism of Antiarrhythmic Action of Calcium Antagonists

4.3.1.1 Mechanism of Action in Supraventricular Arrhythmias

The essential mechanism of action of calcium antagonists with antiarrhythmic efficacy (verapamil group, diltiazem) in supraventricular tachycardia centres on the inhibition of anterograde AV conduction. As detailed in Sect. 3.1.2, calcium antagonists of the verapamil type or diltiazem prolong the functional and effective refractory period in the AV node, the AH interval in the His bundle ECG, and the PR interval in the normal ECG. The delayed AV conduction is the consequence of direct inhibition of slow calcium channels ("slow response") in pacemaker cells of the AV node (Sect. 1.1.1 and 1.2.2).

If there are supraventricular tachycardias of the reentry type[17] (arrhythmia type of reentry circuits), which pass mainly through the AV node, the delayed AV conduction either reduces the ventricular rate or even terminates the arrhythmia (Krikler and Rowland 1980). In atrial fibrillation or atrial flutter, ventricular rate also diminishes as a result of delayed AV conduction.

The slow calcium channels are also inhibited in pacemaker cells in the sinoatrial node, but the overall effect in vivo is weaker than at the AV node, since autonomic counter-regulation (vagal withdrawal, activation of the sympathetic nervous system) at the sinoatrial node virtually annuls the direct effect of calcium antagonists.

Retrograde AV conduction ("fast conduction"), intraatrial conduction, intraventricular conduction and impulse formation in normal atrial and ventricular muscle are not affected by calcium antagonists. Furthermore, conduction through accessory pathways (e.g. in the WPW syndrome) is usually not altered

[17] A cardiac arrhythmia of the reentry type (type of reentry circuits, see Fig. 30) presupposes the presence of two parallel conduction pathways with different rates of conduction. In conduction pathway *A* there is normal conduction; in conduction pathway *B* conduction is retarded, and this leads to a so-called unidirectional block (forward block with blocking in one direction only). If the impulse running along conduction channel *A* takes a retrograde course by curving back into conduction channel *B*, the delayed anterograde excitation and the retrograde excitation usually cancel each other. An extrasystole may fortuitously occur at a moment when only one of the two conduction pathways (*A*) is excitable. If there is retrograde conduction here, and if conduction pathway *B* is then no longer refractory, i.e. normally excitable, then a reentry (*R*, from *B* to *A*), that is, reentry of excitation, is possible. A tachycardia of reentry type may then ensue, by anterograde conduction via pathway *A* and retrograde conduction via pathway *B*.

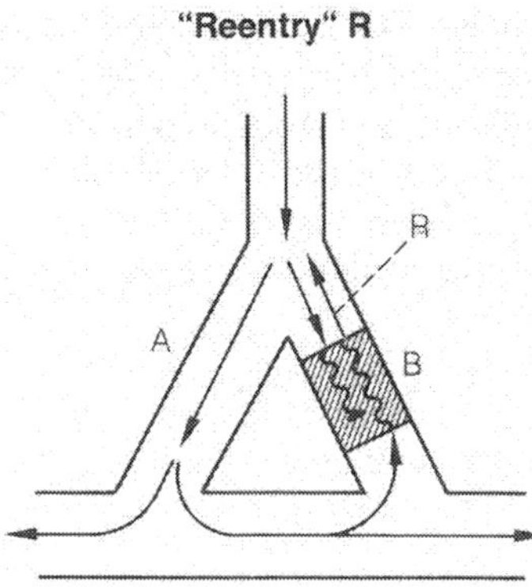

Fig. 30. Principle of "reentry" as a cause of tachycardia. For explanation see text

by calcium antagonists. As stated in Sect. 3.1.2, however, acceleration of anterograde conduction is possible.

If atrial muscle is dilated or ischaemic, individual muscle cells will tend to spontaneous diastolic depolarization, and then supraventricular extrasystoles (ectopic impulse formation) can arise. This spontaneous atrial activity can diminish under calcium antagonists (Braunwald 1982). Furthermore, under certain conditions (e.g. elevation of adrenergic drive) depolarization may occur during early diastole in the vicinity of the mitral valve (so-called after-depolarization); this, too, is inhibited by calcium antagonists (Lydtin und Lohmöller 1980; Braunwald 1982). The mechanism of "slow-response action potentials" and of "triggered automaticity", which is partially present also in supraventricular extrasystoles and explains the effect of calcium antagonists in these arrhythmias, is described further in Sect. 4.3.1.2.

In current electrophysiological models atrial fibrillation is interpreted as being triggered either by several parallel reentry circuits or by multifocal ectopic pacemakers. The rate of fibrillating (atrial) oscillations (undulations) lies between 350 and 600 per minute; at irregular intervals impulses are transferred through the AV node to the ventricles. In atrial flutter there are, in contrast to the irregular undulations of fibrillation, regular flutter waves with a rate, in most cases, between 220 and 350 per minute; here, too, a reentry mechanism is assumed to be the predominant causal factor, or occasionally also a continuously pulsating (oscillating) ectopic atrial centre. Calcium antagonists lower ventricular rate in both forms of arrhythmia by inhibiting AV conduction; the rate of flutter waves is said not to change on administration of calcium antagonists (Schlepper 1985). In some cases atrial fibrillation or atrial flutter is converted to sinus rhythm by their administration; this is explained by suppression of "slow-response action potentials" or early diastolic after-depolarization in diseased atria (Waxman et al. 1981; Theisen and Scheininger 1985; see also Sect. 3.1.2).

4.3.1.2 Mechanisms of Action in Ventricular Arrhythmias

Under certain conditions (e.g. ischaemia) calcium antagonists can also suppress ventricular arrhythmias. If the arrhythmias are based on myocardial ischaemia, an indirect and a direct antiarrhythmic effect can be distinguished. Calcium antagonists exert an indirect antiarrhythmic effect when they diminish myocardial ischaemia or coronary insufficiency (improved coronary blood flow, reduced myocardial oxygen demand, Sect. 4.1.5). Calcium antagonists possibly also have a direct antiarrhythmic effect on slow-response action potentials, and when there is abnormal ectopic impulse formation in the ventricles evoked by triggered repetitive activity ("triggered automaticity").

Ventricular slow-response action potentials arise in ischaemic or infarcted myocardium and are characterized by a low resting potential of −60 to −40 mV, calcium-dependent depolarization, a low action potential amplitude and a low conduction rate (the latter increasing the possibility of reentry arrhythmias). Cardiac arrhythmias in the early phase of myocardial infarction are thought to be partly triggered by slow-response action potentials, and can be, at least in animal experiments, suppressed by calcium antagonists. However, some authors express doubt about the existence of slow-response action potentials, and instead postulate action potentials with "depressed fast response", which are characterized by a partially inactivated fast sodium channel. The mechanism of triggered repetitive activity is based on early diastolic after-depolarizations (delayed after-depolarization), which are also calcium-dependent and have a higher incidence in digitalis intoxication, hypokalaemia, exposure to catecholamines and to theophylline, as well as in hypoxaemia and ischaemia.

Overall, mechanisms of direct antiarrhythmic action of calcium antagonists must still be regarded as hypothetical, since a large proportion of the research results is derived from in vitro or animal experiments and still awaits confirmation in humans (further details in Scholz 1981; Coumel 1983; Opie et al. 1984a, b; Sperelakis 1984; Gülker and Bender 1985).

4.3.2 Supraventricular Arrhythmias

Important atrial arrhythmias, apart from atrial fibrillation and flutter, are summarized in Fig. 31. Classical indications for calcium antagonists (besides digitalis glycosides or β-receptor blockers) are paroxysmal supraventricular tachycardias (mostly atrioventricular nodal reentrant tachycardia or atrioventricular reentrant tachycardia) and atrial fibrillation or flutter with fast AV conduction. Calcium

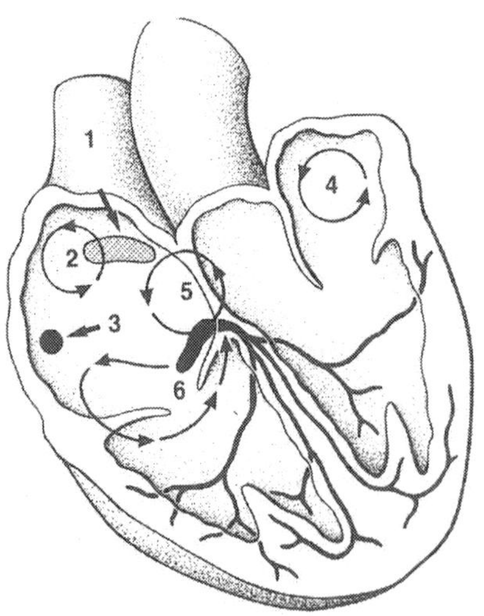

Fig. 31. Atrial tachycardias. *1* Sinus tachycardia (raised automaticity); *2* sinoatrial nodal reentrant tachycardia; *3* atrial tachycardia with ectopic focus; *4* atrial reentrant tachycardia; *5* atrioventricular nodal reentrant tachycardia; *6* atrioventricular reentrant tachycardia (e.g. in WPW syndrome). (After Gülker and Bender 1985)

antagonists are also used in other atrial arrhythmias, although not as agents of primary choice.

4.3.2.1 Sinus Tachycardia

A sinus tachycardia can be considered to be an "arrhythmia" if it is unnecessary for the maintenance of cardiac output (e.g. in physical exertion, lowering of peripheral resistance, volume depletion, and reduced stroke volume on account of heart failure). A sinus tachycardia is regarded as requiring treatment if there are signs of deterioration in the pumping function of the heart, owing to either insufficient diastolic filling (given an adequate blood supply) or ischaemia (Lydtin and Lohmöller 1980). Sinus tachycardia is usually the result of an increased automaticity (activation of sympathetic nervous system, vagal withdrawal). In such cases calcium antagonists exhibit little or no rate-depressing effect. The agents of choice are β-receptor blockers (Lydtin und Lohmöller 1980; Gülker and Bender 1985). Rarely one encounters a sinoatrial nodal reentrant tachycardia with reentry circuits in the sinoatrial node: here the heart rate can be slowed by calcium antagonists and the sinus tachycardia may even be resolved (Antman et al. 1980; Sung et al. 1980; Gülker and Bender 1985; Rowland 1987).

4.3.2.2 Paroxysmal Supraventricular Tachycardia

A large proportion of paroxysmal supraventricular tachycardias (according to Gülker and Bender 1985, about 60%) are based on an AV nodal reentrant tachycardia with reentry circuits in the AV node; in about 30% of cases there is AV reentrant tachycardia (anterograde conduction through the AV node, retrograde conduction through an accessory AV pathway, e.g. in WPW syndrome). Further causes of supraventricular tachycardias are sinoatrial nodal reentry or intraatrial reentry, or an ectopic atrial centre (Fig. 31).

Therapeutically, the first step should always be an attempt to resolve the tachycardia attack by means of vagal stimulation (Valsalva's manoeuvre, unilateral carotid sinus massage, ocular pressure, drinking iced water). If these measures are unsuccessful, then calcium antagonists or β-receptor blockers (besides digitalis glycosides) are agents of primary choice. Parenteral administration of verapamil or diltiazem during an attack has been shown, in up to 95% of cases, to interrupt the tachycardia and restore regular sinus rhythm (Antman et al. 1980; Ellrodt et al. 1980; Krikler and Rowland 1980; Braunwald 1982). In contrast, the success rate of β-receptor blockers is estimated at 50%–70% (Lydtin and Lohmöller 1980; Antman et al. 1980). The following dosages are recommended: for verapamil, 5–15 mg (0.15 mg per kilogram body weight); for diltiazem, 15–25 mg (0.25 mg per kilogram body weight); and for gallopamil, 2–4 mg (Lehmann 1979; Krikler and Rowland 1980; Hung et al. 1984; Waleffe 1985; Schlepper 1985). In some cases calcium antagonists can be combined with β-receptor blockers as oral therapy. In paroxysmal supraventricular tachycardias Yeh et al. (1985) used a single dose of 120 mg diltiazem and 160 mg propranolol, and reported conversion to sinus rhythm after about 20 min in more than 90% of patients. The success rates cited refer to patients with AV nodal or AV reentrant tachycardia.

In intraatrial reentrant tachycardia or tachycardia triggered by an ectopic atrial centre a maximum of two-thirds of patients are thought to convert to sinus rhythm (Braunwald 1982). The pacemaker rate of an ectopic atrial centre is here not affected by calcium antagonists, but ventricular rate decreases, owing to inhibition of AV conduction (Gülker and Bender 1985).

Calcium antagonists have also been successfully employed in the prophylaxis of paroxysmal supraventricular tachycardias. Under a regimen of 240–480 mg verapamil daily or 270–360 mg diltiazem daily the recurrence rate is reported to be markedly reduced in up to 75% of cases (Klein et al. 1982; Hung et al. 1984; Gülker and Bender 1985). Alternatively, in prophylaxis β-receptor blockers or a combination of verapamil and quinidine can also be used (Fleischmann 1985). In contrast to their assured effect during an attack, the role of calcium antagonists in prophylaxis is not undisputed (Ellrodt et al. 1980; Rinkenberger et al. 1980; Rowland 1987).

In the rare cases of AV reentrant tachycardia in the context of WPW syndrome with anterograde conduction via the accessory pathway and retrograde AV conduction, calcium antagonists are without effect, because they do not influence the impulse conduction of the accessory pathway and retrograde AV conduction (Sect. 4.3.1). The drugs of choice in this situation are ajmaline, propafenone, disopyramide or amiodarone (Opie 1984a). (For a discussion of the hazards accompanying treatment of WPW syndrome and atrial fibrillation with calcium antagonists see Sects. 3.2.3 and 4.3.3.)

In individual cases calcium antagonists have been successfully applied in so-called multifocal atrial tachycardia (Levine 1985), for instance, in chronic obstructive lung disease, hypokalaemia, digitalis or theophylline overdosage, as well as in junctional tachycardias initiating from the AV node or the bundle of His (Schamroth 1971).

4.3.2.3 Atrial Fibrillation

In atrial fibrillation with fast AV conduction (i.e. a tachyarrhythmia) digitalis glycosides, β-receptor blockers and calcium antagonists can lower ventricular rate (response). The form of treatment is determined by the underlying heart disease.

In the presence of congestive heart failure, digitalis glycosides are the drugs of first choice, since they exhibit, in contrast to β-receptor blockers and calcium antagonists (with antiarrhythmic action), a positive rather than a negative inotropic effect. If, despite sufficient digitalization, the ventricular rate cannot be satisfactorily reduced, then calcium antagonists can be additionally given. With a dose of 240–480 mg verapamil (Panidis et al. 1983) or 240 mg diltiazem daily (Roth et al. 1986), the heart rate at rest and under stress can usually be reduced to the desired extent. Attention must be given, however, to the possibility of a raised digoxin blood level under chronic verapamil treatment (with gallopamil, too, and perhaps also with diltiazem) owing to reduced digoxin clearance (Belz et al. 1981), and the digitalis dose must therefore be adjusted appropriately (Sects. 4.3.5 and 5.8).

An exception is acute "left heart failure" in mitral stenosis. This is not the result of myocardial disease of the left ventricle, but of impaired left ventricular filling; thus, glycosides are to be used only with caution, since they can increase pulmonary congestion (increased right ventricular contractility). As an alternative to digitalis, therefore, the use of calcium antagonists and/or β-receptor blockers may be considered.

In atrial fibrillation without signs of congestive heart failure, β-receptor blockers or calcium antagonists are preferred as primary therapy, especially if there is hyperthyroidism, mitral stenosis, or hypertrophic obstructive cardiomyopathy. Both forms of treatment satisfactorily lower ventricular rate, but conversion to sinus rhythm is seen in only about 10% of cases receiving either a β-receptor blocker or a calcium antagonist as monotherapy (Schamroth 1971; Lydtin and Lohmöller 1980; Antman et al. 1980; Heuer et al. 1985). Only in newly presenting atrial fibrillation and in patients with a small left atrium (as revealed by chest X-ray or echocardiogram) is the rate of conversion to sinus rhythm higher and can reach around 25% following a single (intravenous) dose of 5 mg verapamil (Tommaso et al. 1983). On the other hand, in 50%–80% of cases atrial fibrillation can be converted to sinus rhythm with a combination of quinidine and a β-receptor blocker or verapamil (Lydtin and Lohmöller 1980; Bender et al. 1984; Heuer et al. 1985; Gülker and Bender 1985); in these studies up to 50% of patients had already been previously digitalized (Heuer et al. 1985). The verapamil–quinidine combination employs doses of 500–1500 mg quinidine and 240–360 (maximum 480) mg verapamil (for a discussion of the quinidine–verapamil interaction see Sects. 4.3.6 and 5.8).

Diltiazem, both in monotherapy and in combination with digitalis glycosides, satisfactorily lowers ventricular rate both in the resting state and under exercise conditions (Roth et al. 1986); it is thought to convert atrial fibrillation only rarely, whether alone or in combination with quinidine (Heuer et al. 1985; Ochs and Eichelbaum 1985; Gülker and Bender 1985). Conversion may of course be attempted with quinidine alone, although this therapy can in certain circumstances induce an accelerated ventricular rate (anticholinergic effect of quinidine) as a consequence of accelerated AV conduction, whereby transition to ventricular fibrillation has also been described (Lydtin and Lohmöller 1980).

Occasionally, calcium antagonists, despite their inability to convert to sinus rhythm, can induce so called "regularization" of heart rate in patients with chronic atrial fibrillation (fluctuations of less than 10% in the RR intervals). This phenomenon is thought to represent AV block of higher degree with a junctional escape rhythm (in the inferior AV node or in the bundle of His; Schamroth et al. 1972; Neuss et al. 1984; Schwartz et al. 1985). Under long-term treatment with verapamil one occasionally observes an AV junctional rhythm (Walker et al. 1983), which, however, is neither registered by the patient nor has any pathological significance.

Continuous treatment with calcium antagonists in chronic atrial fibrillation succeeds, even over longer time periods, in lowering ventricular rate both at rest and under exercise conditions. The recommended doses are: for verapamil, 160–480 mg daily (Ellrodt et al. 1980; Panidis et al. 1983); for diltiazem, 240–360 mg daily (Roth et al. 1986); and for gallopamil, 50–150 mg daily.

A possible alternative in such situations is adequate digitalization; this is equal in respect of exercise tolerance and side effects to treatment with calcium antagonists (verapamil 360 mg daily; Lewis et al. 1987). In some cases, especially in manifest or imminent heart failure, a combination of calcium antagonists and digitalis glycosides may be necessary.

Even in atrial fibrillation of a few years duration, medical cardioversion may be tried. Ochs and Eichelbaum (1985) succeeded in 5 of 11 patients (with on average 2.4 years of atrial fibrillation; 50% with mitral valve disease; all on digitalis) using verapamil (240–480 mg daily) and quinidine (750 mg daily) in restoring sinus rhythm. However, they have published no data on the relapse rate.

In patients with paroxysmal atrial fibrillation, prophylactic treatment with calcium antagonists, even in combination with quinidine, does not significantly reduce the frequency of attacks. Only in animal experiments did Gülker et al. (1982) see a significant rise in the atrial fibrillation threshold under treatment with verapamil and quinidine. β-Receptor blockers provide an alternative prophylactic therapy in atrial fibrillation, either alone or in combination with quinidine (Lydtin and Lohmöller 1980).

In atrial fibrillation occurring in the WPW syndrome calcium antagonists are as a rule contraindicated (Sect. 3.1.2). Calcium antagonists can (like digitalis glycosides and β-receptor blockers), by blocking AV conduction, accelerate anterograde conduction through the accessory pathway into the ventricle and thereby trigger ventricular tachycardias, even to the extent of ventricular fibrillation (Petri et al. 1980; Gulamhusein et al. 1982; Opie 1984a; McGovern et al. 1986). The drugs of choice in the WPW syndrome and atrial fibrillation are ajmaline, disopyramide, propafenone or amiodarone. Only electrophysiological testing can identify those patients presenting with a WPW syndrome and atrial fibrillation who will develop an accelerated ventricular rate with calcium antagonists. We therefore recommend that calcium antagonists be used in the initial treatment of tachyarrhythmias in an unknown patient only if cardioversion or defibrillation is available. This applies especially to parenteral administration of verapamil (Petri et al. 1980, 1982), where continuous close surveillance is mandatory.

4.3.2.4 Atrial Flutter

In atrial flutter calcium antagonists have no effect on the rate of flutter waves, but they depress ventricular rate by inhibiting AV conduction (Fig. 32). This effect was first described and documented by Bender in 1965, and thereafter calcium antagonists were further used in the treatment of arrhythmias. Digitalis glycosides or β-receptor blockers can also be used, the choice of therapy being mainly determined by the underlying heart disease.

In about one-third of cases calcium antagonists (verapamil) can convert atrial flutter to sinus rhythm (Antman et al. 1980; Rinkenberger et al. 1980; Tommaso et al. 1983); in these studies 5–10 mg verapamil was given parenterally. In a further 20% of patients atrial flutter converted to atrial fibrillation (Antman et al. 1980).

In the differential diagnosis of supraventricular tachycardia, parenteral administration of 5 mg verapamil can, besides carotid sinus massage, be of value: a paroxysmal supraventricular reentrant tachycardia will often rapidly change in rate and convert to sinus rhythm; atrial flutter with fast AV conduction will, by contrast, show higher degree AV block, whereby atrial flutter waves are more easily recognized (Singh et al. 1987).

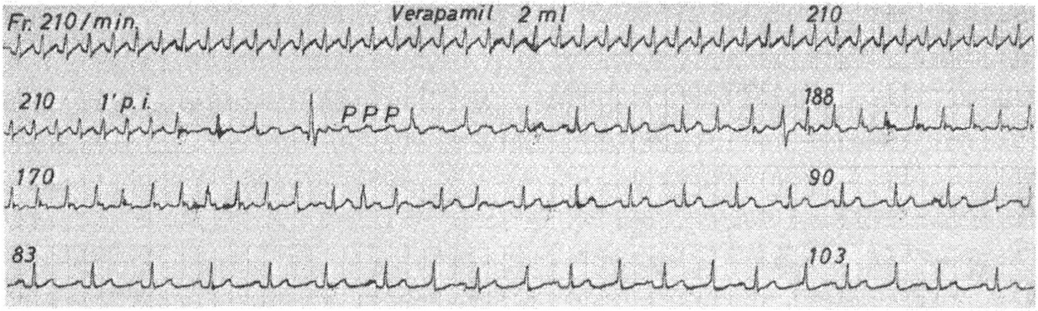

Fig. 32. Effect of verapamil in atrial flutter with 1:1 AV conduction (ventricular rate 210 per minute). Lowering of ventricular rate by AV block; no change in rate of atrial flutter waves. (Original recording by courtesy of Professor Bender, Münster)

4.3.2.5 Supraventricular Extrasystoles

Treatment is only rarely necessary in supraventricular extrasystoles, e.g. in frequent triggering of atrial arrhythmias or in digitalis overdosage (Lydtin and Lohmöller 1980). Calcium antagonists suppress only a small percentage of supraventricular extrasystoles (according to Gülker and Bender 1985, only 20–30%). Drugs of first choice in such cases are antiarrhythmic agents of group Ia, Ic or group III (e.g. quinidine, propafenone, amiodarone; Table 26). Calcium antagonists act principally in extrasystoles arising in the region of the sinoatrial or AV node or in extrasystoles in the dilated or ischaemic atrial myocardium, which are thought to be triggered by so-called slow-response action potentials or by after-depolarization (triggered automaticity; Sect. 4.3.1). The latter seems to play a role in digitalis-induced supraventricular extrasystoles. Calcium antagonists can be used in these situations (Braunwald 1982), or β-receptor blockers may be considered as an alternative (Lydtin and Lohmöller 1980; Sect. 4.3.5).

4.3.3 Wolff-Parkinson-White Syndrome

The WPW syndrome belongs to the so-called preexcitation syndromes (supplementary conduction through an accessory pathway – Kent bundle – from the atria to the ventricles). The ECG characteristically exhibits a short PR interval (less than 0.12 s) and a widened QRS complex with a so-called delta wave. In the context of the WPW syndrome paroxysmal atrial fibrillation or paroxysmal supraventricular tachycardias (AV reentrant tachycardias) occur. The effect of calcium antagonists in such cases has already been discussed in Sects. 4.3.2.2 and 4.3.2.3: in paroxysmal atrial fibrillation and the WPW syndrome calcium antagonists (verapamil, diltiazem, gallopamil) are usually contraindicated on the grounds that accelerated atrioventricular conduction through the accessory pathway can occasionally occur, and a ventricular tachycardia, extending even to ventricular fibrillation, may develop. In this situation the ECG shows widened QRS complexes in irregular sequence.

On the other hand, if electrophysiological tests reveal a paroxysmal supraventricular tachycardia with anterograde conduction through the AV node and retrograde conduction through the accessory pathway, calcium antagonists may be administered, and indeed are, alongside β-receptor blockers, the agents of choice. Until a functional bundle branch block at higher rates occurs, the QRS complexes in the ECG are not widened and the heart rate is regular. Calcium antagonists are ineffective in the rare cases of paroxysmal supraventricular tachycardia with anterograde conduction through the accessory pathway and retrograde conduction through the AV node, and in the ECG widened QRS complexes are seen in regular sequence.

The Lown-Ganong-Levine syndrome (LGL syndrome) also belongs to the preexcitation syndromes. Here there is an accessory pathway between atrium and inferior AV node or the bundle of His (so-called James bundle). In the ECG the PR interval is shortened (less than 0.12 s), the QRS complexes are of normal width. Clinically, paroxysmal reentrant supraventricular tachycardia can also occur in the LGL syndrome, and theoretically calcium antagonists should also be effective in these cases; however, to date no relevant systematic investigations have been conducted.

In general, during the first attack of a paroxysmal supraventricular tachycardia (or atrial fibrillation; Sects. 4.3.2.2 and 4.3.2.3) or in hitherto unknown patients we recommend parenteral administration of calcium antagonists only with continuous surveillance (ideally by monitor). In some cases (in ventricular tachycardia or ventricular fibrillation) electrical cardioversion or defibrillation may become necessary. Further unwanted side effects seen with intravenous administration of calcium antagonists are higher degree AV block, sinus bradycardia, even extending to sinus arrest, and an acute drop in blood pressure. However, these side effects are rare; especially at risk are older patients and those with preexisting diseases of the sinoatrial and/or AV node (Sect. 5.6, Contraindications).

4.3.4 Ventricular Arrhythmias

Ventricular arrhythmias requiring treatment are frequent multifocal ventricular extrasystoles, especially when associated with a bigeminy or R-on-T phenomenon, ventricular salvos, and previous ventricular tachycardia, ventricular flutter or ventricular fibrillation. The treatment of choice is with antiarrhythmic agents of class Ib or III; in some cases β-receptor blockers can be used, and only with this latter form of therapy has a positive influence on mortality been demonstrated in patients with coronary heart disease (reviews by Ryden et al. 1983 and Gülker 1985). Calcium antagonists do not influence the electrical activity (excitability) of normal ventricular myocardium. An antiarrhythmic effect presupposes the occurrence of mainly calcium-dependent excitation processes in the region of diseased ventricular myocardium ("slow-response action potentials" or "repetitive triggered activity"; "triggered automaticity"; Sect. 4.3.1).

On the basis of theoretical electrophysiological considerations (review by Sperelakis 1984; DeLeiris et al. 1984) and of animal experiments (review by Gülker and Bender 1985; Haverkamp et al. 1987), an antiarrhythmic effect has also been postulated for calcium antagonists in humans in the treatment of ventricular arrhythmias. On one hand, a profusion of individual observers have reported a positive effect in ventricular extrasystoles and ventricular tachycardias, extending even to polymorphous ventricular tachycardia in the form of *torsade de pointes* (Schamroth et al. 1972; Arita et al. 1975; Coumel et al. 1983; Bachour 1985; Gülker and Bender 1985; Massumi et al. 1986); furthermore, patients were also treated in the early phase of acute myocardial infarction (Fazzini 1981), whereby ventricular extrasystoles and ventricular tachycardias were suppressed by 5–10 mg verapamil intravenously. On the other hand, a systematic investigation of the efficacy of calcium antagonists in ventricular arrhythmias, especially during or following acute myocardial infarction, has yet to be made. The large-scale intervention studies relating to myocardial infarction (Sect. 4.1.4) have been unable to show any reduction in total mortality or sudden cardiac death. Thus, there is at the present time no proof of an assured antiarrhythmic effect of calcium antagonists in ventricular arrhythmias or in the prevention of ventricular fibrillation in acute myocardial infarction (review by Opie et al. 1984b).

On account of their indirect antiarrhythmic effect (Sect. 4.3.1) calcium antagonists are thought to be more effective in ischaemia-induced cardiac arrhythmias (physical stress, coronary spasm) than in cardiac arrhythmias of other origin (Ellrodt et al. 1980; Gülker and Bender 1985; Nadamanee and Singh 1988). In Prinzmetal's angina pectoris, where malignant ventricular arrhythmias can be triggered by coronary spasm during an attack, calcium antagonists suppress both coronary spasm and arrhythmias (Sect. 4.1.3.1).

Perhexiline and bepridil are both calcium antagonists with an additional action on the fast sodium channel (class I: "membrane stabilizing effect"), which explains their efficacy in ventricular extrasystoles and ventricular tachycardias (review by Antman et al. 1980; Coumel et al. 1983; Davy et al. 1983).

4.3.5 Digitalis-Induced Arrhythmias

Digitalis glycosides and calcium antagonists simultaneously reduce the rate of the sinoatrial node and delay AV conduction. Accordingly, calcium antagonists (verapamil, diltiazem, gallopamil), and β-receptor blockers, are contraindicated in bradycardia and disturbances of AV conduction without supraventricular tachycardia. By contrast, in digitalis overdosage, if supraventricular tachycardias develop (with or without AV block), administration of calcium antagonists and β-receptor blockers, may be tried (Sect. 4.3.2.2 and 4.3.2.5). However, the agent of first choice in digitalis overdosage is phenytoin, which, unlike calcium antagonists and β-receptor blockers, does not inhibit AV conduction (Lydtin and Lohmöller 1980; Lazzara and Scherlay 1980; Braunwald 1982).

4.3.6 Drug Interactions in Antiarrhythmic Combination Therapy

In the combination of calcium antagonists (of the verapamil type, diltiazem) with β-receptor blockers or digitalis glycosides, all three classes of substances can delay AV conduction, and there is thus a possibility of higher degree AV block. AV conduction can be reaccelerated by atropine or β-sympathomimetics (isoprenaline, orciprenaline); intravenous calcium is, however, without effect (Opie 1984a). In combinations of verapamil, gallopamil and diltiazem with β-receptor blockers, the negative inotropic effects can be additive, and in some cases heart failure may be induced or exacerbated. In combinations of calcium antagonists with other antiarrhythmics attention must be given to the frequent negative inotropic properties of the latter.

Maisel et al. (1985) have described several cases of acute hypotension in patients receiving quinidine (orally) and verapamil (i. v.), a phenomenon which they attribute to additive blocking of α-receptors. If verapamil (or gallopamil) is combined with digoxin, as with the quinidine - digoxin combination, an elevation of the digoxin plasma level occurs on account of reduced digoxin clearance (Belz et al. 1981). In some patients, therefore, it is necessary to adjust the dose of digoxin. By contrast, with diltiazem the digoxin plasma level remains unchanged, or is altered only negligibly (Roth et al. 1986; Boden et al. 1986; see also Sect. 5.8). Occasional elevation of the quinidine plasma level has also been observed in combination with verapamil (Trohmann et al. 1986).

With the combination of verapamil with digoxin or quinidine, therefore, close attention must be paid to possible signs of overdose, dosages must be adjusted appropriately, and, if doubts arise, plasma drug levels should be determined (Sect. 5.8).

References

Antman EM, Stone PH, Muller JE, Braunwald E (1980) Calcium channel blocking agents in the treatment of cardiovascular disorders Part I: basic and clinical electrophysiologic effects Ann Intern Med 93: 875–885

Arita M, Ohtake N, Kawasaki K, Takashiba T (1975) Therapeutic effects and mechanism of action of CRD-401 on premature beat unresponsive to procainamide Jap J Clin Exp Med 52: 2430–2439

Bachour G (1985) Therapie der Kammerextrasystolie mit Chinidin-Verapamil In: Bender F, Greeff K (eds) Kombinationstherapie der Herzrhythmusstörungen mit Chinidin und Verapamil Steinkopff Darmstadt pp 127–137

Belz GG, Olesch K, Schmidt Voigt I (1970) Die Behandlung des chronischen Vorhofflimmerns mit einer Kombination von Chinidin und Verapamil Med Welt 38: 1670

Belz GG, Doering W, Munkes R, Aust PE, Belz G (1981) Effects of various calcium-antagonists on blood level clearance of digoxin Circulation 64 (Suppl IV): IV–24

Bender F, Kojima N, Replok HD, Oelmann G (1966) Behandlung tachykarder Rhythmusstörungen des Herzens durch Beta-Rezeptorenblockade des Atrioventrikulargewebes Med Welt 17: 1120–1123

Bender F (1968) Die medikamentöse Behandlung von Herzrhythmusstörungen Therapiewoche 18: 1803

Bender F, Brisse B, Gülker H (1984) Calcium antagonists in supraventricular arrhythmias In: Althaus U, Burckhardt D, Vogt E (eds) Calcium-Antagonismus – Internat. Symposium on Calcium-Antagonism Universimed Frankfurt/Main pp 209–215

Boden WE, More G, Sharma S, Bough EW, Korr KS, Young PM, Shulman RS (1986) No increase in serum digoxin concentration with high-dose diltiazem Am J Med 81: 425–428

Braunwald E (1982) Mechanism of action of calcium-channel-blocking agents N Engl J Med 307: 1618–1627

Coumel P, Maisonblanche P, Cauchemez B (1984) Calcium blockers in the treatment of ventricular arrhythmias In: Althaus U, Burckhardt D, Vogt E (eds) Calcium-Antagonismus – Internat. Symposium on Calcium-Antagonism Universimed Frankfurt/Main pp 190–208

Davy JM, Lainé JF, Sebag C, Motté G, Béclére HA (1983) Bepridil, a new calcium antagonist for the treatment of ventricular tachycardia Circulation 68 (Suppl III): III–310

DeLeiris J, Richard V, Pestre S (1984) Calcium antagonists and experimental myocardial ischemia and infarction In: Opie LH (ed) Calcium antagonists and cardiovascular disease Raven Press New York pp 105–115

Ellrodt G, Chew CYC, Singh BN (1980) Therapeutic implications of slow-channel blockade in cardiocirculatory disorders Circulation 62: 669–679

Fazzini PF, Marchi F, Pucci P, Ledda F, Mugelli A (1981) Treatment of „slow response“ related premature beats in acute myocardial infarction In: Zanchetti A, Krikler DM (eds) Calcium antagonism in cardiovascular therapy: experience with verapamil Excerpta Medica Amsterdam pp 337–341

Fleischmann DW (1985) Therapie rezidivierender supraventrikulärer Tachykardien mit Cordichin In: Bender F, Greeff K (eds) Kombinationstherapie der Herzrhythmusstörungen mit Chinidin und Verapamil Steinkopff Darmstadt pp 139–146

Gülker H, Bender F, Thale J, Heuer H (1982) Vergleichende Untersuchungen zur Wirkung intravenöser Einzel- und Kombinationsgaben von Chinidin, Diltiazem und Verapamil auf elektrisch induzierte atriale repetitive Extrasystolen und Vorhofflimmern In: Bender F, Greeff K (eds) Calciumantagonisten zur Behandlung der Angina pectoris, Hypertonie und Arrhythmie Excerpta Medica Amsterdam pp 278–286

Gülker H (1985) Antiarrhythmische und antifibrillatorische Substanzeigenschaften herzwirksamer Pharmaka In: Bender F, Greeff K (eds) Kombinationstherapie der Herzrhythmusstörungen mit Chinidin und Verapamil Steinkopff Darmstadt pp 1–13

Gülker H, Bender F (1985) Therapie von Herz-Arrhythmien mit Calcium-Antagonisten Schwerpunkt med 8: 28–33

Gulamhusein S, Ko P, Carruthers SG, Klein GJ (1982) Acceleration of the ventricular response during atrial fibrillation in the Wolff-Parkinson-White syndrome after verapamil Circulation 65: 348–354

Haverkamp W, Hindricks G, Kunert M, Thale J, Behrenbeck Th, Gülker H (1987) Wirkungen von Gallopamil und Nifedipin auf ventrikuläre Arrhythmien, Kammerflimmern, epikardiale Leitungsverzögerungen und den Zeitverlauf der ventrikulären Flimmerschwelle während Myokardischämie und Reperfusion In: Bender F, Meesmann W (eds) Therapie mit Gallopamil Steinkopff Darmstadt pp 31–49

Heuer H, Brisse B, Gülker H, Bender F (1985) Antifibrillatorische Wirkung von Chinidin/Verapamil und Chinidin/Pindolol bei Vorhofflimmern In: Bender F, Greeff K (eds) Kombinationstherapie der Herzrhythmusstörungen mit Chinidin und Verapamil Steinkopff Darmstadt pp 115–123

Hung J, Yeh S, Lin F, Fu M, Wu D (1984) Usefulness of intravenous diltiazem in predicting subsequent electrophysiologic and clinical responses to oral diltiazem Am J Cardiol 54: 1259–1269

Klein GJ, Gulamhusein S, Prystowsky EN, Carruthers SG, Donner AP, Ko PT (1982) Comparison of the electrophysiologic effects of intravenous and oral verapamil in patients with paroxysmal supraventricular tachycardia Am J Cardiol 49: 117–124

Krikler DM, Rowland E (1980) The role of calcium-ion antagonists in cardiac arrhythmias In: Fleckenstein A, Roskamm H (eds) Calcium antagonists Springer Berlin Heidelberg New York pp 55–61

Lazzara R, Scherlay B (1980) Treatment of arrhythmias by blocking slow current Ann Intern Med 93: 919–921

Lehmann HU, Taegener-Torge C, Witt E, Beck OA, Hochrein H (1979) Antiarrhythmischer Wirkungsvergleich zwischen Gallopamil (D-600) und Verapamil Angiocardiology 2: 103–111

Levin JH, Michael JR, Guarnieri T (1985) Treatment of multifocal atrial tachycardia with verapamil N Engl J Med 312: 21–25

Lewis R, Lakhani M, Moreland TA, McDevitt DG (1987) A comparison of verapamil and digoxin in the treatment of atrial fibrillation Eur Heart J 8: 148–153

Lydtin H, Lohmöller G (1980) Beta blockers Huber Bern

Maisel AS, Motulsky HJ, Insel PA (1985) Hypotension after quinidine plus verapamil N Engl J Med 312: 167–171

Massumi RA, Marin J, Udhoji VN (1986) Verapamil in parasystolic ventricular tachycardia and other wide QRS rhythms Am Heart J 111: 400–405

McGovern B, Garan H, Ruskin JN (1986) Precipitation of cardiac arrest by verapamil in patients with Wolff-Parkinson-White syndrome Ann Intern Med 104: 791–194

Nademanee K, Singh BN (1988) Control of cardiac arrhythmias by calcium antagonism In: Vanhoutte PM, Paoletti R, Govoni S New (eds) Calcium antagonists New York Academy of Sciences New York pp 536–552

Neuss H, Golling FR, Schlepper M, Thormann J, Weissmüller P, Kindler M (1984) Regularisierung der Kammerintervalle bei Vorhofflimmern – elektrophysiologische Befunde zum zugrundeliegenden Mechanismus Z Kardiol 73: 106–112

Ochs HR, Eichelbaum M (1985) Diltiazem, Verapamil und Chinidin als Mono- und Kombinationstherapie bei chronischem Vorhofflimmern In: Bender F, Greeff K (eds) Kombinationstherapie der Herzrhythmusstörungen mit Chinidin und Verapamil Steinkopff Darmstadt pp 103–111

Opie LH (1984a) Arrhythmias and antiarrhythmic agents In: Opie LH (ed) The Heart Grune & Stratton London pp 314–337

Opie LH, Thandroyen FT, Muller CA, Hamm CW (1984) Calcium channel antagonists as antiarrhythmic agents: contrasting properties of verapamil and diltiazem versus nifedipine In: Opie LH (ed) Calcium antagonists and cardiovascular disease Raven Press New York pp 303–311

Panidis IP, Morganroth J, Baessler C (1983) Effectiveness and safety of oral verapamil to control exercise-induced tachycardia in patients with atrial fibrillation receiving digitalis. Am J Cardiol 52: 1197–1201

Petri H, Kafka W, Hall D, Rudolph W (1980) Potential acceleration of the ventricular rate in WPW with atrial fibrillation after verapamil Circulation 62 (Suppl III) III–262 (abstract)

Petri M, Kafka W, Hansen W, Rudolf W (1982) Discrepant effect of intravenous and oral verapamil on the ventricular rate in WPW syndrome during atrial fibrillation Circulation 66 (4pt2) II–270

Rinkenberger RL, Prystowsky EN, Heger JJ, Troup PJ, Jackman WM, Zipes DP (1980) Effects of intravenous and chronic oral verapamil administration in patients with supraventricular tachyarrhythmias Circulation 62: 996–1010

Roth A, Harrison E, Mitani G, Cohen J, Rahimtoola SH, Elkayam U (1986) Efficacy and safety of medium- and high-dose diltiazem alone and in combination with digoxin for control of heart rate at rest and during exercise in patients with chronic atrial fibrillation Circulation 73: 316–324

Rydén L, Ariniego R, Arnman K, Herlitz J, Hjalmarson A, Holmberg S, Reyes C, Smedgård P, Svedberg K, Vedin A, Waagstein F, Waldenström A, Wilhelmsson C, Wedel H, Yamamoto M (1983) A double-blind trial of metoprolol in acute myocardial infarction N Engl J Med 308: 614–618

Schamroth L (1971) Immediate effects of intravenous verapamil on atrial fibrillation Cardiovasc Res 5: 419–424

Schamroth L, Krikler DM, Garret C (1972) Immediate effects of intravenous verapamil in cardiac arrhythmias Br Med J 1: 660–662

Schlepper M (1985) Antiarrhythmisches Wirkungsprofil von Verapamil und Chinidin In: Bender F, Greeff K (eds) Kombinationstherapie der Herzrhythmusstörungen mit Chinidin und Verapamil Steinkopff Darmstadt pp 27–43

Scholz H (1981) Kalziumantagonisten in der Arrhythmiebehandlung In: Lüderitz B (ed) Ventrikuläre Herzrhythmusstörungen Springer Berlin Heidelberg New York pp 99–101

Schwartz JB, Nielsen AP, Griffin JC (1985) Concentration-dependent enhancement of junctional pacemaker activity by verapamil in man Circulation 71: 450–457

Singh BN, Vaughan Williams EM (1972) A fourth class of antidysrhythmic action? Effect of verapamil on ouabain toxicity, on atrial and ventricular intracellular potentials, and on other features of cardiac function Cardiovasc Res 6: 109–119

Singh BN, Ellrodt G, Peter CT (1978) Verapamil: a review of its pharmacological properties and therapeutic use Drugs 15: 169–203

Sperelakis N (1984) Properties of calcium-dependent slow action potentials: their possible role in arrhythmias In: Opie LH (ed) Calcium Antagonists and Cardiovascular disease Raven Press New York pp 277–291

Sung RJ, Elser B, McAllister RG (1980) Intravenous verapamil for termination of re-entrant supraventricular tachycardias Ann Intern Med 93: 682–689

Theisen K, Scheininger M (1985) Elektrophysiologische Effekte der Kombination Chinidin/Verapamil auf die AV-Leitung des Menschen In: Bender F, Greeff K (eds) Kombinationstherapie der Herzrhythmusstörungen mit Chinidin und Verapamil Steinkopff Darmstadt pp 47–61

Tommaso C, McDonough T, Parker M, Talano JV (1983) Atrial fibrillation and flutter: immediate control and conversion with intravenously administered verapamil Arch Intern Med 143: 877–881
Trohman RG, Estes DM, Castellanos A, Palomo AR, Myerburg RJ, Kessler KM (1986) Increased quinidine plasma concentrations during administration of verapamil: a new quinidine-verapamil interaction Am J Cardiol 57: 706–707
Waleffe A, Hastir F, Kulbertus HE (1985) Effects of intravenous diltiazem administration in patients with inducible tachycardia Eur Heart J 6: 882–890
Walker WS, Winniford MD, Mauritson DR, Johnson SM, Hillis LD (1983) Atrioventricular junctional rhythm in patients receiving oral verapamil therapy JAMA 249: 389–390
Waxman HL, Myerburg RJ, Appel R, Sung RJ (1981) Verapamil for control of ventricular rate in paroxysmal supraventricular tachycardia and atrial fibrillation or flutter Ann Intern Med 94: 1–6
Yeh SJ, Lin FC, Chou YY, Hung JS, Wu D (1985) Termination of paroxysmal supraventricular tachycardia with a single oral dose of diltiazem and propranolol Circulation 71: 104–109

4.4 Cardiomyopathies

The term cardiomyopathy embraces heart diseases of various aetiology and pathogenesis, which primarily involve the myocardium and are not due to coronary heart disease, arterial hypertension, primary pulmonary disease or to valvular or congenital heart disease. Aetiologically, two types of cardiomyopathy are distinguished: primary cardiomyopathies (or idiopathic cardiomyopathies), the cause or causes of which remain unknown; and secondary forms of known origin (e.g. myocarditis caused by defined pathogens, cardiac involvement in sarcoidosis, connective tissue diseases, amyloidosis, haemochromatosis, storage diseases, as well as cardiac lesions caused by alcohol, cobalt or medication). In spite of improved diagnosis, especially via endomyocardial biopsy and immunological parameters, even today only a proportion of cardiomyopathies can be unambiguously aetiologically classified; the greater proportion remains classified as primary cardiomyopathies. For this reason a classification according to pathophysiological, clinical and haemodynamic criteria has been adopted. In the absence of a causal treatment, symptomatic therapy is based on this classification.

4.4.1 Pathophysiological Classification of Cardiomyopathies According to Haemodynamic Criteria

The following forms of cardiomyopathy can be distinguished:

1. Hypertrophic cardiomyopathy, with or without obstruction;
2. Dilated (or congestive) cardiomyopathy;
3. Restrictive (or obliterative) cardiomyopathy (reviews by Goodwin 1972, 1974; Loogen 1977; Wynne and Braunwald 1984).

Figure 33 summarizes the various forms of cardiomyopathy. In considering this classification, one must appreciate that, on one hand, there are transitional forms and, on the other, that the presentation of a cardiomyopathy may be modified by a supplementary cardiovascular disease (e.g. arterial hypertension).

The restrictive form of cardiomyopathy is rare in western Europe and the United States; it includes eosinophilic endocarditis (fibroplastic endocarditis) as

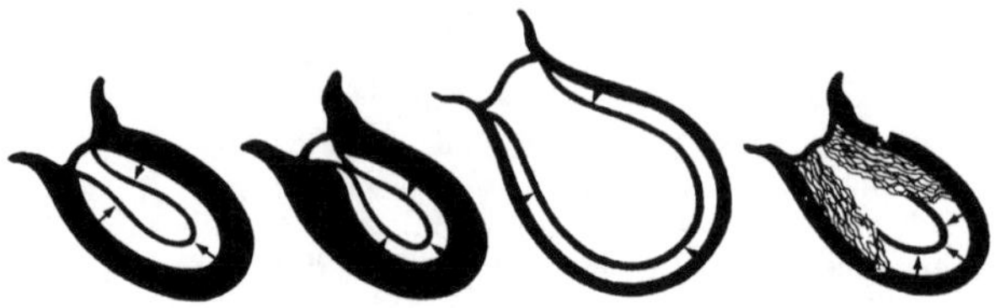

	HCM	HOCM	DCM	RCM
EDV	N (↓)	N (↓)	↑	N (↑)
EF	N (↑)	N (↑)	↓	↓
EDP	↑ (↑)	↑ (N)	↑ (N)	↑ (↑)
Compliance	↓ (N)	↓ (N)	N (↑)	↓ (N)
Obstruction	—	+	—	(+)

Fig. 33. Classification of primary cardiomyopathies. *HCM* hypertrophic cardiomopathy; *HOCM* hypertrophic obstructive cardiomyopathy; *DCM* dilated cardiomyopathy (congestive cardiomyopathy); *RCM* restrictive cardiomyopathy (obliterative cardiomyopathy); *EDV* end-diastolic volume; *EF* ejection fraction; *EDP* end-diastolic pressure; *Compliance* diastolic ventricular elasticity. *Downward arrows* indicate decrease, and *upward arrows* increase; *N* normal. Symbols within parentheses signify rare findings

described by Löffler, as well as cardiac involvement in amyloidosis. In the tropics a so-called endomyocardial fibrosis of unknown origin is also encountered, in which the ventricular cavity is displaced by fibrous tissue and thrombi, and the mitral and tricuspid valves are also often pathologically altered.

Calcium antagonists are used predominantly in hypertrophic cardiomyopathy and in some cases also in dilated cardiomyopathy; these two forms are discussed in the following sections.

4.4.2 Haemodynamics and Diagnosis of Hypertrophic Cardiomyopathy

The characteristic morphological sign of hypertrophic cardiomyopathy is a hypertrophic, unenlarged left ventricle without any indication as to a primary cause of the hypertrophy (e. g. aortic stenosis or arterial hypertension). Histologically, the hypertrophy appears to be the result of a genetically conditioned disturbance of the spatial arrangement of myofibrils; this "textural disorder" is thought to differentiate hypertrophic cardiomyopathy from cardiomyopathies of other origins (Lorell 1985). In addition, connective tissue scars of various sizes are (according to Maron et al. 1987) found in the myocardium, and a thickening and narrowing of small intramural coronary arteries (so-called small vessel disease, without any indication of coronary artery sclerosis). Some authors (Landmark et al. 1983; review by Kaltenbach and Hopf 1984) discuss the possibility, at the molecular level, of intracellular calcium overload in hypertrophic cardiomyopathy.

In hypertrophic cardiomyopathy asymmetrical septal hypertrophy predominantly occurs (according to Wigle 1987, about 90%), which in systole, together with the forward motion of the anterior mitral valve leaflet (mitral valve septum contact), leads to a dynamic obstruction of the left ventricular outflow tract. This form of hypertrophic cardiomyopathy is known as hypertrophic obstructive cardiomyopathy (formerly idiopathic hypertrophic subaortic stenosis IHSS, muscular subaortic stenosis, subvalvular stenosis). As a consequence of the obstruction of the left ventricular outflow tract, a pressure gradient develops between left ventricle and aorta. Both the obstruction and the pressure gradient are dependent on the degree of filling of the left ventricle, contractility and afterload (review by Wynne and Braunwald 1984). The obstruction is increased by reduced filling of the left ventricle owing to nitrates, diuretics or Valsalva's manoeuvre, positive inotropic drugs such as digitalis or β-sympathomimetics, as well as by an abrupt reduction in afterload. In contrast, the obstruction is diminished by negative inotropic substances (β-receptor blockers, calcium antagonists) and by an elevation of the afterload (e.g. resulting from an increase in blood pressure or assumption of the squatting position). An obstruction is found in about 25% of all cases of hypertrophic cardiomyopathy (Maron et al. 1987); but it can also be entirely absent, and there is in such cases frequently a regional hypertrophy in the free cardiac wall or the apex, which spares the ventricular septum.

In all forms of hypertrophic cardiomyopathy (with or without obstruction) diastolic filling of the left ventricle is disturbed. Because of reduced elasticity (compliance) and disturbed relaxation, the stiff and hypertrophied left ventricle fills more slowly and to a lesser extent in diastole than normal (review by Maron et al. 1987). In relation to a normal or reduced end-diastolic volume the end-diastolic pressure in the left ventricle is increased (review by Krayenbühl et al. 1984; Kaltenbach and Hopf 1984; Lorell 1985). Diastolic impairment of left ventricular filling is further increased under stress, since the filling time is further reduced by the increase in heart rate.

After a variable symptomless phase of hypertrophic cardiomyopathy the patient sooner or later seeks medical advice because of dyspnoea, fatigue, anginal symptoms or syncope. The often atypical thoracic pains (pain at rest, protracted, no improvement following administration of nitrates) are thought to be a consequence of a regional myocardial ischemia. Sudden cardiac death resulting from a cardiac arrhythmia (usually ventricular tachycardia) may also be the first manifestation of hypertrophic cardiomyopathy. This is mainly encountered in younger patients and under stress conditions (e.g. sport) (Goodwin and Krikler 1976; Koway et al. 1984; Maron et al. 1987); hypertrophic cardiomyopathy is the commonest cause of sudden cardiac death in young people. Supraventricular and ventricular tachycardias, atrial fibrillation and AV block can occur, and preexcitation syndromes are also observed. If obstruction is present, auscultation yields a systolic murmur, which characteristically increases during Valsalva's manoeuvre. Occasionally a holosystolic murmur is additionally heard at the cardiac apex owing to concomitant mitral insufficiency. In the presence of obstruction, in contrast to valvular aortic stenosis, the carotid pulse curve shows an initial normal rise in the pulse wave, a midsystolic dip, corresponding to stenosis of the outflow tract, and at the end of systole a second peak. This "spike and dome configuration" is pathognomonic for hypertrophic obstructive cardiomyopathy.

Echocardiography (including Doppler echocardiography) is today essential in diagnosing hypertrophic cardiomyopathy: the left ventricle is hypertrophied, but not dilated and there is usually disproportionate hypertrophy of the ventricular septum (so-called asymmetrical septal hypertrophy). Septal hypertrophy constricts the outflow tract of the left ventricle, leading to high outflow velocity in early systole (at this point obstruction is absent). This induces by a Venturi mechanism (details in Wigle 1987; investigations with Doppler echocardiography) a forward motion of the anterior mitral valve leaflet (systolic anterior motion, SAM) with septal contact. Owing to this phenomenon obstruction of the left ventricular outflow tract and perhaps also a reflux via the mitral valve (mitral regurgitation) develops. Rarely a forward motion of the posterior mitral valve leaflet is also seen (Wigle 1987).

If heart catheterization is undertaken, end-diastolic pressure is constantly raised; if there is an obstruction of the left ventricular outflow tract, a pressure gradient is measured between left ventricle and aorta (review by Wynne and Braunwald 1984; Lorell 1985; Maron et al. 1987; Wigle 1987).

4.4.3 Therapy of Hypertrophic Cardiomyopathy with Calcium Antagonists

In the middle of the 1970s, Kaltenbach and his associates, following investigations of hereditary cardiomyopathy in the Syrian hamster (Bajusz and Lossnitzer 1968), used a calcium antagonist (verapamil) for the first time clinically in hypertrophic cardiomyopathy (Kaltenbach et al. 1976).

Cardiomyopathy of the Syrian hamster (review by Lossnitzer et al. 1980) is an autosomal recessive heart disease with histologically demonstrable multifocal myocardial necroses. A disorder of protein synthesis manifests itself not only in the myofibrils, but also in the sarcolemma of cardiac muscle, resulting in an increased uptake of calcium by the myocardium, with consequent calcium overload of myocardial cells. The raised calcium concentration is thought to result in myocardial necroses. Fleckenstein (1971) was able, by administering β-sympathomimetics, to produce a further increase in calcium uptake of myocardial cells; the number of myocardial necroses increased concomitantly. By administration of calcium antagonists (verapamil, gallopamil) or β-receptor blockers, spontaneous or isoproterenol-induced calcium overload in heart muscle cells was prevented (Fig. 34). The incidence of myocardial necroses decreased, and myocardial contractility was maintained by chronic (continuous) administration of verapamil (Rouleau et al. 1982).

Cardiomyopathy in the Syrian hamster and human hypertrophic cardiomyopathy show numerous common features (Lossnitzer et al. 1980). Thus, the use of calcium antagonists in hypertrophic cardiomyopathy has a sound basis in terms of pathophysiology and molecular biology. Nevertheless, it would be premature to explain the effectiveness of calcium antagonists in human hypertrophic cardiomyopathy only by a myocardial mode of action, i.e. by inhibition of calcium overload of myocardial cells alone. Calcium antagonists possibly also lower left ventricular wall tension by reducing afterload and improve coronary blood flow, thereby decreasing subendocardial ischaemia (Lorell 1985).

The haemodynamic effects of calcium antagonists in both acute and long-term therapy have been extensively investigated:

Results are available for verapamil, nifedipine and for a combination of nifedipine with propranolol (review by Rosing et al. 1979a, 1982; Braunwald 1982; Landmark et al. 1983; Kaltenbach and Hopf 1984; Lorell 1985). Calcium antagonists improve impaired left ventricular relaxation and filling, which is measurable as a decrease in the isovolumic relaxation time and as an increase in the left ventricular filling rate (review by Krayenbühl et al. 1984). Beyond improvement in relaxation, calcium antagonists probably have no effect on the passive stiffness

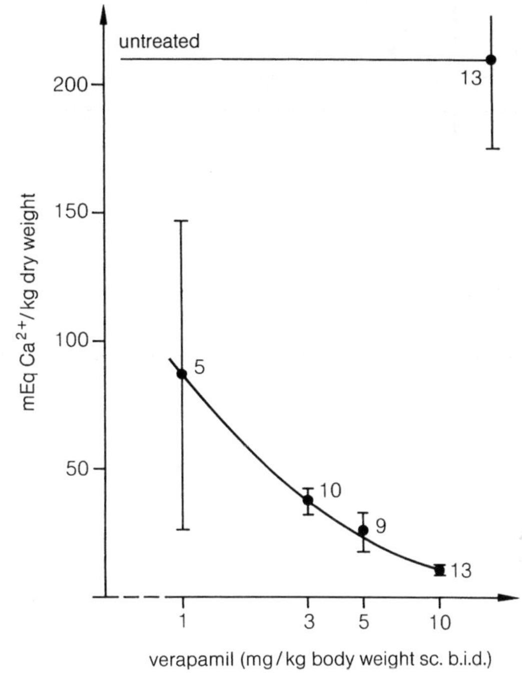

Fig. 34. Decrease of myocardial calcium content in 60-day-old cardiomyopathic hamsters (BIO 8262 strain) after chronic subcutaneous ingestion of different doses of verapamil. Mean values ± standard deviation, and the number of animals are indicated. (After Lossnitzer et al. 1980)

of the left ventricle. Left ventricular end-diastolic pressure decreases, and left ventricular end-diastolic volume remains unchanged, or increases minimally. A dynamic obstruction of the left ventricular outflow tract, if present, is almost always diminished, and the pressure gradient between left ventricle and aorta markedly decreases both at rest and under stress; complete normalization is, however, rare (Kaltenbach et al. 1976, 1984; Rosing et al. 1979a; Krayenbühl et al. 1984; Lorell 1985). In some cases, especially if there is marked reduction in blood pressure or in afterload, the pressure gradient may even increase (Sect. 4.4.2), and this can even induce pulmonary oedema (Epstein and Rosing 1981). Systolic function and contractility of the left ventricle, measurable by cardiac output and ejection fraction, remain largely unchanged (Rosing et al. 1979a; Hanrath et al. 1980; Bonow et al. 1981; Chatterjee et al. 1982; Suwa et al. 1984) and there are only isolated reports (Bonow et al. 1983; Lorell 1985) of a small negative inotropic effect. These predominantly positive results contrast with studies (Ten Cate et al. 1983; Betocchi et al. 1985) in which, following a single dose of verapamil or nifedipine, there is no change in diastolic filling or in the gradient of the left ventricular outflow tract; left ventricular filling pressure even increases to some extent and contractility decreases.

Overall, calcium antagonists improve diastolic relaxation and filling, and they partly reduce obstruction in the left ventricular outflow tract, but normalization is rarely attained. Systolic function of the left ventricle is, in contrast to the negative inotropic effect of β-receptor blockers, usually unchanged. The symptoms of hypertrophic cardiomyopathy (diminished performance, fatigue, dyspnoea, anginal symptoms) are alleviated by calcium antagonists, and parallel to this the patient's exercise tolerance increases (Fig. 35) when measured, for instance, in terms of exercise tolerance (Kaltenbach et al. 1975, 1984; Rosing et al. 1979; Spicer et al. 1983; Bonow et al. 1985). Left ventricular hypertrophy is reported to regress under long-term treatment with calcium antagonists (Kaltenbach et al. 1976, 1984) or to remain unchanged (Rosing et al. 1981), a

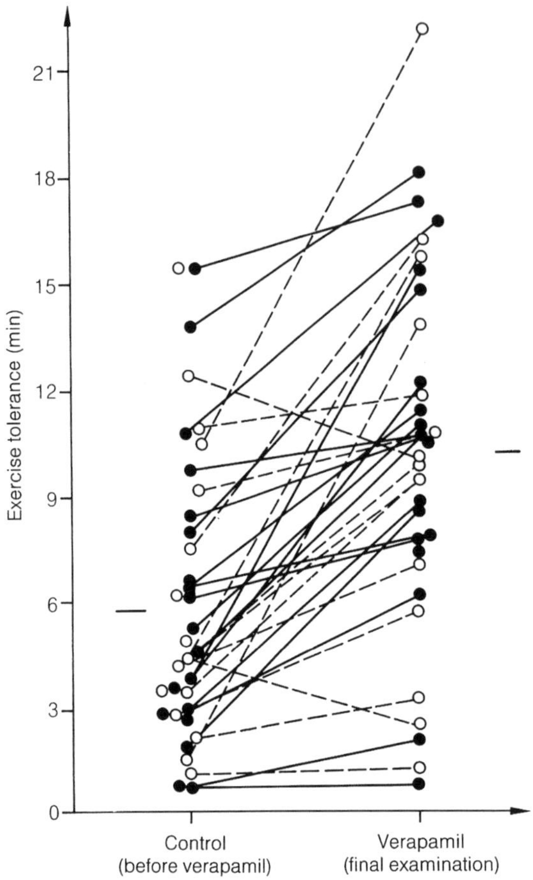

Fig. 35. Exercise tolerance of patients with hypertrophic obstructive cardiomyopathy under long-term treatment with verapamil (*left* values at start of treatment; *right* after a treatment period of 6–25 months). *Transverse lines* indicate mean values (difference significant with $p<0.001$). o - - - o Standard exercise tolerance test; ●——● Improved protocol exercise tolerance test. (After Rosing et al. 1981)

discrepancy which can perhaps be partially explained by methodological differences (e. g. critical assessment of left ventricular hypertrophy by ECG, ventriculography or echocardiography).

Most research and experience of calcium antagonists in this area pertains to the use of verapamil. In two large long-term studies Kaltenbach and Hopf (1984) followed 61 patients for an average of 50 months, and Rosing et al. (1981) studied 68 patients for periods ranging from 6 to 30 months. The daily dose of verapamil was 240–720 mg. In contrast, experience in long-term treatment with nifedipine is still limited (review by Lorell 1985); studies relate to an average daily dose of 40 mg. In a direct comparison (Rosing et al. 1982) exercise tolerance under verapamil treatment was greater than with nifedipine. Verapamil is therefore the calcium antagonist of first choice in hypertrophic cardiomyopathies; the more extensive and longer experience with this substance over a period of 10 years also favours its use. Only in patients with high peripheral vascular resistance and/or little outflow obstruction is nifedipine considered to have advantages (Baughman 1986). Diltiazem has so far not been tested in hypertrophic cardiomyopathy (Maron et al. 1987).

Both verapamil and nifedipine can trigger an acute drop in blood pressure (above all in combination with quinidine) or pulmonary oedema. Hypotheses as to the cause of such effects include an excessive reduction in afterload, with a rise in the left ventricular pressure gradient, decreased coronary blood flow when aortic pressure falls, and a direct negative inotropic effect (Epstein and Rosing 1981; Lorell 1985). Patients whose medical history includes pulmonary oedema or an elevated pulmonary capillary wedge pressure should therefore be treated with calcium antagonists only in exceptional circumstances, and then only on an inpatient basis (at least in the initial phase) to facilitate close surveillance. As in other indications, sinus bradycardia or higher degree AV block can occur during treatment of hypertrophic cardiomyopathy with verapamil; attention must also of course be paid to contraindications (sick sinus syndrome, II or III degree AV block; see also Sect. 5.6).

In Summary. The symptomatic action of calcium antagonists in hypertrophic cardiomyopathy appears to be sufficiently well documented, and these substances are generally well tolerated in spite of the high doses sometimes needed in individual cases (Rosing et al. 1981; Kaltenbach and Hopf 1984). To what extent calcium antagonists positively influence long-term prognosis for patients with hypertrophic cardiomyopathy remains an open question. Kaltenbach and Hopf (1984) cite an annual mortality of 1.3% for their patients, which is distinctly below the rates cited for other patient groups (average mortality 3.3%; β-receptor blocker therapy, surgical treatment or untreated). Randomized prospective studies in uniform patient groups are so far unavailable for any of the therapeutic alternatives.

Patients with hypertrophic cardiomyopathy are at special risk of sudden cardiac death (Goodwin and Krikler 1976; Koway et al. 1984; Maron et al. 1987). Neither calcium antagonists nor β-receptor blockers show a definite action in these patients against malignant cardiac arrhythmias or sudden cardiac death (Lorell 1985; Maron et al. 1987); conversely, the absence of hard, statistically well-founded data does not exclude a positive effect (difficulty of observing sufficiently large groups for sufficiently long periods, an evident necessity, given an annual mortality of 1%–3%). Possibly, the class III antiarrhythmic agent, amiodarone, which in one study prevented sudden cardiac death over a period of 3 years (review by Maron et al. 1987) will in the future provide an effective alternative; there is at present no optimal treatment for arrhythmias in hypertrophic cardiomyopathy.

Since both calcium antagonists and β-receptor blockers are applied in the symptomatic treatment of hypertrophic cardiomyopathy, their combination in this indication has also been studied. Landmark et al. (1982), using 20 mg nifedipine (orally) and 5 mg propranolol (intravenously), observed a more marked reduction in left ventricular pressure gradient than with monotherapy; left ventricular end-diastolic pressure tends (not significantly) to decrease, a sign of improved diastolic filling. Long-term trials with the combination nifedipine–propranolol have not yet been carried out; there are no data on verapamil–propranolol (Lorell 1985).

We recommend the following procedure in patients with hypertrophic cardiomyopathy. Symptomless patients (diagnosed by chance during echocardiography) should be checked at 6-monthly intervals and should avoid more strenuous forms of physical activity (sports, etc.) because of the risk of sudden cardiac death.

We assess risk by exercise tests and Holter ECG. A course of treatment with β-receptor blockers or calcium antagonists is indicated only in individual cases (positive family history of arrhythmias, especially of sudden cardiac death, massive left ventricular hypertrophy or obstruction; Maron et al. 1987). With symptomatic patients either a β-receptor blocker or a calcium antagonist is prescribed. In the case of β-receptor blockers the aim is to attain a heart rate at rest of 60 per minute; propranolol in a daily dose of 160–480 mg (in individual cases up to 640 mg) should first be tried. Digitalis glycosides, β-sympathomimetics and nitrates are to be avoided. If β-receptor blockers have no effect, or more pronounced side effects occur, or if there are contraindications to their use, therapy with verapamil is initiated, in most cases in a daily dose of 240–480 mg (in individual cases up to 720 mg daily).

Rosing et al. (1985) observed a clear improvement with verapamil in 60% of 227 patients who had failed to respond adequately to β-receptor blockers (average daily dose 360 mg verapamil; assessment made 3–56 months following initiation of therapy). In switching a patient from a β-receptor blocker to verapamil, the dose of β-receptor blocker should be slowly reduced to avoid the risk of precipitating tachycardic (including ventricular) arrhythmias (β-receptor blocker withdrawal, raised sympathetic tonus; Wessel et al. 1987).

Supplementary antiarrhythmic treatment must be selected on an individual basis; in spite of a high side effect potential, amiodarone appears to be the drug of choice. It may one day be possible to define and treat specifically groups at high risk of sudden cardiac death (e. g. patients with symptomless episodes of ventricular tachycardia in Holter ECG; Maron et al. 1987). Since atrial arrhythmias (atrial fibrillation, supraventricular tachycardias) have an additional negative effect on diastolic filling, treatment is urgent; verapamil is here the drug of choice, with rapid conversion to sinus rhythm being the therapeutic goal.

In patients with extensive ventricular septal hypertrophy and a large pressure gradient (greater than 50 mmHg) in the left ventricular outflow tract which does not improve with drug therapy, surgical measures (septal myotomy–myectomy) must be considered.

4.4.4 Haemodynamics and Diagnosis of Dilated Cardiomyopathy

Dilated cardiomyopathy is characterized by dilation of the heart with reduced contractility. The compliance of the left ventricle is usually elevated, ventricular filling is unimpaired, and the essential feature is failure of systolic cardiac function. The X-ray shows moderate to extensive enlargement of the heart, and the echocardiogram, which in most cases permits unequivocal diagnosis, exhibits dilated heart cavities without hypertrophy, and wall motion is restricted in all sections. Heart catheterization reveals diminished stroke volume; cardiac output and the ejection fraction are also reduced; the high residual filling of the left ventricle is reflected by increased left ventricular end-diastolic pressure and volume.

Patients mostly seek medical help because of dyspnoea at rest or under stress, diminished physical performance, fatigue, or lower limb oedema, and typical symptoms of congestive heart failure are evident. Further symptoms are cardiac arrhythmias (reportedly, more than a third of patients with dilated cardiomyopathy succumb to sudden cardiac death) and cerebral or peripheral embolism.

Long-term prognosis for dilated cardiomyopathy is poor: the survival rate after 5 years is only 20%–50%. Further details concerning diagnostic methods and prognosis may be found in recent review articles (Johnson and Palacios 1982; Wynne and Braunwald 1984; Hess 1986a, b).

4.4.5 Treatment of Dilated Cardiomyopathy with Calcium Antagonists

Calcium antagonists have been used in dilated cardiomyopathy with congestive heart failure where the latter was not improved by restriction of physical activity, fluid and salt intake, by digitalis and diuretics. As stated in the section on the treatment of heart failure (Sect. 4.5.1), vasodilators are frequently administered in the latter situation in order to achieve recompensation via further reduction of afterload. The effect of calcium antagonists in dilated cardiomyopathy is predominantly dilation of the peripheral resistance vessels with a corresponding reduction in afterload; a direct action on the myocardium and on the coronary arteries plays a comparatively subordinate role.

Further principles of treatment in dilated cardiomyopathy are discussed in recent reviews (other vasodilators such as ACE inhibitors or hydralazine, positive inotropic substances, except digitalis, antiarrhythmic medication, and anticoagulants; Wynne and Braunwald 1984; Hess 1986b). Current opinion in terms of long-term therapy strongly favours ACE inhibitors, the use of which spares diuretics and reduces mortality. Often, particularly in younger patients, heart transplantation remains the last resort. There are published reports on the principles and the outcomes of treatment with β-receptor blockers by, among others, Waagstein et al. (1975), Lydtin and Lohmöller (1980) and Engelmeier et al. (1985).

The acute haemodynamic effects of calcium antagonists in dilated cardiomyopathy have been thoroughly investigated; most studies have employed nifedipine, because, in comparison with verapamil and diltiazem, it has a stronger peripheral vasodilating and a weaker negative inotropic effect (Sect. 3.1.6). Following a dose of 10–20 mg nifedipine, blood pressure, peripheral resistance, left ventricular filling pressure and left ventricular end-systolic volume usually decrease. Parallel with this, ejection fraction, stroke volume and cardiac output are raised. It appears that the greater the initial peripheral resistance, the larger is its reduction (Klugmann et al. 1980; Bellocci et al. 1982; Magorien et al. 1984; Miller et al. 1985; Figulla et al. 1986). A positive influence on left ventricular function is demonstrable both at rest and under stress (Magorien et al. 1984). Hall et al. (1986; 20 mg nifedipine) also observed a decrease in blood pressure and in peripheral resistance, with only a slight increase in stroke volume, effects explained by the negative inotropic effect (of nifedipine) on the heart. In contrast to the ACE inhibitor captopril, nifedipine is therefore thought to be only of limited value in dilated cardiomyopathy without ischaemia.

In contrast to the predominantly favourable effect of nifedipine following a single dose, there is still insufficient experience in the long-term treatment of dilated cardiomyopathy, and the available results are contradictory. Bellocci et al. (1982) observed, over a 2-month treatment period with 80 mg nifedipine daily, the same effect as with a single dose (increase in cardiac output and stroke volume, decrease in blood pressure, peripheral resistance and left ventricular filling pressure); Klugmann et al. (1983), however, reported that, after 3 weeks' therapy, also with 80 mg nifedipine daily, patients showed no improvement in exercise tolerance or in subjective complaints; in comparison with placebo, objective parameters remained unchanged (blood pressure, body weight, echocardiographically determined ventricular diameters). Side effects under nifedipine appear to be rare, although pulmonary oedema or an "unpredictable response" have been described in heart failure (review by Baughman 1986). Dose-dependence of side effects is under discussion.

There are as yet only a few studies concerning the acute effect of verapamil and diltiazem (review by Baughman 1986). Figulla et al. (1986) observed in an acute test with diltiazem an increase in stroke volume and in the ejection fraction, with a parallel decrease in heart rate. Initial positive results of long-term therapy with diltiazem are available (90 mg t.i.d. in 21 patients, 30 "control" patients: reduced mortality and almost abolished haemodynamic deterioration; Figulla et al. 1987).

In Summary. The use of calcium antagonists as vasodilators in dilated cardiomyopathy cannot at present be recommended. Although nifedipine positively influences left ventricular function in the acute test, there is as yet insufficient clinical experience, unambiguous results in long-term treatment are wanting. Neither do the initial positive results with diltiazem permit any general recommendation with regard to treatment. In general, verapamil is cautiously used in heart failure because of its more marked negative inotropic effect (Sect. 4.5.1); in dilated cardiomyopathy the substance has yet to be adequately investigated.

Calcium antagonists in doxorubicin-induced cardiomyopathy are discussed in Sect. 4.11.

References

Bajusz E , Lossnitzer K (1968) Ein neues Krankheitsmodell: Erbliche nicht vaskuläre Myokarddegeneration mit Herzinsuffizienz Münch Med Wschr 31: 1756–1768

Baughman KL (1986) Calcium channel blocking agents in congestive heart failure Am J Med 80 (Suppl 2B) 46–50

Bellocci F, Ansalone G, Santarelli P (1982) Oral nifedipine in the long term management of severe chronic heart failure J Cardiovasc Pharmacol 4: 847–855

Betocchi S, Cannon RO III, Watson RM, Bonow RO, Ostrow HG, Epstein SE, Rosing DR (1985) Effects of sublingual nifedipine on haemodynamics and systolic and diastolic function in patients with hypertrophic cardiomyopathy Circulation 72: 1001–1007

Bonow RO, Rosing DR, Bacharach SL, Green MV, Kent KM, Lipson LC, Maron BJ, Leon MB, Epstein SE (1981) Effects of verapamil on left ventricular systolic function and diastolic filling in patients with hypertrophic cardiomyopathy Circulation 64: 787–796

Bonow RO, Ostrow HG, Rosing DR, Cannon RO III, Lipson LC, Maron BJ, Kent KM, Bacharach SL, Green MV (1983) Effects of verapamil on left ventricular systolic and diastolic function in patients with hypertrophic cardiomyopathy: pressure-volume analysis with a nonimaging scintillation probe Circulation 68: 1062–1073

Bonow RO, Dilsizian V, Rosing DR, Maron BJ, Bacharach SL, Green MV (1985) Verapamil-induced improvement in left ventricular diastolic filling and increased exercise tolerance in patients with hypertrophic cardiomyopathy: short- and long-term effects Circulation 72: 853–864

Braunwald E (1982) Mechanism of action of calcium-channel-blocking agents N Engl J Med 307:1618–1627

Chatterjee K, Raff G, Anderson D, Parmley WW (1982) Hyertrophic cardiomyopathy – therapy with slow channel inhibiting agents Prog Cardiovasc Dis 25: 193–210

Engelmeier RS, O'Connel JB, Walsh R, Rad N, Scanlon PJ, Gunnar RM (1985) Improvement in symptons and exercise tolerance by metoprolol in patients with dilated cardiomyopathy: a double-blind, randomized, placebo-controlled trial Circulation 72: 536–546

Epstein SE, Rosing DR (1981) Verapamil: its potential for causing serious complications in patients with hypertrophic cardiomyopathy Circulation 64: 437–441

Figulla HR, Kreuzer H, Luig H (1986) Verapamil, Diltiazem oder Nifedipin bei schwerer linksventrikulärer Funktionsstörung? Dtsch Med Wschr 111: 11–14

Figulla HR, v. Rechenberg J, Wiegand V, Kreuzer H, Luig H (1987) Beneficial effects of long-term diltiazem treatment in dilated cardiomyopathy Circulation 76 (Suppl 4): 69 (abstract)

Fleckenstein A (1971) Specific inhibitors and promotors of calcium action in the excitation-contraction coupling of heart muscle and their role in the prevention or production of myocardial lesions In: Harris P, Opie L (eds) Calcium and the heart Academic Press London pp 135–188

Fleckenstein A, Janke J, Döring HJ, Leder O (1971) Die intrazelluläre Überladung mit Kalzium als entscheidender Kausalfaktor bei der Entstehung nicht-koronarogener Myokard-Nekrosen Verh Dtsch Ges Kreislaufforsch 37: 345–353

Goodwin JF, Oakley CM (1972) The cardiomyopathies Br Heart J 34: 545–552

Goodwin JF (1974) Prospects and predictions for the cardiomyopathies Circulation 50: 210–219

Goodwin JF, Krikler DM (1976) Arrhythmia as a cause of sudden death in hypertrophic cardiomyopathy Lancet II: 937–940

Hall D, Froer KL, Kraus F, Stautner C, Rudolph W (1986) Effects of nifedipine in congestive heart failure due to dilated cardiomyopathy: A randomized, double-blind, cross-over study In: Bussmann WD, Just H (eds) Vasoaktive Substanzen bei Herzinsuffizienz Springer Berlin pp 160–163

Hanrath P, Mathey DG, Kremer P, Sonntag F, Bleifeld W (1980) Effect of verapamil on left ventricular relaxation time and regional left ventricular filling in hypertrophic cardiomyopathy Am J Cardiol 45: 1258–1264

Hess OM (1986a) Diagnostik der dilatativen Kardiomyopathie Dtsch Med Wschr 111: 219–221

Hess OM (1986b) Therapie der dilatativen Kardiomyopathie Dtsch Med Wschr 111: 222–225

Johnson RA, Palacios I (1982) Dilated cardiomyopathies of the adult N Engl J Med 307: 1051–1058/1119–1126

Kaltenbach M, Hopf R, Keller M (1976) Calciumantagonistische Therapie bei hypertroph-obstruktiver Kardiomyopathie Dtsch Med Wschr 101: 1284–1287

Kaltenbach M, Hopf R (1984) Langzeitbehandlung der hypertrophischen Myokardiopathie mit Calcium-Antagonisten In: Althaus U, Burckhard D, Vogt E (eds) Calcium-Antagonismus – Internat. Symposium on Calcium-Antagonism Universimed Frankfurt/Main pp 276–288

Klugmann S, Fioretti P, Salvi A, Camerini F (1980) Afterload reducing agents in congestive cardiomyopathy; a study with a calcium antagonist drug: nifedipine Eur Heart J 1 (Suppl B): 49–52

Klugmann S, Salvi A, Camerini F (1983) Calcium antagonists in heart failure Herz 8: 88–92

Kowey PR, Eisenberg R, Engel TR (1984) Sustained arrhythmias in hypertrophic obstructive cardiomyopathy N Engl J Med 310: 1566–1569

Krayenbühl HP, Hess OM, Senn M, Ritter M (1984) Wirkungen von Calcium-Antagonisten auf die linksventrikuläre diastolische Funktion bei hypertrophischer Kardiomyopathie In: Althaus U, Burckhardt D, Vogt E (eds) Calcium-Antagonismus – Internat. Symposium on Calcium-Antagonism Universimed Frankfurt/Main pp 266–275

Landmark K, Siré S, Thaulow E, Amlie JP, Nitter-Hauge S (1982) Haemodynamic effects of nifedipine and propranolol in patients with hypertrophic obstructive cardiomyopathy Br Heart J 48: 19–26

Landmark K, Sire S, Thaulow E, Amlie JP, Nitter-Hauge S (1983) Acute influences of nifedipine, propranolol and the combined administration of the two drugs on haemodynamics in patients with hypertrophic obstructive cardiomyopathy In: Kaltenbach M, Neufeld HN (eds) 5th International Adalat Symposium Excerpta Medica Amsterdam pp 81–90

Loogen F, Kuhn H (1977) Classification and natural history of primary cardiomyopathies In: Riecker G, Weber, Goodwin JF (eds) Myocardial failure Springer Berlin pp 232
Lorell BH (1985) Use of calcium channel blockers in hypertrophic cardiomyopathy Am J Med 78 (Suppl 2B): 43–54
Lossnitzer K, Konrad A, Jakob M (1980) Kardioprotektion durch Kalziumantagonisten bei erblich kardiomyopathischen Hamstern In: Fleckenstein A, Roskamm H (eds) Calcium-Antagonismus Springer Berlin pp 151–171
Lydtin H, Lohmöller G (1980) Beta blockers Huber Bern
Magorien RD, Leier CV, Kolibash AJ, Barbush TJ, Unverferth DV (1984) Beneficial effects of nifedipine on rest and exercise myocardial energetics in patients with congestive heart failure Circulation 70: 884–890
Maron BJ, Bonow RO, Cannon RO, Leon MB, Epstein SE (1987) Hypertrophic cardiomyopathy – interrelations of clinical manifestations, pathophysiology, and therapy N Engl J Med 316: 780–789 und 844–852
Miller AB, Conetta DA, Bass TA (1985) Sublingual nifedipine: acute effects in severe chronic congestive heart failure secondary to idiopathic dilated cardiomyopathy Am J Cardiol 55: 1359
Rosing DR, Kent KM, Borer JS, Seides SF, Maron BJ, Epstein SE (1979 a) Verapamil therapy: a new approach to the pharmacologic treatment of hypertrophic cardiomyopathy I. Hemodynamic effects Circulation 60: 1201–1207
Rosing DR, Kent KM, Maron BJ, Epstein SE (1979 b) Verapamil therapy: a new approach to the pharmacologic treatment of hypertrophic cardiomyopathy II. Effects on exercise capacity and symptomatic status Circulation 60: 1208–1213
Rosing DR, Condit JR, Maron BJ, Kent KM, Leon MB, Bonow RO, Lipson LC, Epstein SE (1981) Verapamil therapy: a new approach to the pharmologic treatment of hypertrophic cardiomyopathy. Effects of long-term administration Am J Cardiol 43: 545–553
Rosing DR, Epstein SE (1982) Verapamil in the treatment of hypertrophic cardiomyopathy Ann Intern Med 96: 670–672
Rosing DR, Cannon RO, Watson RM (1982) Comparison of verapamil and nifedipine – effects on symptoms and exercise capacity in patients with hypertrophic cardiomyopathy Circulation 66 (Suppl II): II-24
Rosing DR, Idänpään-Heikkilä U, Maron BJ, Bonow RO Epstein SE (1985) Use of calcium-channel blocking drugs in hypertrophic cardiomyopathy Am J Cardiol 55 (Suppl): 185B-195B
Rouleau JL, Chuck LHS, Hollosi G, Kidd P, Sievers RE, Wikman-Coffelt J, Parmley WW (1982) Verapamil preserves myocardial contractility in the hereditary cardiomyopathy of the syrian hamster Circ Res 50: 405–412
Spicer RL, Rocchini AP, Crowley DC, Vasiliades J, Rosenthal A (1983) Hemodynamic effects of verapamil in children and adolescents with hypertrophic cardiomyopathy Circulation 67: 413–420
Suwa M, Hirota Y, Kawamura K (1984) Improvement in LV diastolic function during intravenous and oral diltiazem therapy in patients with hypertrophic cardiomyopathy: an echocardiographic study Am J Cardiol 54: 1047–1053
Ten Cate FJ, Serruys PW, Mey S, Roelandt J (1983) Effects of short-term administration of verapamil on left ventricular relaxation and filling dynamics measured by a combined hemodynamic-ultrasonic technique in patients with hypertrophic cardiomyopathy Circulation 68: 1279–1283
Waagstein F, Hjalmarson A, Varnauskas E, Wallentin I (1975) Effect of chronic beta adrenergic blockade in congestive cardiomyopathy Br Heart J 37: 1022
Wessel A, Seiffert P, Dettmann R, Heintzen PH (1987) Verapamil-induced ventricular tachycardia in hypertrophic cardiomyopathy Eur Heart J 8: 313–315
Wigle ED (1987) Hypertrophic cardiomyopathy: a 1987 viewpoint Circulation 75: 311–322
Wynne J, Braunwald E (1984) The cardiomyopathies and myocarditides In: Braunwald E (ed) The Heart – a textbook of cardiovascular disease Saunders Philadelphia pp 1399–1446

4.5 Other Cardiac Indications

4.5.1 Congestive Heart Failure

4.5.1.1 Fundamentals of Therapy with Vasodilators

Chronic heart failure is traditionally treated by avoidance of undue physical exertion, a low-salt diet, and digitalis glycosides. Besides glycosides (positive inotropic effect), substances are increasingly used which unload the heart. Diuretic therapy is also generally accepted; indeed in some parts of the English-speaking world such treatment is more common than glycosides. Diuretics unload the heart by reducing blood volume, and thereby also left ventricular filling pressure, without increasing cardiac output. Unwanted effects of diuretic therapy include electrolyte disturbances (e.g. hypokalaemia) and increased blood viscosity (rise in haematocrit). An alternative approach to unloading the failing heart is offered by vasodilators. These substances can either increase the capacity of the low-pressure system (so-called preload reduction, "preload" being interpreted as the left ventricular filling pressure = left ventricular end-diastolic pressure) and/or lower peripheral vascular resistance (so-called afterload reduction, left ventricular "afterload" being interpreted as the wall tension, which depends on heart size, left ventricular muscle mass, and peripheral vascular resistance). As a consequence of preload and/or afterload reduction cardiac output increases, and cardiac oxygen demand can decrease.

In every form of congestive heart failure, there is compensation for the diminution of left ventricular (pumping) function via activation of the sympathetic nervous system or of the renin–angiotensin aldosterone system. This results in an elevation of afterload; since the peripheral resistance vessels (arterioles) constrict, total peripheral resistance increases and preload simultaneously also increases. Increased afterload raises myocardial oxygen demand, and the disproportion between limited oxygen availability and increased oxygen demand further impairs left ventricular (pumping) function. In severe chronic heart failure this vicious circle can be broken by reducing afterload with vasodilators, and to a smaller extent also by reducing preload.

The effects of diuretics, positive inotropic drugs (e.g. digitalis) and also of vasodilators which can reduce preload and/or afterload are summarized in Fig. 36, which shows the interrelationship between left ventricular end-diastolic pressure and ventricular performance (cardiac output). It is of course possible to combine the various therapeutic principles and thereby produce an augmented integrated effect.

In discussing vasodilators, one can distinguish between substances which act directly on the arterial or venous vascular cells (so-called vasodilators in the narrower sense) and those which act by inhibiting the angiotensin-converting enzyme and decreasing formation of angiotensin II and aldosterone (so-called ACE inhibitors). The first group includes, among others, nitrates (which predominantly affect the venous system), hydralazine and the calcium antagonists (which predominantly affect the arterial system) as well as prazosin (a postsynaptic α_1-receptor blocker which affects both the venous and arterial vascular compartments). The second group includes the ACE inhibitors (e.g. captopril and

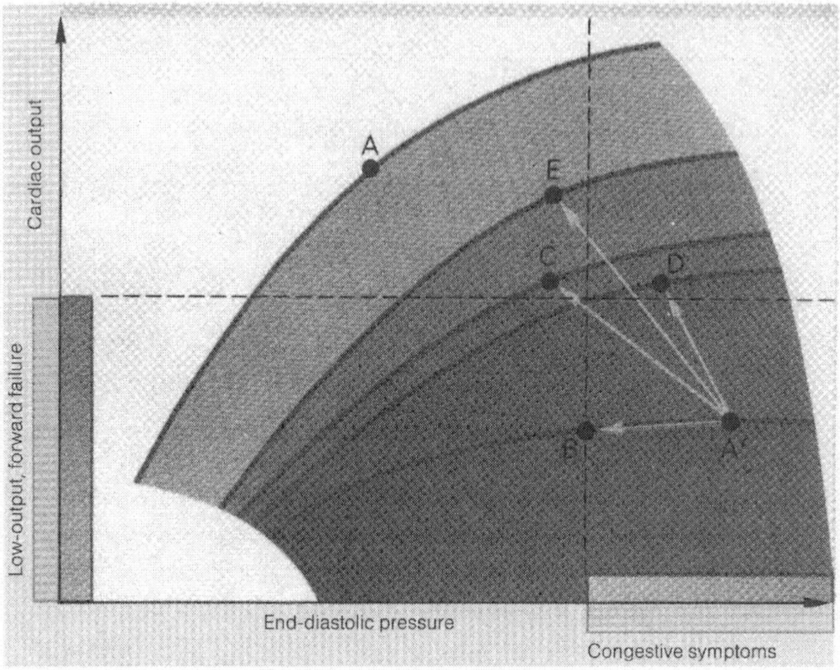

Fig. 36. Interrelationship between left ventricular end-diastolic pressure (preload) and cardiac output, depending on contractility- and afterload-modifying measures. Heart failure is expressed as a rightward bend in the left ventricular performance curve (*A–A'*). Diuretics or nitrates diminish preload and congestive symptoms (backward failure). Cardiac output declines (*A'–B*). Peripheral vasodilators bring about an increase in cardiac output and a reduction in preload (*A'–C*). Digitalis acts in a similar manner, but more weakly (*A'–D*). The combination of vasodilators and a positive inotropic drug acts additively on elevation of cardiac output and diminution of congestive symptoms (*A'–E*). (After Feder and Lydtin 1983)

enalapril), which also influence both the venous and arterial vascular compartments.

Calcium antagonists act in heart failure only as arterial vasodilators; they lower peripheral vascular resistance and thereby afterload and are thus able to raise cardiac output; cardiac preload, or left ventricular filling pressure, is only slightly affected. In long-term treatment most vasodilators show a gradual reduction in efficacy, i.e. development of tolerance (especially with nitrates, prazosin, hydralazine); this phenomenon is thought not to occur with ACE inhibitors and calcium antagonists, or only to a very small degree.

As in every other symptomatic treatment, the clinician must consider the effects not only on the patient's symptoms and on the quality of life, but also on mortality. To date there have been only a few studies concerning treatment of heart failure with vasodilators which also make statements as to prognosis. While treating severe heart failure with hydralazine and isosorbide dinitrate, Cohn et al. (1986) observed an improvement in left ventricular function and a decline in mortality; all patients were concomitantly receiving digitalis and diuretics. The

CONSENSUS study[18] (1987) with the ACE inhibitor enalapril showed similar results; administration of enalapril in addition to standard treatment reduced mortality after 6 or 12 months significantly. Early results (Hess 1986) in dilated cardiomyopathy also show that an afterload-reducing therapy (ACE inhibitors or true vasodilators) favourably influences prognosis. Open questions remain, however: whether milder forms of heart failure profit from such treatment; and whether vasodilators provide not only a complementary therapy to digitalis and diuretics, but possibly also an alternative to these forms of treatment (Opie 1984a; Cowley et al. 1986). More detailed literature on vasodilator therapy is provided by Cohn and Franciosa (1988), Mason (1978), Feder and Lydtin (1983), Bussmann (1984), Braunwald and Colucci (1984), Zähringer (1986).

4.5.1.2 Therapy of Congestive Heart Failure with Calcium Antagonists

Calcium antagonists have been used both in acute heart failure (mostly left-sided heart failure with pulmonary oedema) and in chronic heart failure (due to arterial hypertension, coronary heart disease, cardiomyopathy or valvular heart disease). In addition to their action as arterial vasodilators, calcium antagonists also improve coronary blood flow and diastolic relaxation of the left ventricle. These properties can benefit patients with ischaemic heart disease or cardiomyopathy (Braunwald 1982).

In *acute pulmonary oedema* Polese and co-workers (1979) observed that there was, following administration of 10 mg nifedipine sublingually, a drop in peripheral vascular resistance and arterial blood pressure, while contractility of the left ventricle and cardiac output increased; as a consequence of improved left ventricular function, left ventricular filling pressure also decreased and pulmonary congestion improved. A decrease in left ventricular filling pressure could, however, not be demonstrated by Bussmann in 1984, although he also observed an increase in cardiac output.

From a pathophysiological standpoint, reduction of preload with nitrates and diuretics should be the first priority in acute pulmonary oedema: low cardiac output, raised peripheral vacular resistance, and especially high-normal or elevated blood pressure levels favour the use of calcium antagonists of the dihydropyridine type (e.g. felodipine, nifedipine, nisoldipine). This applies in particular to acute left-sided heart failure in hypertensive crisis (see also Sect. 4.2.5). Here nifidepine, like nitrates, can be administered parenterally, which makes treatment readily controllable (review by Bussmann 1984). In cases of pulmonary oedema under treatment with nitrates and calcium antagonists, arterial blood pressure must be closely controlled; a systolic reduction below 100 mmHg, which would decrease coronary and renal blood flow, is to be avoided. Verapamil should not be used in acute pulmonary oedema. No results for diltiazem are available.

Experience with calcium antagonists in the treatment of *chronic heart failure* is as yet very limited. Although many studies have been published showing the positive action of calcium antagonists in the acute situation, there are still no systematic studies of long-term treatment available.

---

[18] CONSENSUS: Cooperative North Scandinavian Enalapril Survival Study

Nifedipine and the newer dihydropyridine derivatives: nitrendipine, felodipine and nisoldipine, and also diltiazem have been investigated in the acute situation with the following doses (dihydropyridine group): 10–30 mg nifedipine, 5–15 mg felodipine, 20 mg nitrendipine, 6 mg/kg nisoldipine i.v. (Hanrath and Kremer 1983; Leier et al. 1984; Olivari et al. 1984; Klepzig and Strauer 1985; Timmis and Jewitt 1985; for results in dilated cardiomyopathy see Sect. 4.4.5). Workers are not unanimous in their assessment of the positive acute haemodynamic effects of the dihydropyridine derivatives in chronic heart failure. Elkayam et al. (1983), Fifer et al. (1986) and Hall et al. (1986) point out that afterload reduction under nifedipine is frequently accompanied by a negative inotropic effect and that total cardiac output (or stroke volume) changes only slightly. In direct comparison nifedipine is inferior to treatment with captopril, hydralazine, sodium nitroprusside or a combination of sodium nitroprusside and milrinone (a phosphodiesterase inhibitor – positive inotropic effect) (Elkayam et al. 1984a, b; Fifer et al. 1986; Hall et al. 1986).

Diltiazem is said (like nifedipine) to lower afterload, raise cardiac output and to leave left ventricular filling pressure unchanged; but with this substance heart rate is decreased. Dose: 0.2–0.5 mg/kg intravenously, or 270–360 mg daily (Walsh et al. 1984; Materne et al. 1984).

Long-term treatment of heart failure with calcium antagonists cannot yet be conclusively assessed, for in most cases duration of treatment has been too short (maximum: 2 months), the studies were carried out without control groups and in part they present contradictory results (Colucci et al. 1985; Timmis and Jewitt 1985; see also Sect. 4.4.5). The following daily doses were given: 60–80 mg nifedipine (Colucci et al. 1985), 20–30 mg felodipine (Timmis and Jewitt 1985; Dunselman et al. 1987; Kassis et al. 1987), 20–40 mg nisoldipine (Reicher-Reiss et al. 1987; Kowski et al. 1987), 270–360 mg diltiazem.

Since calcium antagonists differ from other vasodilators in their effectiveness and in their main action, the positive results for treatment of heart failure with vasodilators in general (Sect. 4.5.1.1) cannot be directly extended to calcium antagonists. At the present time, therefore, treatment of heart failure with calcium antagonists cannot be generally recommended. In individual cases with favourable haemodynamic parameters (high peripheral vascular resistance, adequate blood pressure, low cardiac output, normal left ventricular filling pressure) and with supplementary indications (arterial hypertension, angina pectoris, hypertrophic cardiomyopathy) calcium antagonists can be a meaningful complement in the therapy of heart failure. Theoretical considerations favour dihydropyridine derivatives with high vascular selectivity in this context. Treatment with calcium antagonists (and other vasodilators) unloads the heart, but also suppresses (in part positively) somatic counter-reactions. It must always be borne in mind that therapy remains purely symptomatic, and that "the patient dies from the heart, not from the periphery" (Sonnenblick).

4.5.2 Aortic Regurgitation

Aortic regurgitation produces a volume load on the left ventricle, leading to left ventricular dilation with hypertrophy. Only in advanced cases do symptoms of coronary insufficiency (diminished coronary perfusion through low diastolic aortic pressure) or heart failure develop. The occurrence of such symptoms (dyspnoea, angina pectoris, etc.) is almost invariably an indication for surgery, i.e. aortic valve replacement. Echocardiography facilitates noninvasive observation of the course of the aortic regurgitation: if the left ventricular end-systolic diameter exceeds 55 mm and if left ventricular function deteriorates (fractional shortening (FS) less than 25%), these measured data alone should provide the impetus for a discussion as to the need for valve replacement (Opie 1984).

Conservative treatment of aortic regurgitation consists of digitalis glycosides, diuretics and vasodilators. In this situation the rationale for the use of vasodilators centres on diminishing the regurgitant fraction by lowering peripheral vascular resistance and elevating the effective stroke volume (forward stroke volume). Left ventricular filling pressure declines, and long-term there is thought to be regression of left ventricular dilation and hypertrophy (review by Greenberg and Rahimtoola 1980). Vasodilators used to date in the treatment of aortic regurgitation include prazosin, hydralazine, sodium nitroprusside and nifedipine.

The acute effect of 20 mg nifedipine sublingually has been investigated by Fioretti et al. (1982) and by Shen et al. (1984). Both authors observed decreases in peripheral vascular resistance, arterial blood pressure and left ventricular filling pressure, as well as increases in heart rate, cardiac output/stroke volume and the ejection fraction during exercise. The change in peripheral vascular resistance was found to be proportional to the change in cardiac output and regurgitant fraction. The overall short-term result was a reduction in afterload which thus brought about an improvement in left ventricular function.

In long-term treatment with 30–60 mg nifedipine daily, however, this favourable haemodynamic effect was only partially demonstrable. Wilson et al. (1984) studied patients after 1 month of treatment; Shen et al. (1986) investigated a group receiving a 3-month course of therapy. Both authors reported that although systolic blood pressure was lower than with placebo, heart rate, stroke volume and the ejection fraction at rest showed no difference in comparison with the placebo group. Only in stress conditions did nifedipine improve left ventricular function; stroke volume and the ejection fraction increased, but exercise tolerance during exercise testing showed no change. In contrast to nifedipine, digitalis glycosides increased the left ventricular ejection fraction, even at rest (Wilson et al. 1984). The results so far available do not permit any definite conclusion concerning the effect of nifedipine in aortic regurgitation, since the duration of the trials was rather short (1 and 3 months, respectively) and mainly patients with no or only slight symptoms were treated.

In Summary. At the present time the treatment of aortic regurgitation with nifedipine as a vasodilator cannot be generally recommended; although the drug acutely improves left ventricular function by afterload reduction, confirmation of this positive haemodynamic effect in the long-term situation is not yet available. Other calcium antagonists have not yet been investigated in aortic regurgitation. In this condition drug therapy should never be allowed to mask the optimal moment for surgical intervention when there is progressive deterioration in left ventricular function.

4.5.3 Application of Calcium Antagonists in Heart Surgery

Most cardiac operative procedures such as correction of valvular defects or aortocoronary bypass grafting can be carried out only during heart arrest. This state is induced by the use of cardioplegic solutions, most commonly a cold potassium chloride solution (St. Thomas' Hospital method) or a procaine–aspartate–magnesium solution (Elert 1983). Systemic circulation of blood and oxygen is taken over by a "heart–lung machine" (so-called cardiopulmonary bypass). During this phase oxygen supply to the heart is severely restricted, and measures are therefore adopted to reduce resting metabolism and thereby oxygen demand, for example, by hypothermia.

Experimental data provided by Fleckenstein (Sect. 3.1.4, Cardioprotection) and animal experiments by Henry et al. (1980) which showed that calcium antagonists (verapamil, nifedipine) can prevent a cardiac contracture ("stone heart") triggered by ischaemia and subsequent reperfusion, have recently led to the use of calcium antagonists as cardioplegic agents. It was postulated that calcium antagonists by virtue of their direct negative inotropic effect can further lower cardiac energy demand, and in addition, can reduce or even prevent deleterious calcium overload of the heart muscle cells (especially of mitochondria). By this means consumption of ATP is said to be reduced and resynthesis following cessation of the ischaemia improved.

No definite conclusions can yet be drawn concerning the application of calcium antagonists in heart surgery in humans, mainly in aortocoronary bypass procedures or valve replacement. Calcium antagonists have been given intravenously both before heart surgery and transfer to the heart–lung machine, and added to the intracoronarily injected cardioplegic solution. Randomized studies using placebo or established cardioplegic solutions yielded, on one hand, positive results with a diminished release of cardiac enzymes, a smaller reduction in ATP in the myocardial cell and reductions in arrhythmias, myocardial ischaemia and postoperative disturbances of contractility (Clark et al. 1981, 1983, 1986 intracoronary nifedipine; Elert 1983 intracoronary nifedipine; Hicks et al. 1984 intracoronary verapamil; Flameng et al. 1983 intravenous lidoflazine; Donegani et al. 1986 intravenous diltiazem). On the other hand, several studies showed no significant difference from the control group (Vouhe 1984 intracoronary diltiazem; Flameng and Stalpaert 1986 intracoronary nifedipine) or even more frequent complications (postoperative bradycardias and reduced contractility with intracoronary verapamil; Kaplan et al. 1984).

In Summary. The use of calcium antagonists in heart surgery, either intravenously before the procedure (perioperatively) or intracoronarily during the procedure as an additive to the cardioplegic solution, while based on convincing pathophysiological arguments, has yet to be supported by conclusive evidence of improvement in cardiac tolerance of ischaemia and in postoperative myocardial function.

Calcium antagonists are also used in cardiac surgery to treat hypertensive episodes (Sect. 4.2.4); most investigations carried out so far have employed nifedipine intravenously. In these situations antagonists appear to provide a true alternative to nitroglycerine, sodium nitroprusside, β-receptor blockers or urapidil (Hempelmann 1982; Boldt et al. 1986; Tarnow 1986).

4.5.4 Calcium Antagonists in Coronary Arteriography and Percutaneous Transluminal Coronary Angioplasty

Coronary arteriography is today regarded as the "gold standard" in the diagnosis or exclusion of coronary heart disease. Furthermore, in known coronary heart disease it enables precise assessment of the extent and severity of vascular lesions and facilitates selection of therapy: balloon catheter dilatation (PTCA), bypass surgery or conservative drug treatment.

Calcium antagonists are able to prevent or resolve coronary spasm during coronary arteriography. In addition they permit differentiation of functional stenoses (increased coronary vasomotor tone) and organic (fixed) stenoses, since intracoronary administration produces maximal dilation of coronary arteries, enabling recognition of anatomically fixed stenoses on the coronary arteriogram. If in suspected Prinzmetal's angina pectoris a provocation test is undertaken with ergonovine or methylergometrine, calcium antagonists (or nitrates) may be employed should the induced coronary spasm not resolve spontaneously.

Today coronary arteriography is routinely carried out following sublingual administration of nifedipine and/or nitroglycerine. As alternatives other calcium antagonists (verapamil, gallopamil, diltiazem) are also used. Only in individual cases (occurrence of coronary spasm, provocation test in Prinzmetal's angina pectoris) are calcium antagonists additionally employed by intracoronary administration (review by Meyer 1986).

Today increasing use is made of PTCA developed by Grüntzig, also known as coronary balloon dilation. Although originally only proximal circumscribed stenoses were dilated in single-vessel disease, more recently patients with multiple-vessel disease, unstable angina pectoris and even acute myocardial infarction are also dilated. During the dilation procedure the selected coronary vessel is occluded, producing myocardial ischaemia, which after a certain time results in angina pectoris, ischaemic ST segment changes in the ECG, disturbed wall motion (measurable, for instance, by echocardiography), and/or cardiac arrythmias.

Intracoronary administration of nifedipine or nitroglycerine before the dilation procedure can improve cardiac ischaemic tolerance. Angina pectoris and signs of ischaemia in the ECG are clearly delayed. A combination of both drugs (0.2 mg nifedipine and 0.2 mg nitroglycerine intracoronarily) leads to an approximate 100% increase in ischaemic tolerance. The improved ischaemic tolerance derives from decreases in left ventricular filling pressure, or end-systolic and/or end-diastolic volume, which diminishes wall tension and ultimately myocardial oxygen consumption (Erbel 1986). Serruys et al. (1984) counterpostulate a "regional cardioplegia" under intracoronarily administered nifedipine; owing to a regionally reduced contractility, ischaemia and thereby anaerobic metabolism can be prevented. Before, during and after catheter dilation, Kaltenbach and co-workers (Bussmann and Kaltenbach 1987) use a combination of isosorbide dinitrate (60–120 mg daily), verapamil (360 mg daily) or gallopamil (100 mg daily), and acetylsalicylic acid (1.5 g daily).

In Summary. Calcium antagonists are today routinely administered during PTCA – either intracoronarily (nifedipine), or intravenously (verapamil, gallopamil and perhaps also diltiazem); supplementary nitroglycerine is also given to produce an augmented effect. This treatment improves cardiac ischaemic tolerance during dilation and prevents coronary spasm evoked by the dilatation process (review by Erbel 1986; Bussmann and Kaltenbach 1987).

Following both coronary bypass surgery and PTCA, there exists the possibility of reocclusion of the recanalized coronary vessel. Drug therapy is used in an attempt to prevent this: some working groups (Meyer, Mainz; de Feyter, Rotterdam: personal communications, 1986) are employing 3×20 mg nifedipine and 500 mg acetylsalicylic acid daily. This treatment is continued for at least 6 months. Here, too, the Kaltenbach group uses the combination of nitrates, verapamil or gallopamil, and acetylsalicylic acid in the dosage already stated (Bussmann and Kaltenbach 1987). Although the regimen appears pathophysiologically justifiable on the basis of inhibition of platelet aggregation by acetylsalicylic acid and (in vitro) by nifedipine, and of coronary dilation by calcium antagonists and nitrates, there are to date no comparative studies with placebo or other forms of treatment (heparin, oral anticoagulants, molsidomine) to substantiate any advantage for combination treatment with calcium antagonists (or nitrates) and acetylsalicylic acid.

4.5.5 Cardiac Involvement in Systemic Sclerosis: Hyperkinetic Heart Syndrome: Cardioprotection in Tocolysis

In progressive systemic sclerosis a focal interstitial fibrosis of the heart develops (Weiss et al. 1942; Leroy 1984) which is probably due to thickening and narrowing of the very small coronary arteries. By analogy with Raynaud's syndrome in systemic sclerosis (Sect. 4.6.2), an organic (anatomical) and a functional component are thought to be involved in the coronary artery stenosis. For this reason nifedipine has been employed in systemic sclerosis, and a reduction in perfusion deficits has been shown by thallium perfusion scintigraphy in the acute test (Kahan et al. 1986). Cold-induced disturbances of wall motion (demonstrated by means of radionuclide ventriculography) also declined under nifedipine treatment (Ellis et al. 1986). Both results favour a potentially reversible disturbance of cardiac microcirculation in progressive sclerosis. The extent to which long-term therapy with calcium can prevent progressive myocardial fibrosis, the cause of death in 15% of scleroderma patients, remains an open question.

The hyperkinetic heart syndrome, the most exaggerated of hyperdynamic cardiovascular regulation disorders, is characterized by increases in cardiac output, cardiac contractility, heart rate, blood pressure, and skeletal muscle blood flow. Exercise capacity is reduced; physical exertion leads to excessive increases in heart rate and blood pressure. The differential diagnosis includes mitral valve prolapse, incipient cardiomyopathy, hyperthyroidism, temporal lobe epilepsy and anxiety syndromes. The treatment of choice is β-receptor blockers and physical exercise (review by Lydtin and Lohmöller 1980). Calcium antagonists have been used in a few isolated cases of hyperkinetic heart syndrome; these studies were carried out with verapamil (Tkaczewski et al. 1980; Rameis and Magometschnigg 1982) which proved less effective in comparison with β-receptor blockers. In hyperkinetic heart syndrome we recommend β-receptor blockers as the agents of choice: their use has both a pharmacologically sound rationale and the widest experience in this indication.

A discussion of the use of calcium antagonists in the prevention of cardiac side effects (cardioprotection) in tocolytic treatment with β-sympathomimetics is to be found in Sect. 4.10.3.2; their application in cytostatic chemotherapy (doxorubicin group) is reviewed in Sect. 4.11.

References

Boldt J, v Bormann B, Hempelmann G (1986) Koronarchirurgische Eingriffe – Steuerung des Blutdrucks durch intravenöse Adalat-Infusionen In: Adalat parenteral Bayer Levekusen pp 105–113

Braunwald E (1982) Mechanism of action of calcium-channel-blocking agents N Engl J Med 307: 1618–1627

Braunwald E, Colucci WS (1984) Vasodilator therapy of heart failure N Engl J Med 310: 459–461

Bussmann WD (1984) Akute und chronische Herzinsuffizienz Springer Berlin

Bussmann WD, Kaltenbach M (1987) Ballondilatation von Herzkranzgefäßen und Herzklappen: Ergebnisse und neuere Entwicklungen Intensivmed 24: 115–119

Clark RE, Christlieb IY, Ferguson TB (1981) The first American clinical trial of nifedipine in cardioplegia: a report of the first 12 months' experience J Thorac Cardiovasc Surg 82: 848–859

Clark RE, Magovern GJ, Christlieb IY, Boe S (1983) Nifedipine cardioplegia experience: results of a 3-year cooperative clinical study Ann Thorac Surg 36: 654–663

Clark RE, Christlieb IY, Clark BK (1986) The use of nifedipine in cardioplegic solutions for high-risk surgical patients In: Lichtlen PR (ed) 6th International Adalat Symposium Excerpta Medica Amsterdam pp 406–413

Cohn JN, Franciosa JA (1977) Vasodilator therapy of cardiac failure N Engl J Med 297: 27–31

Cohn JN, Archibald DG, Phil M, Ziesche S, Franciosa JA, Harston WE, Tristani FE, Dunkman WB, Jacobs W, Francis GS, Flohr KH, Goldman S, Cobb FR, Shah PM, Saunders R, Fletcher RD, Loeb HS, Hughes VC, Baker B (1986) Effect of vasodilator therapy on mortality in chronic congestive heart failure N Engl J Med 314: 1547–1552

Colucci WS, Fifer MA, Lorell BH, Wynne J (1985) Calcium channel blockers in congestive heart failure: theoretic considerations and clinical experience Am J Med 78 (Suppl 2B): 9–17

CONSENSUS Trial Study Group (1987) Effects of enalapril on mortality in severe congestive heart failure. N Engl J Med 316: 1429–1435

Cowley AJ, Stainer K, Wynne RD, Rowley JM, Hampton JR (1986) Symptomatic assessment of patients with heart failure: double-blind comparison of increasing doses of diuretics and captopril in moderate heart failure Lancet 2: 770–772

Donegani E, Costa P, DePaulis R, di Summa M, Poletti GA, Pignatelli MG, Verdecchia C, Morea M (1986) Myocardial protection by perioperative diltiazem drip: a clinical evaluation Thorac Cardiovasc Surg 34: 168–171

Dunselman PHJM, Kuntze CEE, van Bruggen A, Lie KI (1987) Felodipine in heart failure: a double-blind, randomised, placebo-controlled parallel-group study in 23 patients Drugs 34 (Suppl 3): 79–80

Elert O (1983) Myocardial protection with nifedipine in open heart surgery In: Kaltenbach M, Neufeld HN (eds) 5th International Adalat Symposium Excerpta Medica Amsterdam pp 100–106

Elkayam U, Weber L, Torkan B, Berman D, Rahimtoola SH (1983) Acute hemodynamic effect of oral nifedipine in severe chronic congestive heart failure Am J Cardiol 52: 1041–1045

Elkayam U, Weber L McKay CR, Rahimtoola SH (1984a) Differences in hemodynamic response to vasodilation due to calcium channel antagonism with nifedipine and direct-acting agonism with hydralazine in chronic refractory congestive heart failure Am J Cardiol 54: 126–131

Elkayam U, Weber L, Torkan B, McKay CR, Rahimtoola SH (1984b) Comparison of hemodynamic responses to nifedipine and nitroprusside in severe chronic congestive heart failure Am J Cardiol 53: 1321–1325

Ellis WW, Baer AN, Robertson RM, Pincus T, Kronrnberg MW (1986) Left ventricular dysfunction induced by cold exposure in patients with systemic sclerosis Am J Med 80: 385–392

Erbel R (1986) Perkutane transluminale koronare Angioplastie – Verbesserung der Ischämietoleranz durch Adalat intrakoronar In: Adalat parenteral Bayer Leverkusen pp 52–55

Feder E, Lydtin H (1983) Therapie der chronischen Herzmuskelinsuffizienz: Stimulation und/oder Entlastung? Med Klin 78: 110–116

Fifer MA, Lorell BH, Jaski BE, Wright RF, Barry WH, Colucci WS (1986) Nifedipine and milrinone: vasodilators with negative and positive inotropic effects in the treatment of heart failure In: Lichtlen PR (ed) 6th International Adalat Symposium Excerpta Medica Amsterdam pp 124–132

Fioretti P, Benussi B, Scardi S, Klugmann S, Brower RW, Camerini F (1982) Afterload reduction with nifedipine in aortic insufficiency Am J Cardiol 49: 1728–1732

Flameng W, Borgers M, Van der Vusse GJ, Demeyere R, Vandermeersch E, Thoné F, Suy R (1983) Cardioprotective effects of lidoflazine in extensive aorta-coronary bypass grafting J Thorac Cardiovasc Surg 85: 758–768

Flameng W, Stalpaert G (1986) Herzoperation – Myokardschutz durch Adalat intrakoronar In: Adalat parenteral Bayer Leverkusen pp 62–66

Greenberg BH, Rahimtoola SH (1980) Long-term vasodilator therapy in aortic insufficiency Ann Intern Med 93: 440–442

Hall D, Froer KL, Kraus F, Stautner C, Rudolph W (1986) Effects of nifedipine in congestive heart failure due to dilated cardiomyopathy: a randomized, double-blind, cross-over study In: Bussmann WD, Just H (eds) Vasoaktive Substanzen bei Herzinsuffizienz Springer Berlin pp 160–163

Hanrath P, Kremer O (1983) Acute effects of nifedipine on left ventricular performance at rest and during exercise in patients with left ventricular dysfunction In: Just H, Bussmann WD (eds) Vasodilators in chronic heart failure Springer Berlin pp 154–163

Hempelmann G (1982) Prämedikatien, Anästhesie und postoperative Phase bei Hypertonikern und Patienten mit koronarer Herzerkrankung Med Welt 33: 739–745

Henry PD, Clark RE, Williamson JR (1980) Protection of the globally ischemic heart with calciumantagonists: improved recovery after cardiopulmonary bypass In: Fleckenstein A, Roskamm H (eds) Calcium-Antagonismus Springer Berlin pp 181–190

Hess OM (1986) Therapie der dilatativen Kardiomyopathie Dtsch Med Wschr 111: 222–225

Hicks Jr GL, Salley RK, De Weese JA (1984) Calcium channel blockers: an intraoperative and postoperative trial in women Ann Thorac Surg 37: 654–663

Kahan A, Devaux JY, Amor B, Menkés CJ, Weber S, Nitenberg A, Venot A, Guérin F, Degeorges M, Roucayrol JC (1986) Nifedipine and thallium-201 myocardial perfusion in progressive systemic sclerosis N Engl J Med 314: 1400–1401

Kaplan AV, Guffin AV, Jones EL, Kates R, Holbrook GW (1984) Verapamil und Strukturerhaltung des Myokards bei Patienten mit koronarer Bypassoperationen In: Althaus U, Burckhardt D, Vogt E (eds) Calcium-Antagonismus – Internat. Symposium on Calcium-Antagonism Universimed Frankfurt/Main pp 242–253

Kassis E, Amtorp O, Waldorff S, Fritz-Hansen P (1987) Efficacy of felodipine in chronic congestive heart failure: a placebo controlled haemodynamic study at rest and during exercise and orthostatic stress Br Heart J 58: 505–511

Kiowski W, Erne P, Pfisterer M, Müller J, Bühler FR, Burkhart F (1987) Hemodynamic effects of nisolipine in patients with severe chronic congestive heart failure In: Hugenholtz PG, Meyer J (eds) Nisoldipine 1987 Springer Berlin Heidelberg New York pp 329–342

Klepzig M, Strauer BE (1985) Nisoldipin bei Herzinsuffizienz? Z Kardiol 74 (Suppl 3): 8

Leier CV, Patrick TJ, Hermiller J, Dalpiaz Pacht K, Huss P, Magorien RD, Unverferth DV (1984) Nifedipine in congestive heart failure: effects on resting and exercise hemodynamics and regional blood flow Am Heart J 108: 1461–1468

Le Roy EC (1984) The heart in systemic sclerosis N Engl J Med 310: 188–189

Mason DT (1978) Afterload reduction and cardiac performance – physiologic basis of systemic vasodilators as a new approach in treatment of congestive heart failure Am J Med 65: 106–125

Materne P, Legrand V, Vandormael M, Collignon P, Kulbertus HE (1984) Hemodynamic effects of intravenous diltiazem with impaired left ventricular function Am J Cardiol 54: 733–737

Meyer J (1986) Fortschritte in der invasiven Diagnostik und Therapie der koronaren Herzkrankheit In: Adalat parenteral Bayer Leverkusen pp 39–46

Olivari MT, Levine TB, Cohn JN (1984) Acute hemodynamic effects of nitrendipine in chronic congestive heart failure J Cardiovasc Pharmacol 6 (Suppl 7): 1002–1005

Opie LH (1984a) Congestive heart failure In: Opie LH (ed) The Heart Grune & Stratton London pp 301–313
Opie LH (1984b) Valvular heart disease and cardiomyopathy In: Opie LH (ed) The Heart Grune & Stratton London pp 287–300
Polese A, Florentini C, Olivari MT, Guazzi MD (1979) Clinical use of a calcium antagonistic agent (Nifedipine) in acute pulmonary edema Am J Med 66: 825–830
Rameis H, Magometschnigg D (1982) Verapamil: an alternative therapy in hyperkinetic heart syndrome (HHS) – a comparison with oxprenolol In: Intern. Symposium on calcium modulators Venedig: pp 28
Reicher-Reiss H, Vered Z, Goldbourt U, Hertz I, Battler A, Neufeld HN (1987) Influence of chronic nisoldipine therapy upon severe congestive heart failure – comparison with nitrates In: Hugenholtz PG, Meyer J (eds) Nisoldipine 1987 Springer Berlin Heidelberg New York pp 343–346
Serruys PW, van den Brand M, Brower RW, Hugenholtz PG (1983) Regional cardioplegia and cardioprotection during transluminal angioplasty, which role for nifedipine? Eur Heart J 4 (Suppl C): 115–121
Shen WF, Roubin GS, Hirasawa K, Uren RF, Hutton BF, Harris PJ Fletcher PJ, Kelly DT (1984a) Abnormal left ventricular response to isometric exercise in pure, isolated aortic regurgitation: benefical effects of nifedipine Am J Cardiol 54: 605–609
Shen WF, Roubin GS, Hirasawa K, Uren RF, Hutton BF, Harris PJ Fletcher PJ, Kelly DT (1984b) Noninvasive assessment of acute effects of nifedipine on rest and exercise hemodynamics and cardiac function in patients with aortic regurgitation J Am Coll Cardiol 4: 902–907
Shen WF, Roubin GS, Choong CYP, French P, Harris PJ, Fletcher PJ, Kelly DT (1986) Favourable effects of chronic nifedipine therapy on left ventricular response to exercise in chronic aortic regurgitation In: Lichtlen PR (ed) 6th International Adalat Symposium Excerpta Medica Amsterdam pp 110–115
Tarnow J (1984) Koronarchirurgische Eingriffe – Steuerung des Blutdrucks durch intravenöse Adalat-Infusionen In: Adalat parenteral Bayer Leverkusen pp 114–117
Timmis AD, Jewitt DE (1985) Studies with felodipine in congestive heart failure Drugs 29 (Suppl 2): 66–75
Tkaczewski W, Goch JH, Wichan P, Kowalski J (1980) Ergometric test in subjects with sympathicotonus after administration of propranolol, pindolol and oxazepam Pol Tyg Lek 35: 867–869
Vouhé PR, Loisance DY, Aubry P, Piszker G, Cazor JL, Cachera JP (1985) Double-blind evaluation of the intraoperative use of diltiazem during coronary artery surgery In: Just H, Schröder JS (eds) Advances in clinical applications of calcium antagonist drugs Excerpta Medica Amsterdam pp 91–104
Walsh RW, Porter CB, Starling MR, O'Rourke RA (1984) Beneficial hemodynamic effects of intravenous and oral diltiazem in severe congestive heart failure J Am Coll Cardiol 3: 1044–1050
Weiss S, Stead EA, Warren JV, Bailey OT (1943) Scleroderma heart disease with a consideration of certain other visceral manifestations of scleroderma Arch Intern Med 71: 749–776
Wilson RS, Crawford MH, Starling MR, Chaudhuri TK, O'Rourke RA (1984) Improvement in left ventricular function with digoxin but not nifedipine in patients with chronic aortic regurgitation Circulation 70 (Suppl 2): II–144
Zähringer J, Schultheiß HP (1986) Therapie der chronischen Herzinsuffizienz mit Vasodilatantien Internist 27: 315–321

4.6 Other Vascular Indications

4.6.1 Pulmonary Hypertension

In pulmonary hypertension, as with arterial hypertension (Sect. 4.2), a differentiation is made between a primary or idiopathic form and secondary forms. Primary pulmonary hypertension is an uncommon condition which mainly affects younger women. Pulmonary artery pressure and pulmonary vascular resistance are markedly elevated, without evidence of the known causes of secondary pulmonary hypertension: restrictive or obstructive primary pulmonary disease; heart diseases such as valvular heart disease; left-to-right shunts; coronary heart disease; cardiomyopathy; or multiple pulmonary emboli. The secondary forms of pulmonary hypertension usually exhibit marked arterial hypoxia with or without hypercapnia (raised PCO_2).

Patients with primary pulmonary hypertension consult a doctor because of dyspnoea, diminished exercise capacity, anginal symptoms, and/or syncope. Usually, in the course of only a few years, increasing right-sided heart failure develops, which in the final stages of the condition progresses into global heart failure ("right–left heart failure"). The clinical picture can only suggest the diagnosis, which must be confirmed by exclusion of the secondary forms of pulmonary hypertension, usually by means of pulmonary function tests, radionuclide scanning, and heart catheterization.

Although the initiating cause has yet to be conclusively established, an at least partially reversible pulmonary vasoconstriction is postulated to play the decisive role (review by Packer 1985). This hypothesis is supported by evidence of medial hypertrophy of pulmonary arterioles, the absence of structural pulmonary changes, the occurrence of pulmonary hypertension with other vasospastic conditions (Raynaud's syndrome) and also by decreases in pulmonary vascular resistance under treatment with vasodilators and/or oxygen.

In contrast to secondary forms of pulmonary hypertension, where treatment centres on the underlying disorder, therapy of advanced primary pulmonary hypertension has only been feasible since the introduction of combined heart–lung transplantations. Since this procedure is accompanied by manifold risks and can only seldom be performed, an aggressive treatment plan must be adopted in order to restrict the progression of primary pulmonary hypertension by means of oral anticoagulants, diuretics and vasodilators (reviews of primary pulmonary hypertension by Ross 1983; Packer 1985).

The therapy of primary pulmonary hypertension with vasodilators ideally attempts (by analogy with their use in heart failure and arterial hypertension) to adjust the abnormal haemodynamic parameters (raised pulmonary vascular resistance, raised pulmonary artery pressure), to alleviate symptoms and improve exercise tolerance, and thus ultimately to prolong the patient's life. These goals have, however, yet to be fulfilled by any vasodilating medication used to date, including calcium antagonists (review by Packer 1985).

Experimental findings which demonstrated that calcium antagonists (verapamil, nifedipine) can prevent hypoxic pulmonary vasoconstriction (McMurtry et al. 1976) were the stimulus at the close of the 1970s for the first clinical application of verapamil (Landmark et al. 1978), and of nifedipine (Camerini et al. 1980) in the treatment of pulmonary hypertension (primary and secondary forms).

In the short term nifedipine lowers pulmonary artery pressure and pulmonary vascular resistance; cardiac output increases both at rest and during exercise. Verapamil also lowers pulmonary artery pressure, but does not significantly alter cardiac output and pulmonary vascular resistance. In some patients verapamil shows a distinct negative inotropic effect with deterioration of right ventricular function and increase in pulmonary vascular resistance. These initial results were subsequently confirmed by various authors; in addition to nifedipine and verapamil, diltiazem and felodipine were also used (Kambara et al. 1981; Bucca et al. 1982; Crevey et al. 1982; Gaszner et al. 1983; Rubin et al. 1983; Arnman et al. 1984; Packer 1985; Singh et al. 1986). Generally speaking, on account of the acute test results, nifedipine, and in certain circumstances diltiazem or felodipine, are preferable to verapamil in view of their stronger vasodilating and weaker negative inotropic effects (Krebs 1984; Packer 1985). Nearly all the studies in this area involve only small numbers of patients; there is no differentiation between primary and secondary forms of pulmonary hypertension, and no control groups. The doses used were 10–20 mg nifedipine sublingually, 10 mg verapamil intravenously, and 20 mg diltiazem intravenously.

Side effects of treatment with calcium antagonists are described in all forms of pulmonary hypertension. Verapamil (and to a far lesser extent nifedipine) can impair right ventricular function and lead to severe hypotension, even to cardiogenic shock. On account of autonomic counter-regulation with raised cardiac output, nifedipine can even induce an increase in pulmonary artery pressure; furthermore, nifedipine, like felodipine, can induce an increase in arterial hypoxia due to increased perfusion of poorly ventilated alveoli (disturbed ventilation: perfusion ratio; see also Sect. 3.1.5.2; reviews of side effects by Packer et al. 1984, 1985; and Melot et al. 1984).

Experience with calcium antagonists in long-term treatment of primary pulmonary hypertension has so far been limited to extremely small groups of patients (less than ten). Studies were carried out mainly with nifedipine in a daily dose of 40–120 mg or with felodipine 30–40 mg. Although a favourable influence on haemodynamic parameters can be demonstrated even in long-term therapy (decline in pulmonary vascular resistance, increase in cardiac output), no influence can be shown on mortality (Rubin et al. 1983; Klugmann et al. 1983; Arnmann et al. 1984; Rich et al. 1985). Patients with pulmonary hypertension who respond positively to nifedipine at the onset of treatment appear to have a better overall prognosis, which is possibly associated with a partially functional, i.e. potentially reversible, pulmonary hypertension (Rich 1985).

In Summary. In short-term and probably also in long-term therapy nifedipine improves the abnormal haemodynamic parameters in patients with pulmonary hypertension, by lowering pulmonary artery pressure and pulmonary vascular resistance, and raising cardiac output (initial positive results with diltiazem and felodipine have yet to be confirmed by further studies). Randomized long-term studies have yet to be conducted, and thus no statement can be made as to improvement in prognosis as a result of treatment with calcium antagonists. Since calcium antagonists can occasionally worsen a patient's cardiopulmonary condition (increase in systemic hypoxia, drop in blood pressure, right-sided heart failure and cardiogenic shock), initiation of therapy should always be accompanied by monitoring of haemodynamic parameters and by repeated arterial blood gas analysis.

In one form of secondary pulmonary hypertension, namely, hypoxic pulmonary hypertension in the context of chronic obstructive lung disease, mucoviscidosis, pickwickian (or sleep apnoea) syndrome, studies indicate that calcium antagonists, particularly nifedipine, possibly augment, or even provide an alternative for, long-term oxygen treatment. The initial positive results (Bucca et al. 1982; Kennedy et al. 1985; Bratel et al. 1988a, b) require confirmation by randomized, controlled long-term trials.

4.6.2 Raynaud's Syndrome

Raynaud's syndrome as an intermittent "numbness of the digits" was first described by Maurice Raynaud in 1862. In response to exposure to cold or psychic stress, ischaemia develops in one or more fingers (or toes); as the syndrome progresses the digits become cyanotic and, if the circulatory disturbance improves, a reactive hyperaemia arises (consecutive colour triad: white/blue-grey/red). The patient experiences paraesthesia, numbness or pain and during an attack there is considerable functional disability.

Aetiologically, two forms of the syndrome are differentiated: primary Raynaud's syndrome (morbus Raynaud) without any underlying primary basic disease; and secondary Raynaud's syndrome caused by a known underlying primary disease or by medication (e.g. connective tissue diseases such as systemic sclerosis or lupus erythematosus; arterial occlusive diseases such as vasculitis or thoracic outlet syndrome; arterial trauma due to vibration; cryoglobulinaemia; exposure to vinyl chloride; or treatment with ergotamine preparations or oral contraceptives). The terms Raynaud's phenomenon with or without primary disease are also used (review by Diehm et al. 1983; Smith and Rodeheffer 1985).

Pathophysiologically, blood flow through the extremities is reduced in Raynaud's syndrome. This can be a consequence of lowered perfusion pressure (e.g. proximal arterial occlusion), of raised blood viscosity (e.g. cryoglobulinaemia) or of an anatomical (e.g. intimal and adventitial vascular fibrosis in scleroderma) or functional (arterial vasospasm) stenosis of the end-arteries (review by Smith and Rodeheffer 1985). Vascular spasm possibly produces an intense activation of platelets, whereby thromboxane A_2 synthesis is increased and local prostacyclin formation in the vascular wall (endothelium) is reduced.

In most cases, the tendency to arterial vasospasm is not limited to the fingers; many patients suffer from migraine or from Prinzmetal's angina pectoris (Miller et al. 1981). All three diseases are characterized by reversible vasospasm, and local trigger mechanisms are postulated.

The various therapies of Raynaud's syndrome span a spectrum from recognition and avoidance of trigger mechanisms (e.g. exposure to cold) through medication (calcium antagonists, α-receptor blockers such as prazosin, ACE inhibitors such as captopril, prostaglandins, nitrates, etc.) to plasmaphaeresis and thoracic sympathectomy (reviews by Diehm et al. 1983; Smith and Rodeheffer 1985). In secondary Raynaud's syndrome treatment is aimed at correction of the underlying disorder. Since calcium antagonists inhibit vascular smooth muscle and prevent or resolve vasospasm, they have frequently been used in Raynaud's syndrome in recent years. In addition, calcium antagonists are thought to affect

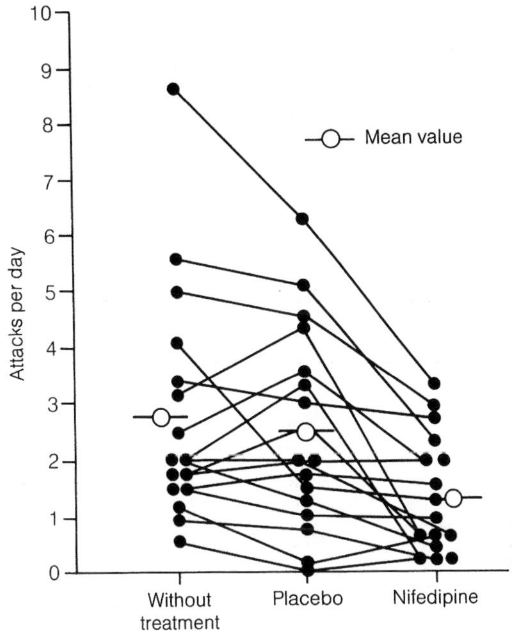

Fig. 37. Frequency of attacks in Raynaud's syndrome under nifedipine and placebo. Difference statistically significant ($p = 0.001$). (After Smith and McKendry 1982)

platelet function, in particular to inhibit platelet activation (Sect. 3.3.3; Malamet et al. 1985).

Randomized studies (in the main double-blind, crossover trials) have been carried out with nifedipine, felodipine, diltiazem and verapamil. During a course of treatment lasting several weeks, with 4 × 10 mg to 3 × 20 mg nifedipine (Kahan et al. 1981; Smith and McKendry 1982; Rodeheffer et al. 1983; Malamet et al. 1985; White et al. 1986), 3 × 30 mg to 3 × 120 mg diltiazem (Vayssairat et al. 1981; Matoba et al. 1982; Kahan et al. 1985) or felodipine 5–20 mg daily (Nilsson et al. 1989), both the frequency and severity of attacks declined (Fig. 37) and patients generally assessed the treatment favourably; under verapamil there appeared to be no essential change in subjective symptoms. The investigations involved patients with primary and secondary Raynaud's syndrome; in general, patients with primary Raynaud's syndrome seem to respond slightly better to treatment with calcium antagonists (Kahan et al. 1985; Müller-Bühl 1985). This is possibly associated with the mainly functional stenosis due to vasospasm in primary Raynaud's syndrome. In contrast to the unequivocal symptomatic improvement under nifedipine, felodipine or diltiazem, measurement of objective vascular parameters yielded inconsistent results.

While Rodeheffer et al. (1983) could demonstrate no change in perfusion pressure in the digital arteries, Creager et al. (1984) observed a decrease in fingertip vascular resistance and an increase in fingertip blood flow. Müller-Bühl (1983) also reported an increase in digital blood flow following exposure to cold. White et al. (1986) measured the digital skin temperature recovery time after immersion of the patient's hand in ice water for 20 s. With nifedipine, the recovery time was significantly shorter than with placebo but did not reach the levels of normal volunteers.

The studies so far carried out with verapamil (dose 3 × 40 mg to 3 × 120 mg daily) revealed no improvement in subjective symptoms or in objective measurable parameters (review by Müller-Bühl 1985); however, in these investigations only patients with secondary Raynaud's syndrome were treated.

In Summary. Nifedipine (dose 40–60 mg daily), felodipine (dose 5–20 mg daily) and diltiazem (doses 3 × 40 to 3 × 120 mg daily) are effective as symptomatic treatment in primary and secondary Raynaud's syndrome. However, results so far refer to studies carried out only over weeks, or at most for a few months. There are no available data concerning long-term treatment of Raynaud's syndrome, and thus it is not yet established whether calcium antagonists can improve prognosis in secondary Raynaud's syndrome (e.g. in systemic scleroderma as the underlying primary disease; Smith and McKendry 1982).

4.6.3 Ergotism: Mesenteric Ischaemia: Arterial Occlusive Disease: Chronic Venous Insufficiency: Accidental Intraarterial Injection of Drugs

Employment of calcium antagonists in the indications discussed in this section can be justified on pathophysiological grounds. However, although reports of individual cases are available, no large-scale investigations have been conducted, and the effect of calcium antagonists cannot be conclusively assessed for any of the following situations.

Epidemics of ergotamine poisoning (ergotism) occurred in earlier times owing to contamination of cereals with the fungus, *Claviceps purpurea*. Today, ergotism is usually caused by overdosage of drugs containing ergotamine or by hypersensitivity to the drug. This most often occurs during migraine therapy and, in rare instances, also in the prophylaxis of thrombosis with dihydroergotamine–heparin. Besides cerebral symptoms, accompanied by nausea and emesis, arterial vasospasm is of the greatest clinical importance. Spasm of the cerebral arteries, coronary vessels, mesenteric and renal arteries and, very commonly, of the arteries in the extremities has been described. Treatment aims to resolve vasospasm, to decrease blood viscosity, and to prevent further clot formation. On account of their effectiveness in vasospastic conditions calcium antagonists should also show a positive effect in ergotism, the strong vasodilating substances of the dihydropyridine group being preferred. As alternatives hydralazine, captopril, prazosin, sodium nitroprusside, nitroglycerine or prostacyclin may be considered (review by Creutzig and Alexander 1985).

Ischaemia of the mesenteric arteries may lead to intermittent pain (angina abdominalis, abdominal angina), a malabsorption syndrome, or in the extreme situation to acute mesenteric infarction. As in angina pectoris, a combination of organic stenoses and functional changes in vasomotor tone and/or vasospasm is thought to occur in abdominal angina. The use of calcium antagonists in the patient with abdominal angina can therefore be justified.

During digitalis therapy, and particularly in digitalis overdosage, vasospasm of the mesenteric arteries can occur. Calcium antagonists resolve or even prevent such vasospasm in animal experiments (Sect. 3.1.5.2) and perhaps also in humans (Kauffmann et al. 1983).

In peripheral arterial occlusive disease, mainly of the lower extremities, exercise (walking), improvement of microcirculation by isovolaemic or hypervolaemic haemodilution as well as the use of platelet aggregation inhibitors form the basis of treatment. Along with other vasodilating substances, calcium antagonists (verapamil, nifedipine, flunarizine) are also prescribed in this indication. Calcium antagonists are thought not only to increase blood flow in the extremities, but also to increase the walking distance in intermittent claudication (stage II of arterial occlusive disease, according to La Fontaine; Solti et al. 1987; Di Perri et al. 1988). Besides vasodilation, an improvement of the flow properties of the blood (erythrocyte fluidity and deformability, platelet aggregation) is postulated. As with other vasodilating substances, however, calcium antagonists can also dilate mainly noncompromised vascular areas and lead to the so-called steal phenomenon. Randomized, controlled long-term studies in arterial occlusive disease have been carried out chiefly with flunarizine (10–20 mg daily; Dormandy, 1988). In the main the success of treatment (increase in walking distance) was statistically significant; however, the study involved only a small number of patients and the duration of treatment was relatively short. As stated in Sect. 4.6.4, a protective effect for calcium antagonists on the progression of atherosclerosis has not yet been proven.

Calcium antagonists (chiefly nifedipine) are successfully used in the treatment of digital ulcers in the context of Raynaud's syndrome, e.g. in scleroderma (Sect. 4.6.2). In some cases nifedipine is said to be effective in refractory diabetic ulcers, even to the point of preventing amputation (Lazarus et al. 1983).

Also in varicosis and chronic venous insufficiency a calcium antagonist (flunarizine) has been applied. In several placebo-controlled studies subjective symptoms improved and venous ulcers healed completely (review by Roeckaerts and Vanden Bussche 1980). Unfortunately, no study has included objective haemodynamic or rheological parameters. Calcium antagonists of the dihydropyridine type (prototype nifedipine) appear to be less suitable in these situations, since they can induce ankle oedema as a side effect (Sect. 5.1). A discussion of the administration of calcium antagonists in the treatment of patients presenting with accidental intraarterial injection of drugs is to be found in Sect. 4.11.

4.6.4 Atherosclerosis

On the basis of numerous animal trials, an antiatherogenic effect of calcium antagonists has also been postulated for humans. Section 3.4 contains a critical analysis of the role of calcium antagonists in human atherosclerosis; a protective effect has yet to be conclusively proven.

References

Arnman K, Ryden L, Smedgard P, Thorhallsson E (1984) Felodipine in primary pulmonary hypertension: report of two cases Acta Med Scand 215: 275–280

Bratel T (1988) Pulmonary hypertension in chronic obstructive lung disease. Effects of a calcium antagonist, felodipine, on central haemodynamics and gas exchange Thesis Stockholm

Bratel T, Hedenstierna G, Lundquist H, Nyquist O, Ripe E (1988) Cardiac function and central haemodynamics in severe chronic obstructive lung disease. Acute and long-term effects of felodipine Eur Respir J 1: 262–268

Bucca C, Scalvo M, Rolla G, Borello G, Polizzi S (1982) Verapamil versus oxygen in pulmonary hypertension complicating chronic obstructive lung diseases Bull Physiopath Resp 18 (Suppl 4): 87–93

Camerini F, Alberti E Klugmann S, Salvi A (1980) Primary pulmonary hypertension: effects of nifedipine Br Heart J 44: 352–356

Craeger MA, Pariser KM, Winston EM, Rasmussen HM, Miller KB Coffman JD (1984) Nifedipine-induced fingertip vasodilation in patients with Raynaud's phenomenon Am Heart J 108: 370–373

Creutzig A, Alexander K (1985) Ergotismus Dtsch Med Wschr 110: 1420–1422

Crevey BJ, Danitzker DR, Bower JS, Popat KD, Walker SD (1982) Hemodynamic and gas exchange effects of intravenous diltiazem in patients with pulmonary hypertension Am J Cardiol 49: 578–583

Diem C, Müller-Brühl U, Mörl H (1983) Neue Aspekte der Pharmakotherapie des Raynaud-Phänomens Med Klin Prax 78: 33–37 u 55

Di Perri T, Laghi Pasini F, Pecchi S, Domini L, Ralli L, Materazzi M, De Franco V, Monaci A, Ceccatelli L (1988) Pharmacological basis and therapeutic efficacy of flunarizine in peripheral artery disease In: Vanhoutte PM, Paoletti R, Govoni S (eds) Calcium antagonists New York Academy of Sciences New York pp 491–501

Dormandy J (1988) Clinical use of calcium antagonists in peripheral circulatory disease In: Vanhoutte PM, Paoletti R, Govoni S (eds) Calcium antagonists New York Academy of Sciences New York pp 611–620

Gaßner A, Pichler M, Fridrich L, Sykora J, Tizek H, Lenz K (1983) Einfluß von Nifedipin auf die Hämodynamik bei präkapillärer pulmonaler Hypertension in Ruhe und unter Belastung Dtsch Med Wschr 108: 1790–1794

Kahan A, Weber S, Amor B, Saporta L, Hodara M (1981) Nifedipine and Raynaud's phenomenon Ann Intern Med 94: 546

Kahan A, Foult JM, Weber S, Amor B, Menkes CJ, Degeorges M (1985) Nifedipine and alpha 1-adrenergic blockade in Raynaud's phenomenon Eur Heart J 6: 702–705

Kambara H, Fujimoto K, Wakabayasli A, Kawai C (1981) Primary pulmonary hypertension: beneficial therapy with diltiazem Am Heart J 101: 230–231

Kauffmann GW, Friedburg H, Anger P, Rückauer K (1982) Diagnose, Differentialdiagnose und Behandlung der sogenannten „non-occlusive disease" Chirurg 53: 641–654

Kennedy TP, Micheal JR, Summer W (1985) Calcium channel blockers in hypoxic pulmonary hypertension Am J Med 78 (Suppl 2B): 18–26

Klugmann S, Salvi A, Camerini F (1983) Calcium antagonists in heart failure Herz 8: 88–92

Krebs R (1984) Calcium antagonists: new vistas in theoretical basis and clinical use In: Opie LH (ed) Calcium antagonists and cardiovascular disease Raven Press New York pp 347–357

Landmark K, Refsun AM, Simonsen S, Storstein O (1978) Verapamil and pulmonary hypertension Acta Med Scand 204: 299–302

Lazarus RJ, Onor A, Hameed K (1983) Nifedipine and a diabetic ulcer Ann Intern Med 98: 414

Malamet R, Wise RA, Ettinger WH, Wigley FM (1985) Nifedipine in the treatment of Raynaud's phenomenon Am J Med 78: 602–608

Matoba T, Ogata M, Kuwahara H (1982) Diltiazem and Raynaud's Syndrome Ann Intern Med 97: 455

McMurtry IF, Davidson AB, Reeves JT, Grover RF (1976) Inhibition of hypoxic pulmonary vasoconstriction by calcium antagonist in isolated rat lungs Circ Res 38: 99–104

Melot C, Hallemans R, Naeije R, Mols P, Lejeune P (1984) Deleterious effect of nifedipine on pulmonary gas exchange in chronic obstructive pulmonary disease Am Rev Respir Dis 130: 612–616

Miller D, Waters DD, Warnica W, Szlachcic J, Kreeft J, Théroux P (1981) Is variant angina the coronary manifestation of a generalized vasospastic disorder? N Engl J Med 304: 763–766

Müller-Brühl U (1985) Raynaud-Syndrom Behandlung mit Kalzium-Antagonisten Münch Med Wschr 127: 1033–1034

Nilsson H, Blychert E, Jonasson T, Leppert J, Ringqvist I (1989) The effect of felodipine on cold-induced digital vasospasm J Cardiovasc Pharmacol (in press)

Packer M, Medina N, Yushak M (1984) Adverse hemodynamic and clinical effects of calcium channel blockade in pulmonary hypertension secondary to obliterative pulmonary vascular disease J Am Coll Cardiol 4: 890–901
Packer M (1985) Vasodilator therapy for primary pulmonary hypertension Ann Intern Med 103: 258–270
Rich S, Brundage BH, Levy PS (1985) The effect of vasodilator therapy on the clinical outcome of patients with primary pulmonary hypertension Circulation 71: 1191–1196
Rodeheffer RJ, Rommer JA, Wigley F, Smith CR (1983) Controlled double-blind trial of nifedipine in the treatment of Raynaud's phenomenon N Engl J Med 308: 880–883
Roeckaerts F, Vanden Bussche G (1980) Double-blind placebo-controlled studies with flunarazine in venous insufficiency Angiology 31: 833–845
Ross J (1983) Vascular disease of the lung: primary pulmonary hypertension In: Harrison's Principles of Internal Medicine McGraw Hill New York pp 1559–1561
Rubin LJ, Nicod P, Hillis LD, Firth BG (1983) Treatment of primary pulmonary hypertension with nifedipine Ann Intern Med 99: 433–438
Singh H, Ebejer MJ, Higgins DA, Henderson AH, Campbell IA (1986) Chronic cor pulmonale. Acute haemodynamic and neurohumoral effects of nifedipine at rest and on maximal exercise In: Lichtlen PR (ed) 6th International Adalat Symposium Excerpta Medica Amsterdam pp 537–542
Smith CD, McKendry RJR (1982) Controlled trial of nifedipine in the treatment of Raynaud's phenomenon Lancet 2: 1299–1301
Smith CR, Rodeheffer RJ (1985) Treatment of Raynaud's phenomenon with calcium channel blockers Am J Med 78 (Suppl 2B): 39–42
Solti F, Iskum M, Szabó Z, Czakó E (1978) Die Wirkung des Ca^{++}-Antagonisten Isoptin auf die periphere Durchblutung bei obliterierenden Angiopathien Therapiewoche 28: 5993–5994
Vayssairat M, Capron L, Fiessinger JN, Mathieu JF, Housset E (1981) Calcium channel blockers and Raynaud's disease Ann Intern Med 95: 243
White CJ, Phillips WA, Abrahams LA, Watson TD, Singleton PT (1986) Objective benefit of nifedipine in the treatment of Raynaud's phenomenon Am J Med 80: 623–625

4.7 Calcium Antagonists in Neurological Disease

Calcium antagonists of the verapamil type, of the dihydropyridine type, piperazine derivatives, and diltiazem are employed in the treatment of central nervous system disorders. They are currently indicated in migraine prophylaxis and anti-vasospastic therapy following subarachnoid haemorrhage. Further therapeutic possibilities include the treatment of cerebral ischaemia, organic brain syndrome, vestibular vertigo and epilepsy.

4.7.1 Migraine

4.7.1.1 Classification and Definition

Most of the clinical studies which will be cited in Sect. 4.7.1.3 define migraine according to the "Classification of Headache" of the Ad-hoc Committee on Classification of Headache (1962). Although this describes and classifies the concept of migraine, there are ambiguities, with the result that direct comparisons are not always possible (Blau 1984). The Anglo-American classification of common, classical and complicated migraine is internationally accepted (Soyka 1978, 1985a; Gelmers 1985; Table 27), but will be replaced by a new international classification in 1991 (Committee of the International Headache Society 1988). Cluster headache is a unilateral, orbital and periorbital localized, piercing or

Table 27. Classification of migraine (After Soyka 1985a)

Common migraine	Attacks of headache Autonomic concomitants Absence of neurological disorders
Classical migraine	Attacks of headache, usually with phasic course Transient neurological prodromes (particularly scintillating scotomas, paraesthesias)
Complicated migraine (includes hemiplegic and ophthalmoplegic migraine)	Attacks of headache Autonomic concomitants Prolonged disturbances of neurological function (lasting a few hours to days)

throbbing headache accompanied by reddening and lacrimation of the affected eye, possibly also by anisocoria (Saller et al. 1982) and sweating abnormality (Sjaastad and Fredriksen 1986). Attacks tend to occur at night for a number of successive days, followed by attack-free intervals (review by Soyka 1985a).

4.7.1.2 Pathogenetic Factors

Although a plethora of individual factors and sequential pathophysiological cascades concerning the pathogenesis of migraine have been proposed and discussed, experimental simulation of an attack in animals has not been achieved. Thus, discussion of the various triggering factors and resultant characteristics of migraine can be described only hypothetically.

Vascular Factors

According to Wolff (1963), a migraine attack requires an initial cerebral vasoconstriction with resultant secondary cerebral ischaemia. This phase embraces the neurological deficits of complicated migraine, and also the visual disturbances of classical migraine, as well as the prodromes of common migraine. This theory proposes that pain derives from the subsequent phase of vasodilation in which there is stimulation of pain receptors in overdistended and pulsating blood vessels. The initial hypoperfusion can be demonstrated by measurement of regional cerebral blood flow (rCBF) during an attack (Olesen et al. 1981; Anthony 1984). Rheoencephalographic studies (Soyka 1978) also confirm this initial intracranial vasoconstriction. Olesen et al. (1981), however, were unable to show such hypoperfusion in common migraine. More recent investigations by Olesen and Lauritzen (1984) with the ^{131}Xe inhalation technique, on the other hand, showed a focal hypoperfusion at the start of a classical migraine attack, spreading from the occipital lobe to the sulcus Sylvii at the rate of 2–3 cm/min. This form of propagation corresponds to another neuronal phenomenon of the central nervous system: the "spreading depression" which was first described by Leao in 1944 as an electroencaphalographically demonstrable general reduction of cortical activity, spreading as a wave at a rate of about 3 cm/min and resulting in local hypoperfusion. Lauritzen et al. (1982) were able to demonstrate this neuronal phenomenon with a subsequent reduction in cerebral blood flow of 20%–25% in the rat cortex, and surmised a connection with the pathophysiology of migraine. The more recent hypothesis of a primary neuronally conditioned cortical hypoxia as the releasing mechanism of the migraine attack is based on these publications.

In addition to cerebral disturbances, migraine patients often exhibit peripheral vasomotor disturbances which are postulated to be caused by an increased activation of the sympathetic system, demonstrable by a rise in dopamine β-hydroxylase activity. A reinforced sympathetic activity in the muscular nerve supply (SAMN) has not been confirmed (Fagius et al. 1985).

Serotonin and Plasma Factors

The occurrence of fluctuations in serotonin concentrations during a migraine attack has long been recognized. Initially serotonin levels increase, and then in the course of an attack they decrease to below normal levels (Sicuteri et al. 1961). In this context an increased release of thrombocytic serotonin may be of significance. Furthermore, serotonin plays a significant role as a neurotransmitter in central pain processing (so-called pain modulation). Thus, changes in serotonin levels could be directly responsible for pain during a migraine attack (Soyka 1985a).

According to a further hypothesis, foods can trigger a migraine attack (e.g. in association with tyramine or phenylethylamine). Grant (1979) has reported successful therapeutic attempts at the recognition and avoidance of noxae. Furthermore, an increased release of prostaglandin E_1 is postulated and an increase in plasma free fatty acids has been demonstrated. Both these factors could, as part of a chain reaction, result in an increased release of serotonin. Soyka (1985a) proposes a "releasing factor" to which the "key role in the pathogenesis of the migraine attack" is attributed. This factor, the author suggests, may be "effective peripherally on the level of the thrombocytes as well as centrally at the neuronal synapses".

Thrombocytes

Changes in thrombocyte aggregation appear to be of pathogenetic significance. Platelet sensitivity (reactivity of thrombocytes towards aggregation-triggering substances) in migraine patients is raised both in the prodromal phase and in the phase following headache; thrombocyte aggregation peaks in the post-headache phase, concomitant with increased cerebral blood flow (Grötz 1985).

In Summary. A migraine attack develops owing to the interplay of peripheral and central vascular reactions, focal cerebral hypoxia and increased serotonin release by thrombocytes. In spite of a profusion of known releasers (alcohol, effort, psychic and hormonal changes, pregnancy, foodstuffs) and of hereditary factors, on the molecular level a "common factor" has yet to be identified.

4.7.1.3 Mechanism of Action of Calcium Antagonists: Experimental Studies in Animals

Based on the hypothesis (Sect. 4.7.1.2) that hypoxia is a precondition, if not the actual trigger factor, of a migraine attack, the following experimental studies assume a primarily antihypoxic effect for centrally acting calcium antagonists.

During a brief period of ischaemia in the rat cortex, previous administration of flunarizine improved cortical function (measured as cerebral electrical activity and somatosensory evoked potentials; Tolu et al. 1983). Flunarizine reduced the hypoxia-induced electrophysiological functional deficit in the rat cortex (Clincke and Wauquier 1982) as well as the subsequent damage to short-term memory (Clincke and Wauquier 1984). In contrast to verapamil, flunarizine provided in rat brain tissue protection against cytotoxic hypoxia (induced by potassium cyanide; Dubinsky et al. 1984); furthermore, it diminished morphologically detectable cellular damage following severe hypoxia both in vivo (van Reempts et al. 1983) and in vitro (in hippocampus cells; Wauquier et al. 1984). White et al. (1982) showed normal cortical cerebral blood flow (cCBF) in the dog following a brief period of cardiac arrest if flunarizine was promptly injected after the start of reperfusion.

A possible explanation of these results is offered by an antivasoconstrictive effect of flunarizine, which is not synonymous with a vasodilating effect (van Nueten and Janssen 1973 showed no influence on myogenic activity of portal and mesenteric veins).

Without influencing normal reactions of vascular smooth muscle to vasoactive substances or myocardial function, flunarizine specifically reduces the ischaemia-induced increase in transmembrane calcium permeability, thus blocking calcium-dependent vascular contraction as a so-called calcium overload blocker. Vasospasm in the early phase of a migraine attack can thus perhaps be prevented by flunarizine (Amery et al. 1981; Wauquier 1982; Sect. 1.2.5), the effect being particularly marked in the carotid and basilar arteries (van Nueten and Van Houtte 1981). The specific antihypoxic effect in nerve cells is probably due to a membrane-stabilizing effect under hypoxic conditions (Amery et al. 1981). This could be explained by a maximum effect of flunarizine on so-called fast calcium channels of the cell membrane, which in hypoxia induce rapid calcium influx into the cell, so conditioning the intracellular consequences of hypoxia. In this context, nimodipine is said to be the most selective of calcium antagonists (Peroutka et al. 1984; Gelmers 1985). Flunarizine is further held to impede, on the neuronal level, complete depletion of cerebral "energy reservoirs" during epileptic seizures, thereby maintaining minimal cellular function (Amery 1981).

In "spreading depression" (triggered in rats by potassium chloride), the administration of flunarizine resulted in a significant shortening of corticoelectrical recovery time, and thereby also inhibited the propagation of "spreading depression": this could be a further explanation of the mode of action of flunarizine in the prophylactic treatment of migraine (Amery 1981).

Furthermore, flunarizine is said to inhibit excessive calcium inflow into erythrocytes during hypoxia and thereby to counteract reduction of erythrocyte deformability (Holmes et al. 1984; Sect. 3.3.3). The consequent reduction in blood viscosity can improve microcirculation.

Increased histamine release can trigger migraine or, in particular, attacks of cluster headache (inter alia, mast cells are said to release calcium-dependent histamine; Amery 1981). In this context, flunarizine acts not only as a calcium antagonist, but also as a (relatively weak) histamine-1 receptor blocker.

In Summary. Calcium antagonists can intervene in the pathogenesis of migraine via the following mechanisms: protection against hypoxic cellular damage within the central nervous system; inhibition (antagonization) of cerebral vasoconstriction; protection of cerebral "energy reservoirs"; reduction of viscosity owing to maintenance of erythrocyte deformability and weak histamine-1 receptor blockade.

4.7.1.4 Prophylactic Treatment of Migraine: Clinical Studies

To date the calcium antagonists flunarizine, nimodipine, and verapamil have been used in the prophylactic treatment of migraine. For these substances placebo-controlled studies are available; so far only open studies have been conducted with diltiazem. In part, the calcium antagonists were compared directly with other drugs well established in migraine prophylaxis, such as pizotifen and β-receptor blockers.

Drillisch and Girke (1981) reported the first clinical trial (double-blind, not placebo-controlled) with flunarizine (3×3 mg/day) versus cinnarizine in which both substances reduced the frequency and duration of attacks in 30 patients with classical migraine. Flunarizine also lowered the intensity of pain during residual migraine attacks and, based on the patients' opinions, was better tolerated than cinnarizine. Maximum efficacy in migraine prophylaxis appears to be achieved only after regular ingestion over a 3-month period.

In two placebo-controlled, double-blind trials (Louis 1981; Frenken 1982) flunarizine (10 mg daily) was shown to reduce the frequency of migraine attacks by more than half in 83 (72%) patients. While Louis (1981) recommends a treatment duration of 2 months, Frenken (1982) sees a positive effect within 1 month. Neither author demonstrated for flunarizine any influence on the duration or intensity of residual attacks. Louis (1981) observes a tendency for younger patients to respond more favourably to flunarizine than older patients. In comparative studies flunarizine shows the same prophylactic effects as pizotifen (Louis and Spierings 1982; Caers 1982; Rascol 1986).

In all three studies there is a consensus that flunarizine lowers the frequency of attacks: a reduction of 54% in patients under flunarizine, 45% under pizotifen (Louis and Spierings 1982); of 86% under flunarizine and of 60% under pizotifen (Caers 1982); and a reduction of 60% under flunarizine and of 45% under pizotifen (Rascol 1986).

Only Louis and Spierings (1982) report a decrease in the intensity of residual migraine attacks after 2 months' flunarizine treatment. On the other hand, there is unanimous agreement that the minimal duration of treatment should be 2–3 months.

A comparison of 160 mg propranolol with 10 mg flunarizine per day in 54 patients showed significant reduction in the frequency of attacks with both substances after treatment periods of 2 and 4 months, respectively, but no influence on the intensity of attacks (Andersson 1985). According to Smith and Schwartz (1984) diltiazem (dosage 4×60 to 4×90 mg/day) is said to reduce the frequency of attacks in therapy-refractory (despite treatment with β-receptor blockers) migraine. Nimodipine in a dose of 40 mg/day in 60 migraine patients reduced the frequency of attacks by more than half in 69% of the patients (Gelmers 1983; double-blind, placebo-controlled). Here too patients reported milder attacks. Verapamil reduced in 49% and 56% of patients, respectively, the number of attacks by more than half with a dosage of 4×80 mg/day (Solomon et al. 1983; Markley et al. 1984), and a shorter duration of individual attacks was observed. Severe forms of classical and complicated migraine were successfully treated with up to 4 × 160 mg/day. In particular those migraine patients who also suffer from Raynaud's syndrome are said to benefit from nifedipine prophylaxis (3 × 10 mg/day) in respect of both the number and intensity of attacks (Kahan et al. 1983).

In patients with systemic lupus erythematosus and migraine (mostly with visual disturbances) nifedipine prophylaxis led to a reduction in attacks; a dose of 10 mg nifedipine sublingually at the start of an attack relieved pain within 10 min (Miller and Satoro 1984).

In Summary. In the prophylaxis of migraine with calcium antagonists flunarizine (10 mg daily) has been proven effective and can thus take its place alongside well-established medication in prophylactic therapy such as β-receptor blockers, pizotifen, methysergide, amitriptyline and dihydroergotamine (Gelmers 1983). To what extent calcium antagonists, beyond reduction of the frequency of attacks, also reduce the intensity of attacks remains uncertain (Amery 1983). There is minimal evidence for age-dependent effectiveness and tolerance of flunarizine. Nimodipine, cinnarizine and verapamil are indeed also effective, but are associated with a far greater incidence of side effects, in particular affecting the gastrointestinal tract and the cardiovascular system (Chap. 5). Nifedipine is the agent of choice if further vasospasm, e.g. Raynaud's syndrome or Prinzmetal's angina pectoris, is present. In cluster headaches verapamil (up to 3 × 80 mg/day) is said to provide an effective prophylaxis following ingestion over at least 14 days (Pfaffenrath 1988).

At present flunarizine and β-receptor blockers compete with one another in migraine therapy. Among the β-receptor blockers the most effective are propranolol, timolol, metoprolol, nadolol and atenolol, all without ISA (intrinsic sympathomimetic activity; Anthony 1985). They are particularly indicated in migraine prophylaxis when "the stress-protective and anxiolytic effect could be useful" (Soyka 1985 b). Gradual increment in dosage and the extension of treatment over several months are important. The profile of side effects of β-receptor blockers is generally known and the frequency of substance-specific side effects (e.g. practolol syndrome) is extremely low, at least among first-generation β-receptor blockers. These are thus arguments for primary application of β-receptor blockers in migraine prophylaxis. When there are contraindications to the use of β-receptor blockers or when tolerance problems develop, flunarizine should be considered. However, serious side effects have recently been observed with this drug, such as extrapyramidal motor disturbances and depressive reactions (Chouza et al. 1986; Micheli et al. 1989; Sect. 5.3), apparently predominantly among older patients (Sect. 5.3). Flunarizine may also cause slight weight gain, dizziness (avoidable by taking medication before retiring) and constipation. In our opinion, drugs of second choice include pizotifen, methysergide and amitriptyline.

The indication for prophylaxis is at present a controversial issue. Most authors of the studies listed in Sect. 4.7.1.4 institute prophylaxis if there have been at least six attacks in the preceding 6 months, or two attacks per month or one attack per month lasting longer than 4 days with β-receptor blockers (preferably metoprolol) (Pfaffenrath 1988). Although we too regard this as sound, both the duration and intensity of attacks and the patient's subjective burden of suffering must also be taken into account.

4.7.1.5 Prophylactic Treatment in Childhood

Migraine in childhood is not a rare syndrome; 30% of adult migraine sufferers have already experienced symptoms by their tenth year of life (Paulson 1980). Attacks in childhood which last no more than 12 h (Bille 1984) are treated, if at all, with mild analgesics such as acetylsalicylic acid or paracetamol together with an antiemetic (Hockaday 1984). Often, rest and sleep are enough to bring about recession of an attack. Ferriere (1984, personal communication, open study) as well as Sorge and Marano (1985, randomized, placebo-controlled, double-blind trial) tested the effectiveness of 5–10 mg flunarizine daily as migraine prophylaxis in children (average age 10–12 years), in both studies for a period of 3 months. Both investigations demonstrated that flunarizine reduced the frequency of attacks significantly (by 73% in Ferriere, and by 66% in Sorge and Marano 1985), and in some children shortened the duration of residual attacks. Both authors also emphasized that, apart from tiredness and a slight weight gain, no "dangerous" side effects were observed.

In our own view, the indication for prophylactic migraine treatment in children requires rigorous, critical consideration (Hockaday 1984), especially in view of the recently reported extrapyramidal motor side effects with flunarizine (Chouza et al. 1986) and also since long-term results are lacking. Furthermore, it is questionable whether migraine in childhood requires drug treatment at all, since Sorge and Marano (1985) also describe a mean duration of attack of only 3–4 h. If such attacks cannot be tolerated by the child, we recommend a nonmedical treatment such as biofeedback (Werder and Sargent 1984) wherever possible, and if drug prophylaxis is required, treatment with β-receptor blockers should be attempted.

References

Ad Hoc Committee on Classification of Headache (1962) Classification of Headache JAMA 179: 717–718

Amery WK, Wauquier A, van Nueten JM, de Clerck F, van Reempts JV, Janssen PAJ (1981) The anti-migrainous pharmacology of flunarizine (R 14 950), a calcium antagonist Drugs Exptl Clin Res 7: 1–10

Amery WK (1983) Flunarizine, a calcium-channel blocker: A new prophylactic drug in migraine Headache 23: 70–74

Andersson PG (1985) Flunarizine and propranolol in the treatment of migraine – a double-blind study J Neurol (Suppl 232): 220

Anthony M (1984) Was ist Migräne? Münch Med Wschr 126: 1131–1134

Anthony M (1985) Der Einsatz von Betablockern in der Migränebehandlung. Eine Übersicht In: Pfaffenrath V, Sjaastad O, Carroll JD (eds) Migräne und Betablockade Banaschewski München-Gräfelfing pp 70–80

Bille B (1984) Migräne bei Kindern Münch Med Wschr 126: 1149–1151

Blau JN (1984) Towards a definition of migraine headache Lancet 1: 444–445

Cears LJ (1983) Kalziumblocker bei der Behandlung gewöhnlicher und klassischer Migräne In: Soyka D (ed) Migräne Pathogenese-Pharmakologie Therapie Enke Stuttgart pp 141–153
Chouza C, Scaramelli A, Coamano JL, de Medina O, Aljanati R, Romero S (1986) Parkinsonism, tardive dyskinesia,akathisia and depression induced by flunarizine Lancet I: 1303–1304
Clincke GHC, Wauquier A (1982) Protective effects of flunarizine versus verapamil on water intake attenuated by normobaric hypoxia In: Wauquier A et al. (eds) Protection of tissues against hypoxia Elsevier Biomedical Press Amsterdam pp 287–290
Clincke GHC, Wauquier A (1984) Pharmalogical protection against hypoxia-induced effects on medium-term memory in a two-way avoidance paradigm Behavioural Brain Research 14: 139–142
Committee of the International Headache Society (1988) Classification and diagnostic criteria for headache disorders, cranial neuralgia, and facial pain Cephalalgia 8/7 Suppl
Drillisch C, Girke W (1980) Ergebnisse der Behandlung von Migräne-Patienten mit Cinnarizin und Flunarizin Med Welt 31: 1870–1872
Dubinsky B, Sierchio JN, Temple DE, Ritchie DM (1984) Flunarizine and verapamil: effects on central nervous system and peripheral consequences of cytotoxic hypoxia in rats Life Sciences 34: 1299–1306
Fagius J, Sundlöf G, Wallin BG (1985) Die Sympathikusaktivität der muskelversorgenden Nerven (SAMN) unter besonderer Berücksichtigung von Betablockade und Migräne In: Pfaffenrath V, Sjaastad O, Carroll JD (eds) Migräne und Betablockade Banaschewski München-Gräfelfing pp 62–69
Frenken CWGM (1984) Flunarizine, a new preventive approach to migraine. A double-blind comparison with placebo Clin Neurol Neurosurg 86: 17–20
Gelmers HJ (1983) Nimodipine, a new calcium antagonist in the prophylactic treatment of migraine Headache 23: 106–109
Gelmers HJ (1985) Calcium channel blockers in the treatment of migraine Am J Cardiol 55: 139 B–143 B
Grant ECG (1979) Food allergies and migraine Lancet 1: 966–968
Grötz J (1985) Migräne In: Hossmann V, Grötz J, Schrör R (eds) Kalziumantagonisten und zerebrale Erkrankungen Springer Berlin pp 117–147
Hockaday JM (1984) Behandlung der Migräne im Kindesalter Münch Med Wschr 126: 1152–1153
Holmes B, Brogden RN, Heel RC, Speight TM, Avery GS (1984) Flunarizine- a review of its pharmacodynamic and pharmaco-kinetic properties and therapeutic use Drugs 27: 6–44
Kahan A, Weber S, Amor B, Guerin F, Degeorges M (1983) Nifedipine in the treatment of migraine in patients with Raynaud's phenomen N Engl J Med 308: 1102–1103
Karasawa A, Kumada Y, Yamada K, Shato K, Nakamizo N (1982) Protective effect of flunarizine against cerebral hypoxiaanoxia in mice and rats J Pharm Dyn 5: 295–300
Lauritzen M, Jorgensen MB, Diemer NH, Gjedde A, Hansen AJ (1982) Persistent oligemia of rat cerebral cortex in the wake of spreading depression Ann Neurol 12: 469–474
Leao AAP (1944) Spreading depression of activity in cerebral cortex J Neurophysiol 7: 359–390
Louis P (1981) A double-blind placebo-controlled prophylactic study of flunarizine (Sibelium) in migraine Headache 21: 235–239
Louis P, Spierings ELH (1982) Comparison of flunarizine (Sibelium) and pizotifen (Sandomigran) in migraine treatment. A double-blind study Cephalalgia 2: 197–203
Markley HG, Cheronis JCD, Piepho RW (1984) Verapamil in prophylactic therapy of migraine Neurology 34: 973–976
Meyer JS (1985) Calcium channel blockers in the prophylactic treatment of vascular headache Ann Intern Med 102 :395–397
Miller FW, Santoro TJ (1984) Nifedipine in the treatment of migraine headache and amaurosis fugax in patients with systemic lupus erythematosus N Engl J Med 311: 921
Van Nueten JM, Vanhoutte PM (1981) Calcium entry blockers and vasospasm In: Vanhoutte PM, Keusen I (eds) Vasodilatation Raven Press New York pp 459–468
Olesen J, Tfelt-Hansen P, Henriksen L, Larsen B (1981) The common migraine attack may not be initiated by cerebral ischaemia Lancet 2: 438–440

Olesen J, Lauritzen M (1984) Vaskuläre Aspekte in der Pathophysiologie der Migräne Münch Med Wschr 126: 1135–1137
Paulson GW (1980) Migraine headaches in children Pediatr Nurs 6: 41–42
Pfaffenrath V (1988) Was ist gesichert in der Therapie? Der chronische Kopfschmerz – Spannungskopfschmerz und Schmerzmittelmißbrauch Arcis München
Rascol A, Montastruc JL, Rascol O (1986) Flunarizine versus pizotifen : a double-blind study in the prophylaxis of migraine Headache 26: 83–85
Saller R, Berger T, Hellenbrecht D (1982) Behandlung chronischer Kopfschmerzen Der informierte Arzt 3: 33–38
Sicuteri F, Testi A, Anselm B (1961) Biochemical investigation in headache: increase in hydroxyindolacetic acid excretion during migraine attacks Int Arch Allergy 19: 55–58
Sjaastad O, Fredriksen T (1986) Cluster headache syndrome: interrelationship of autonomic phenomena and pain – an editorial. Cephalalgia 6: 3–5
Smith R, Schwartz PHD (1984) Diltiazem prophylaxis in refractory migraine N Engl J Med 310: 1327–1328
Solomon GD, Steel JG, Spaccavento LJ (1983) Verapamil prophylaxis of migraine JAMA 250: 2500–2502
Sorge F, Marano E (1985) Flunarizine versus placebo in childhood migraine. A double-blind study Cephalalgia Suppl 2: 145–148
Soyka D (1978) Entstehungsbedingungen der Migräne Dtsch Med Wschr 103: 1342–1345
Soyka D (1985a) Hypothesen der Migräneentstehung Nervenheilkunde 4: 167–174
Soyka D (1985b) Beta-Rezeptorenblocker bei Migräne Dtsch Med Wschr 110: 185–186
Tolu E, Mameli O, Caria MA, Melis F, Anania V, Desole MS (1983) Protective effects of Ca^{++}-antagonist agents on the cerebral electrical activity during anoxia in the rat Pharmacological Research Communications 15: 833–842
Van Reempts J, Borgers M, van Dael L, van Eyndhoven J, van de Ven M (1983) Protection with flunarizine against hypoxic-ischaemic damage of the rat cerebral cortex: A quantitative morphologic assessment Arch int Pharmacodyn 262: 76–88
Wauquier A (1982) Brain properties of etomidate and flunarizine J Cereb Blood Flow Metabol 2 Suppl 1: S 53–S 56
Wauquier A, van Nueten JM, Ashton D (1984) Die Wirkung von Flunarizin bei tierexperimentellen Ischämie-und Hypoxieversuchen Therapiewoche 34: 5642–5658
Wauquier A, Clincke G, Ashton D, Edmonds HL (1984) Protective and resuscitative properties of flunarizine In: Carciagli F, Giacobini E, Paoletti R (eds) Developmental neuroscience: physiological-pharmacological and clinical aspects Elsevier Science Publishers B V pp 119–127
Werder DS, Sorgent JD (1984) A study of childhood headache using biofeedback as a treatment alternative Headache 24: 122–126
White BC, Gadzinski DS, Hoehner PJ, Krome C, Hoehner T, White JD, Trombley JH (1982) Effect of flunarizine on canine cerebral cortical blood flow and vascular resistance post cardiac arrest Ann Emergency Med 11: 119–126
Wolff HG (1963) Headache and other head pain Oxford University Press New York

4.7.2 Vasospasm Following Subarachnoid Haemorrhage

4.7.2.1 Pathogenesis and Experimental Studies

Subarachnoid haemorrhage, which in most cases results from a ruptured intracranial aneurysm, causes a brief immediate vasospasm of the cerebral vessels, as has been reproduced in animal experiments (Hossmann 1985; Tanaka et al. 1982). This spasm involves in particular the large arteries as well as the pial vessels and resolves spontaneously after a few minutes (Hossmann 1985). Far more important is delayed vasospasm, which can produce secondary deterioration and delayed ischaemic deficits and/or severe disturbances of consciousness (grades I–V, according to Hunt and Hess 1968). These secondary vasospasms occur between the 4th and 14th day in 30% of patients with subarachnoid haemorrhage (Allen 1985a), the greatest incidence occurring on the 7th day (Fisher et al. 1977; Allen and Bahr 1979; Editorial Lancet 1985). Further factors favouring the development of cerebral ischaemia due to vasospasm following subarachnoid haemorrhage include: hydrocephalus; cerebral oedema; electrolyte imbalance; dehydration; raised blood viscosity; and respiratory disturbances. Neurosurgical interventions during this period increase the risk to the patient (Editorial Lancet 1985). As a result of this permanent tonic vasoconstriction of the cerebral arteries, morphological changes develop in the vascular smooth muscle cells within a few days (Allen 1985a; Weir et al. 1985); these can induce permanent vascular constriction without any additional influence by vasoactive substances.

Animal experimental data (Tanaka et al. 1982; Weir et al. 1985) as well as clinical findings (computerized tomography data: Bell 1980; Allen 1985a) consistently attest that the volume of blood in the basal subarachnoid cisterns and the degree of intraventricular haemorrhage due to rupture of an aneurysm determine the extent of subsequent secondary vasospasm. This supports the hypothesis that one or several substances from extravasated blood cause this secondary vasospasm (animal experimental investigation: Simeone and Vinall 1975; Tanaka et al. 1982; Weir et al. 1985). Further studies deal in vitro with potential releasing factors of vasospasm, such as catecholamines, serotonin, prostaglandins and metabolites of haemoglobin, and of their effect on the cerebral vessels. In this context, oxyhaemoglobin produces a particularly potent vasoconstrictive effect (Suzuki 1979), especially in association with serotonin (Hossmann 1985).

Tanaka et al. (1982) postulate that the haemorrhage triggers regulatory dysfunction in noradrenergic nerve endings of cerebrovascular smooth muscle.

Allen et al. (1976) demonstrated a strong contraction in the canine basilar artery when the tissue was incubated with CSF taken from patients a few days following subarachnoid haemorrhage. Here, by means of the chromatographic separation technique, dose-dependent vasoconstriction by serotonin was observed.

Towert (1981) was able to demonstrate in isolated basilar arteries of the rabbit that a serotonin-induced contraction could be annulled by nimodipine. This antivasoconstrictive effect of nimodipine does not apply to the femoral artery: nimodipine reacts selectively with the special receptors of cerebral vascular smooth muscle cells (Allen 1985a). Using nimodipine, as well as nifedipine in equivalent dosage, Allen and Bahr (1979) were able to prevent, or at least to resolve, early and late vasospasm following experimentally induced subarachnoid haemorrhage in the dog. Here nimodipine was effective even in nanomolar concentrations.

Weir et al. (1985) were unable to confirm these results in animal experimental subarachnoid haemorrhage. They ascribed this to the fact that they had given nimodipine only every 8 h, in contrast to the 4-h intervals adopted by Allen (1985a).

In Summary. Animal experimental studies indicate the efficacy of dihydropyridines in cerebral vasospasm following subarachnoid haemorrhage.

4.7.2.2 Clinical Studies and Therapy of Vasospasm

The first important clinical investigation demonstrating the efficacy of nimodipine in the prophylaxis of vasospasm following subarachnoid haemorrhage was conducted by Allen et al. (1983). This study (prospective, randomized, placebo-controlled, five centres) involved 125 patients without focal neurological deficits following aneurysmal haemorrhage (diagnosed by computerized tomography or lumbar puncture). Within 96 h of the haemorrhage they were treated with nimodipine (initial dosage 0.7 mg per kilogram body weight, thereafter 0.35 mg/kg

Table 28. Results of Allen et al. (1983): Classification of patients with deficits from spasm, according to neurologic outcome

Outcome	Placebo ($n=60$)	Nimodipine ($n=56$)
Normal	6	8
Neurological deficits		
Mild or moderate	2	4
Severe	8 (3 deaths)	1 (death)

every 4 h for 21 days). Neurological status was assessed after 21 days in both patient groups (placebo group and nimodipine group). The results are summarized in Table 28.

Allen et al. conclude that under the given conditions nimodipine provides an effective prophylaxis against vasospasm. No side effects attributable to nimodipine were observed. However, the authors emphasize that the validity of the study is somewhat marred by the small number of patients with delayed ischaemic deficits in both groups. The dosage, which was established from experimental data (determination of density of dihydropyridine receptors; Cohen and Allen 1980), could, in the opinion of the authors, be further increased, in order to prevent residual severe vasospasm. To test their hypothesis, Allen et al. (1985b) are planning a new and more ambitious study, with the same criteria as in the preceding one, with the exception of the dosage (new study: patient groups receive 30, 60 or 90 mg nimodipine every 4 h). Critical comments on the study by Allen et al. (1983) emphasize, for instance, that the time at which neurosurgical interventions were undertaken in both placebo and treatment groups is not indicated, that initial neurological status is not precisely defined and that the small number of cases involved does not permit an unambiguous conclusion to be drawn (Editorial Lancet 1983; Vermeulen et al. 1983).

In a further prospective (but not placebo-controlled) study (Koos et al. 1985) 91 patients with manifestations of neurological deficits (84% of patients in grades III–V, according to Hunt and Hess 1968) were treated with nimodipine. In comparison with the clinical data of earlier studies, efficacy was demonstrated in already symptomatic vasospasm, earliest possible initiation of treatment appearing to be important.

Other open studies confirm these results (efficacy of nimodipine in the prophylaxis of vasospasm and in established vasospasm: Kazner 1985; Auer et al. 1985). Recent studies (placebo-controlled, double-blind, randomized) have shown that nimodipine can be effective in preventing delayed ischaemic deficits due to cerebral vasospasm, but not in preventing vasospasm per se (Philippon et al. 1986; Jan et al. 1988; Petruk et al. 1988; Öhman and Heiskanen 1988). Petruk and co-workers (1988) found in 188 poor-grade aneurysm patients (grades III–V according to Hunt and Hess), a significant improvement in the successful outcome rate and a reduction in delayed ischaemic deficits in both grade III and IV patients of the nimodipine group; no significant difference between nimodipine and placebo in grade V patients, and no significant difference in the incidence of moderate or severe diffuse vasospasm. Thus, the authors conclude that the effect of nimodipine results through a mechanism other than prevention of large vessel spasm.

Grossman et al. (1984) showed in a placebo-controlled, double-blind study involving 58 patients with subarachnoid haemorrhage that treatment with 4×80 mg verapamil daily results in fewer vasospasms and improved outcome of neurological deficits. Two patients manifested bradycardia which led to interruption of therapy.

In Summary. The efficacy of nimodipine in secondary vasospasm following subarachnoid haemorrhage has so far been documented in placebo-controlled and open studies with varying preconditions: in patients with different grading according to Hunt and Hess; for prophylaxis of vasospasm and in established vasospasm. In open studies demonstration of the efficacy of nimodipine is based entirely on comparison with previously published data.

Nimodipine possibly reduces ischaemic deficits without influencing the occurrence of vasospasm. In randomized studies significant reduction of ischaemic deficits was proven only in subgroups (patients with early surgical intervention, grades IV–V according to Hunt and Hess, evaluated after 3 and 6 months, diminished mortality following vasospasm). Verapamil, because of its pharmacological profile and because of scant available data, is only an agent of second choice in this indication.

References

Allen GS, Gross CJ, French LA, Chou SN (1976) Cerebral arterial spasm. Part 5. In vitro contractile activity of vasoactive agents including human CSF on human basilar and anterior cerebral arteries J Neuro Surg 44: 594–600

Allen GS, Bahr AL (1979) Cerebral arterial spasm. Part 10. Reversal of acute and chronic spasm in dogs with orally administered nifedipine Neurosurgery 4: 43–47

Allen GS, Ahn HS, Preziosi TJ, Battye R, Boone SC, Chou SN, Kelly DL, Weir BK, Crabbe RA, Lavik PJ, Rosenbloom SB, Dorsey FC, Ingram CR, Mellits DE, Bertsch LA, Boisvert DPJ, Hundley MB, Johnson RK, Strom JA, Transou CR (1983) Cerebral arterial spasm – a controlled trial of nimodipine in patients with subarachnoid hemorrhage N Engl J Med 308: 619–624

Allen GS (1985a) Role of calcium antagonists in cerebral arterial spasm Am J Cardiol 55: 149 B-153 B

Allen GS (1985b) Cerebral arterial spasm – controlled trials of nimodipine in subarachnoid hemorrhage patients In: Betz E, Deck K, Hoffmeister F (eds): Nimodipine – pharmacological and clinical properties Schattauer Stuttgart pp 397–407

Auer LM, Reulen HJ, Gilsbach J, Oppel F, Gröger U, Harders A (1985) Prevention of brain infarction from arterial spasm by acute aneurysmal surgery and preventive nimodipine treatment in patients with subarachnoid hemorrhage In: Betz E, Deck K, Hoffmeister F (eds): Nimodipine – pharmacological and clinical properties Schattauer Stuttgart pp 739–407

Bell BA, Kendall BE, Symon L (1980) Computerized tomography in aneurysmal subarachnoid hemorrhage J Neurol Neurosurg Psychiatry 43: 522–524

Editorial Lancet (1983) Calcium antagonists and aneurysmal subarachnoid hemorrhage Lancet 2: 141–143

Fisher CM, Roberson GH, Ojemann RG (1977) Cerebral vasospasm with ruptured saccular aneurysm: the clinical manifestations Neurosurgery 1: 245–248

Glossmann H, Goll A, Rombusch M, Ferry DR (1985) Molecular pharmacology of Ca^{++}-channels: receptor binding studies In: Betz E, Deck K, Hoffmeister F (eds) Nimodipine pharmacological and clinical properties Schattauer Stuttgart pp 57–73

Grossmann W, Beck O, Gerstenbrand F, Paal G, Rumpl E, Voth D (1984) Doppelblinduntersuchung über den prophylaktischen Einsatz eines Kalziumantagonisten (Verapamil) gegen Vasospasmen nach Subarachnoidalblutung (vorläufige Ergebnisse) In: Grossmann W (ed) Symposium „Der zerebrale Angiospasmus" 4.–6. Okt Mainz 1983 de Gruyter Berlin pp 501–507

Hossmann V (1985) Subarachnoidalblutung In: Hossmann V, Grötz J, Schrör K (eds) Kalziumantagonisten und zerebrale Erkrankungen Springer Berlin pp 41–62

Hunt WE, Hess RM (1968) Surgical risk as related to time of intervention in the repair of intracranial aneurysms J Neurosurg 28: 14–20

Jan M, Buchheit F, Tremoulet M (1988) Therapeutic trial of intravenous nimodipine in patients with established cerebral vasospasm after rupture of intracranial aneurysms Neurosurgery 23: 154–157

Janis RA, Bellemann P, Sarmiento JG, Triggle DJ (1985) The dihydropyridine receptors In: Fleckenstein A, van Breemen C, Groß R, Hoffmeister F (eds) Cardiovascular effects of dihydropyridine-type calcium antagonists and agonists Springer Berlin pp 140–155

Kazner E (1985) Clinical experience with nimodipine in the prophylaxis of neurological deficits after subarachnoid hemorrhage Neurochirurgia 28: 110–113

Koos WT, Auer L, Böker DK, Fuhrmeister U, Gaab M, Jaksche H, Kostron H, Meinig G, Muizelaar JP, Perneczky A, Seibert HK, Sprung CH, Ulrich F, van der Werf AJM, Dycka J, Tettenborn D (1985) Treatment of neurological deficits from cerebral vasospasm after subarachnoid hemorrhage with nimodipine (Nimotop) In: Betz E, Deck K, Hoffmeister F (eds) Nimodipine – pharmacological and clinical properties Schattauer Stuttgart pp 387–395

Öhman J, Heiskanen O (1988) Effect of nimodipine on the outcome of patients after aneurysmal subarachnoid hemorrhage and surgery J Neurosurg 69: 683–686

Petruk KC, West M, Mohr G, Weir KA, Benoit BG, Gentili F, Disney LB, Khan MI, Grace M, Holness RO, Karwon MS, Ford RM, Cameron GS, Tucker WS, Purves GB, Miller DR, Hunter KM, Richard MT, Durity FA, Chan R, Clein LJ, Maroun FB, Godon A (1988) Nimodipine treatment in poor-grade aneurysm patients J Neurosurg 68: 505–517

Philippon J, Grob R, Dagreou F, Guggiari M, Rivierez M, Viars P (1986) Prevention of vasospasm in subarachnoid haemorrhage. A controlled study with nimodipine Acta Neurochir 82: 110–114

v.Planta M, Ritz R (1986) Konservative Prinzipien in der Therapie der Subarachnoidalblutung Dtsch Med Wschr 111: 1333–1337

Simeone FA, Vinall P (1975) Mechanism of contractile response of cerebral artery to externally applied fresh blood J Neurosurg 43: 37–47

Suzuki J (1979) Cerebral vasospasm: Prediction, prevention and protection In: Pia HW, Langmaid C, Zierski J (eds) Cerebral aneurysms. Advances in diagnosis and therapy Springer Berlin pp 203–208

Tanaka K, Gotoh F, Muramatsu F, Fukuuchi Y, Okayasu H, Suzuki N, Kobari M (1982) Effect of nimodipine, a calcium antagonist, on cerebral vasospasm after subarachnoid hemorrhage in cats Arzneimittelforschung 32 (II): 1529–1534

Towart R (1981) The selective inhibition of serotonin-induced contractions of rabbit cerebral vascular smooth muscle by calcium antagonistic dihydropyridines. An investigation of the mechanism of action of nifedipine Circ Res 48: 650–657

Vermeulen M, van Gijn J, Hijdra A, Wijdicks EFM (1985) Nimodipine for cerebral arterial spasm in subarachnoid hemorrhage N Engl J Med 309: 437

Weir B, Espinosa F, Nosko M, Norris S, Cook D, Krueger C (1985) Development of a primate model of chronic vasospasm and delayed ischemic deficit – attempts at modification using nimodipine In: Betz E, Deck K, Hoffmeister F (eds) Nimodipine – pharmacological and clinical properties Schattauer Stuttgart pp 375–377

4.7.3 Calcium Antagonists in Cerebral Ischaemia and in Cardiopulmonary Resuscitation

4.7.3.1 Pathophysiology and Animal Studies

In the context of cerebral ischaemia, as with other central nervous system syndromes, two mechanisms of calcium antagonist action are discussed:

1. Calcium antagonists are postulated to act as antivasoconstrictors via inhibition of calcium inflow into vascular smooth muscle cells, thereby counteracting the postischaemic hypoperfusion (Ames et al. 1968) that follows cerebral ischaemia (Hossmann and Hossmann 1973; White et al. 1983; Hossmann 1985).
2. Calcium antagonists are further postulated to inhibit the inflow of calcium into damaged nerve cells which, in the final analysis, follows hypoxia, and to protect the cells from the consequences of calcium overload in the extracellular space, mitochondria and endoplasmic reticulum. (Discussions of the possible precise pathobiochemical course of hypoxic damage are provided by White et al. 1983; Borges and van Reempts 1984; Hossmann 1985; see also Sect. 4.7.1.2 and Fig. 38.)

There exists a profusion of animal studies which assess neurological deficits, cerebromorphological changes or cerebral blood flow (CBF). Since different methodologies are employed, these studies are difficult to compare. Most authors conclude, however, that flunarizine improves cerebral blood flow following complete ischaemia (van Nueten 1982; White et al. 1982; Holmes 1984; see also Sect. 4.7.1.2), and perhaps even raises it (White et al. 1983). Despite using the

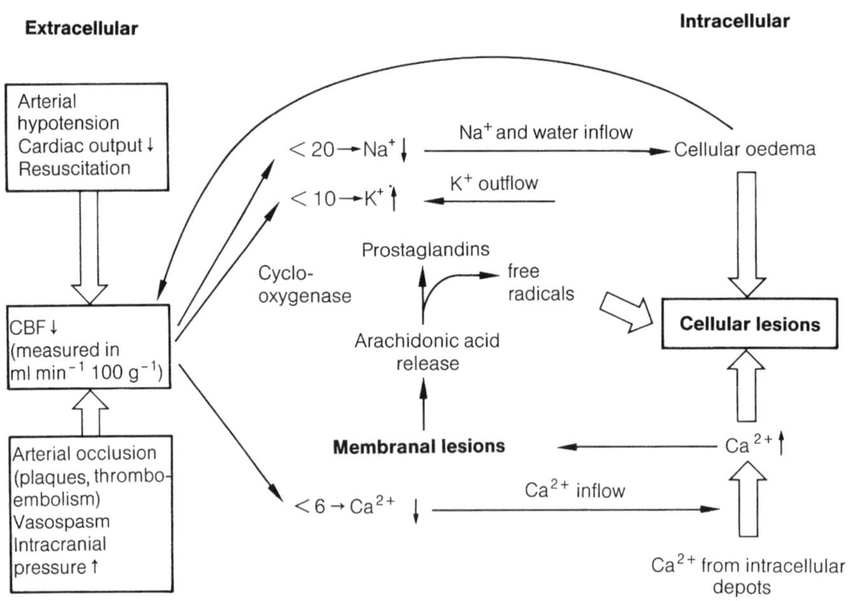

Fig. 38. Simplified scheme of cerebral blood flow (*CBF*) and cellular damage resulting from hypoxia. (Measured values after Dearden 1985)

same methodology as White et al. (1983), Newberg et al. (1984) working in the dog were unable to find any improvement with flunarizine in CBF or in neurological deficits following complete ischaemia (in individual cases flunarizine even led to pulmonary oedema). When flunarizine was replaced by nimodipine, however, significant improvement of neurological deficits was recorded (Steen and Newberg et al. 1983; Kazda et al. 1982). The general conclusion can be drawn that nimodipine probably reduces the postischaemic phase of hypoperfusion (Kazda et al. 1982; Hossmann 1985).

4.7.3.2 Clinical Studies

There are only a few available clinical investigations relating to the use of calcium antagonists in cerebral ischaemia and cardiopulmonary resuscitation. The effect of nifedipine on CBF in various neurological syndromes was first investigated by Schmitz et al. (1975). Here autoregulation was unaffected and there was no occurrence of "steal phenomena". In an open study, CBF (measured by the ^{133}Xe inhalation technique) in patients with acute myocardial infarction was raised following ingestion of nimodipine (15–30 μg/kg i.v.) in comparison with data recorded in the literature (Gelmers 1985). In 60 patients with ischaemic cerebral infarction (31 in the control group, 29 in the nimodipine group; nonrandomized pilot study), nimodipine proved favourable in restoration of neurological function, as measured by the Matthew Scale (Gelmers 1985). This was confirmed in a placebo-controlled double-blind study (multicentre) involving 160 patients with stroke in which there was a mortality of 8% in the nimodipine group (dosage 120 mg daily) against 21% in the placebo group (Gelmers et al. 1987).

Preliminary results in a further placebo-controlled double-blind trial with nimodipine (randomized; Paci et al. 1985) have so far revealed no significant differences between placebo and nimodipine groups (dosage 40 mg nimodipine daily); here patients are treated with nimodipine within 12 h of a cerebral infarction and are neurologically examined at first daily and thereafter at weekly intervals, assessment being based on the Matthew Scale (duration of treatment 28 days). A study of intravenous flunarizine in stroke is in preparation (2 × 25 mg i.v. for 7 days, then 30 mg orally for 3 weeks, then 10 mg each evening for 1 year), where comparison will be made with haemodilution and acetylsalicylic acid (Peters and Sallers 1987).

Despite the postulated theoretical model of action of calcium antagonists in cerebral hypoxia and following resuscitation, and some success in animal studies, clinical effectiveness has not been confirmed in practice. Although the pilot study and the multicentre investigation by Gelmers (1985, 1987) provide positive indicators in favour of nimodipine (improvement of neurological function following cerebral infarction), there has to date been no convincing placebo-controlled double-blind study following either resuscitation or cerebral infarction which, according to precise neurological scaling, demonstrates any significant advantage over placebo. Provisional results from Paci et al. (1983) raise the possibility that calcium antagonists in association with barbiturates, indomethacin or free-radical-binding medicaments may provide greater efficacy.

References

Ames A, Wright RL, Kowada M (1968) Cerebral ischemia II: The no-reflow-phenomenon Am J Pathol 52: 437

Borgers M, Van Reempts J (1984) The distribution of calcium in normal and ischemic brain tissue. Clinical Research Reviews 4: 70–71

Dearden NM (1985) Ischaemic brain Lancet 2: 255–257

Gelmers HJ (1985) Effect of nimodipine on the clinical course of patients with acute ischemic stroke In: Betz E, Deck K, Hoffmeister F (eds) Nimodipine – pharmacological and clinical properties Schattauer Stuttgart pp 467–471

Gelmers HJ, Gortner K, De Weerdt CJ, Wiezer HJA (1987) New aspects in stroke therapy International Symposium on Calcium Antagonists The New York Academy of Sciences New York p 100 (Abstract)

Hossmann V, Hossmann KA (1973) Return of neuronal functions after prolonged cardiac arrest Brain Res 60: 423–438

Hossmann V (1985) Zerebrale Ischämie In: Hossmann V, Grötz J, Schrör K (eds) Kalziumantagonisten und zerebrale Erkrankungen Springer Berlin pp 63–88

Kazda S, Garthoff B, Krause HP, Schlossmann K (1982) Cerebravascular effects of the calcium-antagonist dihydropyridine derivate nimodipine in animal experiments Arzneimittelforschung 32: 331

Newberg LA, Steen PA, Milde JH, Michenfelder JD (1984) Failure of flunarizine to improve cerebral blood flow or neurologic recovery in a canine model of complete cerebral ischemia Stroke 15: 666–671

Paci A, Ottaviano P, Trenta A, Jannone G, Acuto G, Carosi M (1985) Treatment of acute ischemic cerebral infarction with nimodipine (Baye 9736): totally randomized double-blind placebo-controlled study – preliminary results In: Betz E, Deck K, Hoffmeister F (eds): Nimodipine – pharmacological and clinical properties Schattauer Stuttgart pp 473–477

Peters T, Seiler KU (1987) New opportunities with a calcium overload blocker in brain protection? International Symposium on Calcium Antagonists The New York Academy of Sciences New York p 109 (Abstract)

Schmitz H, Schierl W, Beck O, Lydtin H (1975) Über die Wirkung von Nifedipine (Adalat) auf regionale Hirndurchblutung und Unterschenkeldurchblutung In: Schlegel B (ed) Verhandlungen der Deutschen Gesellschaft für Innere Medizin Bergmann München pp 888–890

Steen PA, Newberg LA, Milde JH, Michenfelder JD (1983) Nimodipine improves cerebral blood flow and neurologic recovery after complete cerebral ischemia in the dog J Cereb Blood Flow Metabol 3: 38–43

White BC, Winegar CD, Wilson RF, Hoehner PJ, Trombley JH (1983) Possible role of calcium blockers in cerebral resuscitation: a review of the literature and synthesis for future studies Critical Care Medicine 11: 202–207

4.7.4 Vertigo

The statement that "vertigo is not a defined disease entity, but a multisensory syndrome of varying aetiologies and pathogenetic origins" (Brandt 1985) reflects the central problem in the design and assessment of clinical studies relating to the efficacy of calcium antagonists, in particular flunarizine and cinnarizine, in the treatment of vertigo. (Table 29 provides a diagnostic classification of vertigo and the corresponding forms of nystagmus.)

Preliminary results relating to the effect of flunarizine on vestibular functions stem from Oosterveld (1974). Flunarizine lowers (dose-independently between 20 and 40 mg per kilogram body weight) the nystagmus response by 20% (survey of further experimental studies of vertigo by Holmes et al. 1984).

Both in volunteers and in patients with vertigo (without precise diagnostic classification) several investigators have found a reduction of angular velocity in the slow phase of vestibular nystagmus under flunarizine treatment (Oosterveld 1982). The authors thus propose that flu-

Table 29. Diagnostic classification of vertigo (after Brandt and Daroff 1980; Brandt 1985) and the correlation to the corresponding nystagmus syndromes treated with flunarizine or cinnarizine

	Syndrome	Nystagmus
Peripheral vestibular disorders	Acute unilateral vestibular deficit	Spontaneous or fixation nystagmus towards healthy ear
	Menière's disease	Spontaneous nystagmus towards healthy or deficit ear
Peripheral or central vestibular disorders	Vertebrobasilary – brain stem ischaemia – basilar artery migraine	Peripheral: vestibular deficit Central: vertical gaze nystagmus or spontaneous nystagmus with high frequency, low amplitudes
Central vestibular disorders	Vestibular epilepsy	Contraversive nystagmus

narizine (or cinnarizine) exhibits a "vestibular-depressant" (vestibular-function-inhibiting) action, although the precise mechanism of action remains obscure. As a histamine antagonist, flunarizine could block histamine-specific receptors in the vestibular system, or alternatively influence, via inhibition of calcium inflow, the excitability of vestibular neurones (Hofferberth 1980).

Ell and Gresty (1983) observed no change in the slow nystagmus phase in volunteers and three patients with central positional vertigo under flunarizine treatment, but found a reduction in amplitude in the rapid phase of vestibular nystagmus. Since the rapid phase is thought to be centrally conditioned, the authors surmise a central focus of effect of flunarizine (suppression of type II vestibular neurones in the brain stem by flunarizine, leading to smaller amplitudes, more frequent saccades and thus a relative resolution of vestibular lateral symmetry, with consequent recession of subjective sensations of dizziness (for details see Ell and Gresty 1983). The authors thus regard as justifiable the use of flunarizine in patients exhibiting laterally asymmetrical vestibular response, whether of peripheral or of central origin, in the ENG (electronystagmogram). They criticize studies by other authors on the grounds that dizziness occurring during the first few days of flunarizine treatment is not considered and that an ENG investigation was made at the onset of treatment; at that time, a lower degree of concentration could explain (Ell and Gresty 1983) the reduction in the speed of the slow phase and the shorter duration of nystagmus during treatment of flunarazine (Oosterveld 1974; Boniver 1979).

Several clinical studies (mainly double-blind and crossover) conclude that flunarizine has a positive effect in patients suffering from vertigo (Boniver 1979, 50 patients, 31 patients, 18 patients, initial dosage 40 mg flunarizine, then, after 4th week of treatment, 10 mg flunarizine daily; Hofferberth 1980, 2 patients, dosage 20 mg daily; Oosterveld 1982, 41 patients, dosage 10 mg daily; Materna 1985, 30 patients, dosage 1st week 30 mg, 2nd week 20 mg, 3rd and 4th week 10 mg daily). It must here be noted that patients with vertigo of different aetiologies are compared and that treatment success is inadequately documented (i.e. no ENG control).

Studies with cinnarizine reach the same result with regard to the symptom of vertigo, and the same reservations apply (Mangabeira-Albernaz et al. 1978, 58 patients, dosage 50 mg cinnarizine daily; Amery and Oosterveld 1980, 82 patients, dosage 25 mg cinnarizine daily).

Reviews of clinical studies with flunarizine and/or cinnarizine are provided by Wouters et al. (1983), Holmes et al. (1984), Hofferberth (1984) and Hossmann (1985).

In Summary. Flunarizine (average dosage 2×10 mg daily) or cinnarizine (25 mg daily) favourably influence some, particularly peripheral, forms of vertigo. There are, however, no studies available which record which patients actually benefit from calcium antagonists, as based on precise diagnostic criteria. Such studies, in addition to further animal data, would lead to a clarification of the mechanism of action. It also remains unclear whether flunarizine acts only as a sedative at the onset of therapy, thus reducing the speed of the slow phase of vestibular nystagmus, and has in general a central locus of attack, or whether the frequently mentioned "vestibular-depressant effect" (Oosterveld 1974) plays the decisive role. There is, however, a consensus that, if vertigo symptoms are positively influenced, the effect is dose-unrelated above a daily dose of 20 mg flunarizine.

In general, the establishment of a clear indication for the use of flunarizine or cinnarizine in the treatment of vertigo should be based on rigorous criteria, since side effects such as dizziness, fatigue, gastrointestinal complaints and extrapyramidal motor disturbances (Sect. 5.3) can occur under treatment with these agents.

References

Amery WK, Oosterveld WJ (1980) Cinnarizine in the treatment of vertiginous complaints associated with old age R Soc Med 33: 43–47

Boniver R (1979) Vertigo, particularly of vascular origin, treated with flunarizine (R 14950) Acta Oto-Rhino-Laryngology Belgica 33: 270–281

Brandt T (1985) Differentialdiagnose klinischer Schwindelformen Münch Med Wochenschr 127: 1137–1140

Brandt T, Daroff RB (1980) The multisensory physiological and pathological vertigo syndromes. Ann Neurol 7: 195–203

Ell J, Gresty M (1983) The effects of the "vestibular sedative" drug, flunarizine upon the vestibular and oculomotor system J Neurol Neurosurg Psychiatry 46: 716–724

Hofferberth B (1980) Die Anwendung von Flunarizin bei Patienten mit vertebrobasilärer Insuffizienz Arzneimittelforschung 30: 1817–1819

Hofferberth B (1984) Kalzium-Blocker in der Therapie der vertebrobasilären Insuffizienz Fortschr Med 102: 725–728

Holmes B, Brodgen RN, Heel RC, Speight TM, Avery GS (1984) Flunarizine – a review of its pharmacodynamic and pharmacokinetic properties and therapeutic use Drugs 27: 6–44

Hossmann V (1985) Zerebrale Ischämie. In: Hossmann V, Grötz J, Schrör K (eds) Kalziumantagonisten und zerebrale Erkrankungen. Springer Berlin pp 63–88

Mangabeira-Albernaz PL, Ganaca MM, Novo F, Rodrigues de Paiva E (1978) Flunarizine and cinnarizine as vestibular depressants A statistical study J Oto-Rhino-Laryngology Borderlands 40: 92–100

Materna F (1985) Leitsymptom Schwindel: Vergleichsstudie mit Flunarizin und einem α-Adrenolytikum Med Prax 80: 3–6

Oosterveld WJ (1974) Vestibular pharmacology of flunarizine compared to that of cinnarizine Oto-Rhino-Laryngology Borderlands 36: 157–161

Oosterveld WJ (1980) Cinnarizine in the vertinginous syndrome R Soc Med 33: 29–37

Oosterveld WJ (1982) Flunarizine in vertigo: a double-blind placebo-controlled cross-over evaluation of a constant-dose schedule ORL 44: 72–80

Schwerdtfeger F (1978) Disturbed vestibular function: results with flunarizine in a 2-phase trial against placebo Ärztliche Praxis 30: 515–519

Wouters L, Amery W, Towse G (1983) Flunarizine in the treatment of vertigo J Laryngol Otol 97: 697–704

4.7.5 Epilepsy

Animal experiments concerning the anticonvulsant action of calcium antagonists have been largely carried out with flunarizine. In experimentally induced cerebral spasm in the rat, dihydropyridine derivatives, in addition to flunarizine, also exhibit anticonvulsant properties (nifedipine and nitrendipine, Van Wieringen et al. 1986; nimodipine, Meyer et al. 1987).

4.7.5.1 Mechanism of Action of Flunarizine

In comparing antiepileptic substances most investigators conclude that flunarizine probably acts similarly to phenytoin or carbamazepine (Holmes et al. 1984; Ashton et al. 1985; Wauquier et al. 1985; Van Wieringen et al. 1986).

In this situation flunarizine does not reduce burst potential via inhibitory neurones of the GABA (γ-aminobutyric acid) system, but rather decreases the amplitude, not the frequency, of spontaneous neuronal discharge (Ashton et al. 1985). Furthermore, it increases the threshold of the electrically induced "spreading depression" (Sect. 4.7.1.2) and is said to block the receptors responsible for activating sodium channels in the cell membrane (Van Wieringen et al. 1986). The active sodium and calcium channels are considered to be of importance in the development of spontaneous neuronal discharge, the synchronization of which, in the final analysis, brings about epilepsy-specific potentials; a calcium-dependent potassium potential determines hyperpolarization following a discharge and prevents renewed repolarization, and thereby further discharge (Alger and Nicoll 1980).

In Summary. Flunarizine reduces the tendency to burst firing, which arises via electrical changes in the cell membrane (via the sodium and calcium channels; Ashton et al. 1985). In animal experiments, further seizures could not be induced, even 24 h after a dose of flunarizine (Clincke and Wauquier 1979; in rats, with the metrazol test); these results are in agreement with pharmacokinetic data on flunarizine (see Sect. 2.2.4, details of animal experiments in Ashton et al. 1985; Van Wieringen et al. 1986).

4.7.5.2 Clinical Studies

Clinical data concerning flunarizine as an anticonvulsant are at present limited to studies of baseline antiepileptic medication with the additional administration of flunarizine (mainly in chronic therapy-resistant seizures). A summary of the most important clinical studies is given in Table 30; it must be emphasized that the plasma level of antiepileptic agents is unaffected by flunarizine (Mancia et al. 1986), but that the flunarizine plasma level itself is somewhat lower than in patients without antiepileptic treatment (Overweg et al. 1984; Van Wieringen et al. 1986) owing to hepatic enzyme induction by the antiepileptics (Binnie et al. 1985). (A discussion of the drug interaction between verapamil and carbamazepine is to be found in Sect. 5.8.) Under flunarizine treatment, the aura of complex–partial seizures was completely absent in some patients: there occurred an immediate secondary generalization of the seizures. The reason for this phenomenon is unknown (Van Wieringen et al. 1986; Binnie 1987). Analysis of available data shows that patients with partial epilepsies and complex–partial seizures, in particular, appear to benefit from supplementary treatment with flunarizine, although they may also be at risk by the nonappearance of the aura. In their

Table 30. Long-term studies (3 × 3 months) of anticonvulsant supplementary (add-on) treatment with flunarizine. (Baseline treatment with anticonvulsant agents unchanged)

Study	Method	*n*	Type of seizure	Flunarizine (dose per day)	Results
De Clerck and Wauquier 1987	db pc	20	tr epilepsy (all types of s)	10 mg	Overall fewer s; 3 patients free from s
Overweg et al. 1984	db pc	30	tr epilepsy (particularly complex-partial s)	10 mg	Reduction of complex-partial s in 20 patients (sig)
Binnie et al. 1985	db pc co Open dose	47	Complex-partial s	10–25 mg increasing dosage to normal values	Reduction of s greater than 50% in 16 patients (sig)
Mancia et al. 1986	db pc co	90	Chronic epilepsy complex-partial s (61 patients) grand mal (23 patients)	10 mg (up to 70 kg) 15 mg (over 70 kg)	Slight reduction in s (sig)
Fröscher et al. 1986	db pc co	16	Particularly complex-partial s	20 mg for 3 weeks, then 15 mg	Tendency to reduction in s (not sig)
Gigli et al. 1987	db pc co	16	tr epilepsy	10 mg	No reduction in s

s seizures; co crossover; db double-blind; *n* number of patients; pc placebo-controlled; tr therapy-resistant; sig significant.

studies involving various dosages, Binnie et al. (1985; see Table 30) found that the best results were obtained with a regimen of 15–20 mg flunarizine; however, this dosage was also accompanied by the most side effects (Sect. 5.3).

On the other hand, more recent results by Gigli et al. (1987) have again brought the efficacy of flunarizine treatment into question; in this study no significant difference could be observed in respect to frequency of seizures between the flunarizine and placebo groups. To date there have been no studies designed to test the efficacy of flunarizine as monotherapy in epilepsy. (Reviews on flunarizine as supplementary add-on therapy: Vanden Bussche et al. 1985; Ashton et al. 1985; Van Wieringen et al. 1986; Schmidt and Ried 1989).

References

Alger BE, Nicoll RA (1980) Epileptiform burst after hyperpolarization: calcium dependent potassium potential in hippocampal CAL pyramidal cells Science 210: 1122–1124

Ashton D, Marrannes R, Pauwels PJ, Van Belle H, Reid K, Wauquier A (1985) Possible mechanisms of flunarizine's anti-convulsant activity International Workshop on Flunarizine Epilepsy Beerse Belgium pp 2–31

Binnie CD, de Beukelaar F, Meijer JWA, Meinardi H, Overweg J, Wauquier A, van Wieringen A (1985) Open dose-ranging trial of flunarizine as add-on therapy in epilepsy Epilepsia 26: 424–428

Binnie CD (1987) Clinical trials of flunarizine in epilepsy International Symposium on Calcium Antagonists The New York Academy of Sciences New York pp 97 (abstract)

Clincke G, Wauquier A (1979) Metrazol-produced impairment of passive avoidance retention specifically antagonised by anti-petit mal drugs Psycho Pharmacol 66: 234–236

De Clerck AC, Wauquier A (1978) Double-blind study of the effectiveness of flunarizine in therapy resistant epilepsy in mentally retarded children 10th Epilepsy International Symposium Vancouver pp 169

Fröscher W, Bülau P, Burr W, Kiefer H, Kreiten K, Rao ML, Penin H (1986) Double-blind placebo-controlled trial of flunarizine in therapy-resistant epiletic patients In: Schmidt D, Wauquier A, Meinardi H, Vanden Bussche G (eds) Health Science Review International Workshop on Flunarizine in Epilepsy Beerse Belgium pp 81–88

Gigli GL, Musumeci SA, Colognola RM, Ferri R, Valente M, Bergonzi P, Mazza S (1987) Flunarizine in epilepsy: a worthwhile therapy? International Symposium on Calcium Antagonists The New York Academy of Sciences New York pp 101 (abstract)

Holmes B, Brogden RN, Heel RC, Speight TM, Avery GS (1984) Flunarizine a review of its pharmacodynamic and pharmacokinetic properties and therapeutic use Drugs 27: 6–44

Mancia D, Bergamasco B, Moglia A, di Perri R (1986) Double-blind placebo-controlled crossover study of the efficacy of flunarizine as add-on therapy in epilepsy In: Schmidt D, Wauquier A, Meinardi H, Vanden Bussche G (eds) Health Science Review International Workshop on Flunarizine in Epilepsy Beerse Belgium pp 89–92

Meyer FB, Sundt TM, Anderson RE (1987) Anticonvulsant properties of dihydropyridine Ca^{++}-antagonists International Symposium on Calcium Antagonists The New York Academy of Science New York pp 106 (abstract)

Overweg J, Binnie CD, Meijer JWA, Meinardi H, Nuijten STM, Schmaltz S, Wauquier A (1984) Double-blind placebo-controlled trial of flunarizine as add-on therapy in epilepsy Epilepsia 25: 217–222

Schmidt D, Ried S (1989) Clinical relevance of calcium antagonists in the treatment of epilepsy Arzneimittelforschung/Drug Res 39: 156–158

Vanden Bussche G, Wauquier A, Ashton D, de Beukelaar F (1985) Flunarizine in therapy-resistant epileptic patients In: Meldrum BS, Porter RJ (eds) Current problems in epilepsy: "New anticonvulsant drugs" Janssen Pharmaceutica Beerse Belgium pp 1–13

Wauquier A, Clincke G, Ashton D, Fransen J, Gillardin JM, Van Clemen G (1986) Pharmacological evaluation of anticonvulsant drugs In: Schmidt D, Wauquier A, Meinardi H, Vanden Bussche G (eds) Health Science Review pp 15–23

Van Wieringen A, Ashton D, de Beukelaar F, Binnie CD, Overweg J, Wauquier A (1986) Flunarizine in the treatment of epilepsy In: Speckmann EJ, Schulze H, Walden J (eds) Epilepsy and calcium Urban & Schwarzenberg München pp 387–399

4.7.6 Further Indications

In patients with organic brain syndrome (mainly of vascular origin) and with dementia syndromes, calcium antagonists, in particular flunarizine, cinnarizine and nimodipine, have been tested in numerous studies against placebo. Their efficacy was assessed by means of neuropsychological testing. On the whole patients show an improvement in "cerebral (mental) capacity" after taking calcium antagonists (review by Grötz 1985; nimodipine studies, Bono et al. 1984; Held et al. 1985; Baumel et al. 1987; flunarizine and cinnarizine studies, Staessen 1977; Ronchini et al. 1982). Most of these studies can be criticized on the grounds that they do not define, or aetiologically classify, the diagnosis of "cerebral dysfunction", and thus compare heterogeneous groups, mostly of older patients with multiple disorders. Precisely among such patients flunarizine and cinnarizine frequently exhibit serious side effects such as extrapyramidal motor disorders (Parkinson's syndrome), depressive moods and anxiety states (Sect. 5.3). In conclusion, a general recommendation to treat patients with "cerebral atherosclerosis, dementia of unknown origin, or organic brain syndrome" with flunarizine or cinnarizine cannot be made.

Intractable hiccups have been successfully treated with nifedipine (20 mg every 8 h; Lux 1985; Mukhopadhyay et al. 1986; Sect. 6.8.4). Application of verapamil in Duchenne muscular dystrophy has been described; Gran and his coworkers (1987) used nifedpine with success in myotonic dystrophy.

The efficacy of verapamil in manic disturbances has been investigated by Aldenhoff and Schlegel (1987) and by Dinan and Silverstone (1987). Both studies showed successful therapy with a daily dosage of 120–400 mg for 2 weeks, the success being measured in terms of rating scales and laboratory chemical data (CRF corticotropin-releasing factor, growth hormone and prolactin). The mechanism of action appears to differ considerably from that of the neuroleptics or benzodiazepines, and the degree of clinical improvement in the manic disorder, especially in the state of agitation, is not paralleled by the degree of sedation, as observed with neuroleptics (Dinan and Silverstone 1987). In both studies the clinical improvement is accompanied by a decrease in the plasma cortisol level.

References

Aldenhoff JB, Schlegel S (1987) Therapeutic action of calcium-antagonists in psychic disorder. Conclusions from experimental and clinical studies In: International Symposium on Calcium Antagonists The New York Academy of Sciences New York pp 96 (abstract)

Baumel B, Eisner LS, McNamara R, Korakin M (1987) Nimodipine in the treatment of primary degenerative dementia In: International Symposium on Calcium Antagonists The New York Academy of Sciences New York pp 96 (abstract)

Bono G, Sinforiani E, Trucco M, Cavallini A, Acuto GC, Nappi G (1985) Nimodipine in cCVD patients: Clinical and neuropsychological results of a double-blind cross-over study In: Betz

E, Deck K, Hoffmeister F (eds) Nimodipine – pharmacological and clinical properties Schattauer Stuttgart pp 275–285
Dinan TG, Silverstone T (1987) A pilot study of verapamil in the treatment of mania In: International Symposium on Calcium Antagonists The New York Academy of Sciences New York pp 99 (abstract)
Gran R (1987) Nifedipine in the treatment of myotonia in myotonic dystrophy J Neurol Neurosurg Psychiat 50: 199–206
Grötz J (1985) Hirnorganisches Psychosyndrom und chronische Hirnleistungsschwäche In: Hossmann V, Grötz J, Schrör K: Kalziumantagonisten und zerebrale Erkrankungen Springer Berlin pp 89–116
Held K, Boehme K, Rode CP (1985) Efficacy and tolerability of nimodipine in patients with old-age cerebrovascular dysfunction In: Betz E, Hoffmeister F (eds) Nimodipine – pharmacological and clinical properties Schattauer Stuttgart pp 289–292
Lux G (1985) Leitsymptom Singultus Dtsch Ärztebl 82: 1429–1433
Mukhopadhyay P, Osman MR, Wajima T, Wallace TJ (1986) Nifedipine for intractable hiccups N Engl J Med 314: 1256
Ronchini P, d'Ambrosio E, Anesi E, Pelliconi P (1982) Effetti della flunarizina sull' insuffizienza cerebro vasculare cronica: esperienza di un reparto geriatrico J Clin Med 63: 131–139
Staessen AJ (1977) Treatment of circulation disturbances with flunarizine and cinnarizine VASA 6: 59–71

4.8 Gastrointestinal Tract

Calcium antagonists also inhibit the smooth muscle of the gastrointestinal tract (Sect. 3.2.1); they diminish basal muscular tone and resolve or prevent spasm. Although calcium antagonists have been used to a greater or lesser extent for nearly 10 years in the indications discussed in the following sections, larger, in particular randomized, placebo-controlled trials are the exception. Thus, the place of calcium antagonists in the treatment of disorders of the gastrointestinal tract cannot be precisely assessed. Current reviews of this area are provided by Castell (1985) and Mössner (1987).

4.8.1 Oesophagus

With regard to the oesophagus, calcium antagonists reduce the frequency, amplitude and duration of peristaltic waves; furthermore, they decrease the pressure at the lower oesophageal sphincter. Calcium antagonists have to date been applied in localized or diffuse oesophageal spasm, in radiation oesophagitis and in achalasia (review by Castell 1985).

Oesophageal spasm can occur as either a localized or a diffuse form; a clinical entity "diffuse oesophageal spasm" is recognized (review by Richter and Castell 1984; Editorial Lancet 1987), together with the subtype "nutcracker oesophagus". The main symptoms are retrosternal pains (partially radiating into the arms, neck or back) and dysphagia. Oesophageal spasm is important in the differential diagnosis of angina pectoris and is found relatively often in patients with retrosternal pain and an inconspicuous coronary arteriogram. Coincidence of oesophageal and coronary spasm is a subject of current discussion (review in Editorial Lancet 1986). Verapamil, as well as nifedipine and diltiazem, have been employed in the therapy of oesophageal spasm (Weiser et al. 1978; Lipton et al. 1978; Blackwell et al. 1981; Nasrallah 1982; Richter and Castell 1984) with such daily doses as 3×10 to 3×20 mg nifedipine and 4×60 to 4×90 mg diltiazem.

Under such treatment peristalsis and the incidence of spontaneous contraction of oesophageal smooth muscle declines, and the frequency and intensity of subjective symptoms diminishes. It remains to be clarified whether treatment should be carried out as continuous prophylaxis (3 × 10 to 3 × 20 mg nifedipine, or 4 × 60 to 4 × 90 mg diltiazem before meals), or whether it should be limited to the acute attack (e.g. 10–20 mg sublingually; Editorial Lancet 1987).

In radiation oesophagitis nifedipine (40–60 mg daily) is also said to relieve or reduce symptoms (retrosternal burning and dysphagia; Finkelstein 1986). In this situation nifedipine is used as an adjunct to conventional treatment with antacids, histamine-2 blockers, lidocaine and opiates.

Achalasia belongs to the group of oesophageal motility disorders, a neuromuscular disorder of unknown origin, characterized by absent or incomplete relaxation of the lower oesophageal sphincter and concomitant absence of peristaltic contractions in the caudal oesophagus (Wienbeck and Bergers 1983). The patient typically suffers from dysphagia, regurgitation of food, aspiration and/or retrosternal pain. Therapeutically, calcium antagonists may be used with the aim of reducing the pressure of the lower oesophageal sphincter and thereby the amplitude of the proximal oesophageal contractions. The most extensive experience in this indication to date has been gained with nifedipine (Weiser et al. 1978; Berger and McVallom 1982; Wienbeck and Berges 1983). In a placebo-controlled crossover study Bortolotti and Labo (1981) reported marked symptomatic improvement in 14 of 20 patients. The daily dose of nifedipine used in these studies lies mostly between 30 and 40 mg. If there is no response to medication, further therapeutic possibilities include intraluminal dilation treatment by means of a pneumatic dilator or longitudinal myotomy (Wienbeck and Berges 1983).

Reducing the pressure of the lower oesophageal sphincter by treatment with calcium antagonists can occasionally lead to symptoms of gastro-oesophageal reflux (heartburn, retrosternal pain) in patients with no primary gastrointestinal disorder. Especially in patients with coronary heart disease, care must be taken not to confuse such symptoms with inadequately treated angina pectoris (Sect. 5.1).

4.8.2 Small Intestine: Large Intestine

Calcium antagonists are also prescribed in motility disorders of the small and large intestine which are associated with uncoordinated or excessive smooth muscle contractions. The so-called irritable bowel syndrome is clinically characterized by abdominal pain, constipation and/or diarrhoea. Therapeutically, nifedipine (3 or 4 × 10 mg daily) can be used alongside dietary measures such as bran, *Plantago ovata* and anticholinergics or metoclopramide (Blume and Schuster 1983; Fritsch 1986). Pain and diarrhoea are usually improved by these means, but constipation may be aggravated owing to the relaxant effect of nifedipine on intestinal smooth muscle.

Mössner (1987) discusses the application of calcium antagonists in certain forms of diarrhoea (functional acceleration of stool passage, possibly also secretory diarrhoea), on the grounds that they act by relaxation of smooth muscle and by inhibition of small intestinal secretions.

The syndrome known as proctalgia fugax is characterized by spasm of the rectal muscle. Some authors regard it as a variant of the irritable bowel syndrome; on the other hand, there are certain parallels with migraine. In individual cases a sympathomimetic (salbutamol) or a calcium antagonist (diltiazem 2 × 80 mg daily) has been used with success (Boquet et al. 1986; Jonard and Essamri 1987). In this situation diltiazem is held to lower the basal tone of the inner anal sphincter (which consists of smooth muscle), once again the mechanism of action appearing to pivot on a relaxant effect on smooth muscle (Jonard and Essamri 1987).

4.8.3 Gallbladder: Bile Ducts

Studies with nifedipine have demonstrated that calcium antagonists diminish contractions of the gallbladder induced by cholecystokinin (Schmidt et al. 1986, personal communication; ceruletide-induced contractions of the gallbladder remain unaffected, Porschen et al. 1987). Theoretically, calcium antagonists could therefore provide an alternative to anticholinergic spasmolytics and to nitrates in gallbladder colic. Systematic investigations in this direction have yet to be undertaken. It also remains an open question whether, or to what extent, long-term treatment with calcium antagonists increases the risk of gallstone formation as a result of reduced gallbladder motility.

4.8.4 Other Gastrointestinal Indications

The employment of calcium antagonists in circulatory disorders of the mesenteric artery is dealt with in Sect. 4.6.3. From a theoretical standpoint, portal vein pressure can be lowered by calcium antagonists in patients with portal hypertension; however, since calcium antagonists, like β-receptor blockers and nitrates, simultaneously lower arterial blood pressure, deterioration in arterial perfusion of the liver with negative sequelae for liver function could occur. Systematic investigations with calcium antagonists have not yet been carried out; in individual cases verapamil is said to lower the raised portal vein pressure in patients with liver cirrhosis, while under nifedipine it remains unchanged (review by Mössner 1987).

Intractable hiccups which fail to respond either to so-called household remedies, general measures such as pharyngeal stimulation and use of a stomach probe, or to medicinal measures such as metoclopramide or phenothiazines (review by Lux 1985) can sometimes be interrupted by administration of a calcium antagonist (nifedipine 3 × 10 to 3 × 20 mg daily). The patient may thus be spared more drastic therapeutic measures such as succinylcholine, phrenic nerve block or diaphragmatic pacing (Mukhopadhyay et al. 1986).

References

Berger K, McCallum W (1982) Nifedipine in the treatment of achalasia Ann Intern Med 96: 61–62

Blackwell JN, Holt S, Heading HC (1981) Effect of nifedipine on oesophageal motility and gastric emptying Digestion 21: 50–56

Blume M, Schuster M (1983) Effect of nifedipine on colonic motility in the irritable bowel syndrome John Hopkins University Baltimore: Digestive Disease Week

Boquet J, Moore N, Lhuintre JP, Boismare F (1986) Diltiazem for proctalgia fugax Lancet 1: 1493
Bortolotti M, Labo G (1981) Clinical and manometric effects of nifedipine in patients with oesophageal achalasia Gastroenterology 80: 39–44
Castell DO (1985) Calcium-channel blocking agents for gastrointestinal disorders Am J Cardiol 55 (Suppl): 210 B–213 B
Editorial Lancet (1986) Angina and oesophageal disease Lancet 1: 191–192
Editorial Lancet (1987) Management of diffuse oesophageal spasm Lancet 1: 80–81
Finkelstein E (1986) Nifedipine for radiation oesophagitis Lancet 1: 1205–1206
Fritsch WP (1986) Reizdarmsyndrom Dtsch Med Wschr 111: 1776–1777
Jonard P, Essamri B (1987) Diltiazem and internal anal sphincter Lancet 1: 754
Lipton SA, Markis JE, Pine MB, Paulin S (1978) Cessation of smoking followed by Prinzmetal's variant angina and diffuse esophageal spasm N Engl J Med 290: 775–776
Lux G (1985) Leitsymptom: Singultus Dtsch Ärztebl 82: 1429–1433
Mössner J (1987) Calciumantagonisten in der experimentellen und klinischen Gastroenterologie Dtsch Med Wschr 112: 1092–1099
Mukhopadhyay P, Osman MR, Wajima T, Wallace TI (1986) Nifedipine for intractable hiccups N Engl J Med 314: 1256
Nasrallah SM (1982) Nifedipine in the treatment of diffuse oesophageal spasm Lancet 2: 1285
Porschen R, Pieper S, Bernhardt L, Schade B, Wienbeck M (1987) Beeinflussen Kalziumantagonisten die Gallenblasenkontraktion in vivo? Klin Wochenschr 65 (Suppl IX): 30
Richter JE, Castell DO (1984) Diffuse esophageal spasm: a reappraisal Ann Intern Med 100: 242–245
Weiser HF, Lepsien G Golenhofer K, Siewert R (1982) Clinical and experimental studies on the effect of nifedipine on smooth muscle of the esophagus and LES In: Dutie H (ed) Gastrointestinal motility in health and disease MTP Press Lancaster pp 565–572
Wienbeck M, Berges W (1983) Motilitätsstörungen des Verdauungstraktes Dtsch Ärztebl 80: 17–24

4.9 Respiratory Tract

With regard to the respiratory tract calcium antagonists are applied in the treatment of primary and secondary pulmonary hypertension (Sect. 4.6.1) and may be applied in asthma and chronic obstructive lung disease.

Neither in asthma nor in chronic obstructive lung disease are calcium antagonists drugs of first choice. In contrast to β-receptor blockers, however, calcium antagonists can be given virtually without risk to patients with coronary heart disease or with arterial hypertension and concomitant bronchial obstruction (Editorial Thorax 1983). An increase in bronchial obstruction, as occasionally observed in treatment with β-receptor blockers, has thus far not been described with calcium antagonists.

4.9.1 Asthma

Asthma is commonly defined as a state of reversible bronchial obstruction in which not only bronchial smooth muscle cells, but also mast cells, mucus secretory cells and specific inflammatory cells are involved (Tinkelmann 1985). Since the functions of these four different types of cell are all broadly speaking calcium-dependent, calcium antagonists should theoretically be able to influence asthma positively; they have a proven inhibitory effect upon the bronchial muscle cell. An additional inhibition of mast cell degranulation is postulated; however, a clear effect on the glandular secretion of the respiratory tract and on inflammatory cells has yet to be demonstrated (reviews in Editorial Thorax 1983; Ferlinz 1986).

Although, as discussed in Sect. 3.2.2, basal tone (resting bronchomotor tone) of the bronchial muscles, i.e. bronchial width at rest, is unaffected by calcium antagonists (McFadden 1981; Editorial Thorax 1983), they can lower the reactivity of the bronchial system to bronchoconstrictory stimuli (e.g. effort, hyperventilation, exposure to antigens, or administration of histamine and methacholine). Beyond inhibiting bronchoconstriction, however, they do not induce bronchodilation. To date, the results available for nifedipine, diltiazem and verapamil reflect studies involving only small groups of patients in the acute test, and the study protocols rarely included randomization of patients or crossover design (review by McFadden 1981; Editorial Thorax 1983; Middleton 1983; Hartmann and Magnussen 1983).

Because of the lack of clinical experience both in the short- and long-term treatment of asthma, at the present time calcium antagonists do not offer a genuine alternative to local and systemic therapy with bronchodilators (xanthine derivatives, β-sympathomimetics, anticholinergics) and steroids. Tinkelmann (1985) expresses doubt that the classic calcium antagonists (nifedipine, verapamil) are of benefit in the treatment of chronic asthma, since they lack any specific effect on cells involved in bronchial obstruction.

By contrast, ketotifen and cinnarizine, both substances with calcium antagonistic and antihistaminic activities (histamine-1 receptor blockers), appear to be effective in the prophylaxis of chronic asthma (Emanuel et al. 1979; Tinkelmann 1985). Disodium cromoglycate is said to act similarly and to possess calcium antagonistic activity (Foreman and Garland 1976). However, the latter three substances do not meet the criteria of a classical calcium antagonist delineated by Fleckenstein (Sect. 1.2.1).

In Summary. The classic calcium antagonists (nifedipine, verapamil, diltiazem) are not generally accepted in the treatment of asthma. However, these substances can be given to patients with coronary heart disease, arterial hypertension (or any other indication for calcium antagonists) with concomitant asthma, without risk of increasing bronchial obstruction. Calcium antagonists, with supplementary antihistaminic activity (ketotifen, cinnarizine, disodium cromoglycate) may prove useful in the prophylaxis of chronic asthma.

4.9.2 Chronic Obstructive Pulmonary Disease

In the 1970s it was demonstrated (Schneider et al. 1976) that verapamil, in contrast to β-receptor blockers (although a β-receptor blocking action had originally been postulated for verapamil by Haas 1964), does not induce a raised airway resistance either in healthy volunteers or in patients with chronic obstructive lung disease. In this study a slight, though statistically significant, bronchospasmolytic effect was observed.

Only in the last few years have calcium antagonists (chiefly nifedipine) been systematically investigated in patients with chronic obstructive lung disease. In the acute test, following administration of 10–20 mg nifedipine, there is a drop in airway resistance and an increase in forced expiratory volume (FEV, maximal percentage of vital capacity which can be breathed out in 1 s). The effect was comparable with inhalation of 0.2 mg salbutamol (as aerosol; Emslander et al.

1984, 1986; Ikeda et al. 1985). In one study (Ikeda et al. 1985), despite bronchodilation, decreased oxygenation of the blood together with a rise in the alveolar–arterial oxygen pressure difference was observed. This is explained by the systemic and pulmonary vasodilation which increases cardiac output and thereby partially leads to perfusion of inadequately ventilated areas of the pulmonary vascular system (ventilation–perfusion mismatch).

There are also no long-term studies concerning treatment of chronic obstructive lung disease with calcium antagonists, and thus at the present time their use in this indication cannot be generally recommended. However, in patients with obstructive pulmonary disease who also have coronary heart disease or arterial hypertension, calcium antagonists are the drugs of choice since β-blockers are contraindicated in variable bronchial obstruction (particularly in asthma). The use of calcium antagonists in pulmonary hypertension following chronic obstructive lung disease is discussed in Sect. 4.6. If chronic obstructive lung disease includes cor pulmonale with right-sided heart failure, calcium antagonists are, as in every form of cardiac failure, to be used with caution, and substances with a relatively strong negative inotropic effect (verapamil group) should be avoided.

References

Editorial (1983) Calcium-channel blockers and asthma Thorax 38: 481–485

Emanuel MB, Chamberlain JA, Whiting S, Rigden BG, Craven AH (1979) Cinnarizine in the treatment of chronic asthma Br J Clin Pharmacol 7: 189–195

Emslander HP, Sauer E, Freisinger H, Sebening H, Daum S (1984) Akute Bronchospasmolyse mit dem Kalzium-Antagonisten Nifedipin Atemw-Lungenkrkh 10: 374–382

Emslander HP, Sauer E, Munteanu J, Sebening H, Daun S (1986) The acute effect of nifedipine on airways resistance in patients with chronic obstructive lung disease In: Lichtlen PR (ed) 6th International Adalat Symposium Excerpta Medica Amsterdam pp 497–503

Ferlinz I (1986) Nifedipine in myocardial ischemia, systemic hypertension and other cardiovascular disorders Ann Intern Med 105: 714–729

Foreman JC, Garland LG (1976) Cromoglycate and other antiallergic drugs: a possible mechanism of action Br Med J 1: 820–821

Haas H (1964) Selektive Sympathikolyse und Myokardfunktion Dtsch Med Wschr 89: 2117–2121

Hartmann V, Magnussen H (1983) Zur Wirkung des Kalzium-Antagonisten Dilzem auf den Histamin- und Carbachol-induzierten Bronchospasmus bei Gesunden und Patienten mit Asthma bronchiale Atemw.-Lungenkrkh 9: 294–296

Ikeda H, Takahashi K, Sasaki H, Ito E, Yasui S (1985) Acute effects of nifedipine on patients with chronic obstructive lung disease Arzneimittelforschung 35 (I): 518–521

McFadden JR (1981) Calcium-channel blocking agents and asthma Ann Intern Med 95: 232–233

Middleton E jr (1983) Role of calcium and calcium antagonists in airway function Eur J Respir Dis 64 (Suppl 128): 123–132

Schneider H, Wylicil, Kardos P (1976) Die Wirkung von Verapamil auf den Bronchialwiderstand und auf die Fenoterol-induzierte Bronchospasmolyse bei Patienten mit obstruktiver Lungenerkrankung und bei Gesunden Therapiewoche 26: 5378–5383

Tinkelman DG (1985) Calcium channel blocking agents in the prophylaxis of asthma Am J Med 78 (Suppl 2 B): 35–38

4.10 Urogenital Tract

As described in Sect. 3.2.4, calcium antagonists inhibit the smooth muscle of the urinary tract (ureter, bladder) and of the uterus (Forman et al. 1978, 1979; Maggi and Meli 1984). Inhibition of myometrial contractility can be demonstrated in healthy female volunteers, in women with dysmenorrhoea and in pregnant women with prostaglandin-induced abortion or premature labour (review by Ulmsten et al. 1980; Andersson 1988).

4.10.1 Ureteral Spasm: Ureteral Colic: Renal Colic

Owing to their spasmolytic effect on smooth muscle cells, calcium antagonists have been used in individual cases of colic in the urinary tract. Carrol (1985) observed that nifedipine in a dose of 6 × 10 mg daily produced a rapid recession of ureteral colic pain which had previously only partially responded to anticholinergics and opiates. In a randomized study of 20 patients, however, the effect of 10–20 mg nifedipine sublingually was less than that of intramuscularly administered meperidine (McCormack et al. 1986). We, too, have used nifedipine in isolated cases of ureteral colic, in particular in patients who had not reponded to antispasmodics or analgesics (*N*-butylscopolamine, paracetamol, dipyrone, indomethacin), and we observed an improvement (i.e. relief of pain). Although the place of nifedipine in the treatment of ureteral colic is not yet clear, it can be tried in refractory colic pain, and it is almost always well tolerated.

4.10.2 Detrusor Instability (Detrusor Hyperreflexia – Irritable Bladder – Nervous Bladder)

Urge incontinence with detrusor instability is the commonest form of incontinence in the elderly (Williams and Pannill 1982). It is characterized by uninhibited and spontaneous detrusor contractions of the bladder. Causally, there is either disturbed central nervous system inhibition (motor urge incontinence), hyperexcitability of afferent pathways (sensory urge incontinence) or a disorder of the micturition reflex (so-called deconditioning; Williams and Pannill 1982; May et al. 1986). After the elimination of organic causes (e.g. prostatic hyperplasia in stage II: increased micturition frequency and nocturia, but without retention of urine), the treatment of urge incontinence consists both of pharmacological inhibition of bladder muscle activity and a thorough course of bladder training to extend micturition intervals. Various drugs are used, and the aim of treatment is to raise the stimulation threshold of the detrusor and therewith to improve detrusor instability (detrusor hyperreflexia). Calcium antagonists, indomethacin and bromocriptine are used in addition to anticholinergics (e.g. *N*-butylscopolamine), direct muscle relaxants (e.g. flavoxate), polysynaptic inhibitors (e.g. emepronium), or β-sympathomimetics (e.g. terbutaline) (review by Williams and Pannill 1982; May et al. 1986).

Calcium antagonists can increase bladder capacity and thereby improve both the symptoms and the urodynamic parameters of incontinence. The studies to date have employed nifedipine (Rud et al. 1979) and flunarizine (20 mg daily Palmer et al. 1981). However, the patients were mainly women, the patient groups were small and the trials were of relatively short duration. Overall, side effects were rare; only in isolated instances did a disturbance in micturition develop (variable urine retention, overflow incontinence).

In Summary. Calcium antagonists (flunarizine, nifedipine) supplement the treatment of urge incontinence with detrusor instability. However, since mainly elderly patients are involved, careful attention should be given to the occurrence of possible side effects (e.g. hypotension with vertigo during nifedipine treatment, and extrapyramidal symptoms with flunarizine therapy).

4.10.3 Tocolysis

Tocolytic treatment attempts to inhibit premature labour activity and to delay the moment of birth since premature labour may result in premature delivery or perinatal damage to the foetus. Calcium antagonists can be used either as directly acting tocolytic agents, or as supplementary therapy in tocolysis with β-sympathomimetics in order to diminish cardiotoxic side effects (review by Fleckenstein et al. 1978; Döring et al. 1981; Wischnik 1983, 1985a).

4.10.3.1 Direct Tocolytic Action of Calcium Antagonists

The inhibition of uterine (myometrial) activity by verapamil, described by Fleckenstein et al. (1972), relates to animal experiments. Here extremely high doses of calcium antagonists were given which, because of cardiovascular side effects, would probably exclude their clinical use. Thus, verapamil probably has no direct tocolytic effect in humans (according to Wischnik 1985a the relative efficacy ratio of papaverine to verapamil to the β-sympathomimetic, fenoterol, is 1:10:5000).

In contrast to verapamil, for which no teratogenic effect has been observed in pregnancy (Roundtable 1981), animal experimental data raise the possibility of such an effect for nifedipine. This difference is perhaps explained by the virtual nonpassage of verapamil in normal dosage across the human placental barrier (Strigl et al. 1980). Nifedipine has therefore been applied in pregnancy only in isolated cases. However, in recent investigations which demonstrated satisfactory inhibition of labour in the third trimester with a dose of 6×10 mg nifedipine (Ulmsten et al. 1980; Read and Wellby 1986), no teratogenic side effects were observed. Overall, nifedipine is thought to exhibit a markedly stronger tocolytic effect than verapamil (Wischnik 1985a). Studies comparing nifedipine with β-sympathomimetics (currently the most frequently applied tocolytic substances) are not yet available. At the present time a definitive statement concerning the effect of nifedipine in tocolysis, and in particular the question of its possible teratogenicity, cannot be made.

4.10.3.2 Application of Calcium Antagonists for Cardioprotection in Tocolysis with β-Sympathomimetics

β-Sympathomimetics are today the most frequently used tocolytic agents. The newer β-sympathomimetics of the fenoterol type act mainly on β_2-receptors (β_2-receptor selectivity). Their relaxant effect on the uterus (β_2-receptor stimulation) is, however, hampered by a number of side effects (due mainly to β_1-receptor stimulation). These side effects range from harmless functional disorders such as palpitations or states of unease or anxiety, via cardiac arrhythmias to cardiomyopathy with heart failure. Morphologically, in extreme cases myocardial necroses are exhibited (review by Döring et al. 1981). The latter are thought to be a consequence of excessive β-receptor stimulation (an isoproterenol-like effect) with intracellular calcium overload (Rona et al. 1959; Fleckenstein 1971).

Their early data demonstrating the prevention of isoproterenol-induced myocardial necroses by verapamil stimulated Fleckenstein et al. (1978) to apply verapamil in an attempt to reduce the cardiotoxic effect of tocolytics (fenoterol). In animal experiments (rat heart) fenoterol-induced calcium overload of myocardial cells could be prevented by concomitant administration of verapamil (50 mg per kilogram body weight). These studies also indicated a cardioprotective effect by compensation of a potassium or magnesium deficiency (Fleckenstein et al. 1978; Döring et al. 1981).

On the basis of these results a combination of fenoterol and verapamil was applied clinically in tocolytic therapy; however, the normal dosage of 2 μg verapamil per kilogram body weight per minute showed no effect on subjective dysphoria, no improvement in the cardiac energy balance and thus no cardioprotective effect, as confirmed by measurements during tocolysis and assessment of cardiac function post partum (Strigl et al. 1980; Irmer et al. 1981, 1983; Wischnik et al. 1982; review by Wischnik 1985a). Furthermore, at this dosage verapamil exhibits no direct tocolytic effect as described for higher doses in animal experiments (Sect. 4.10.3.1; Fleckenstein et al. 1972). Overall, there is no proof for cardioprotection by verapamil in tocolysis with β-sympathomimetics (verapamil dosage according to Weidinger 1983, 40–100 μg/min i.v., 60-240 mg/day orally).

It remains to be clarified to what extent a higher dose, as used, for instance, in hypertrophic obstructive cardiomyopathy (up to 960 mg daily) can be given in pregnancy (risk of cardiodepression and hypotension). Irmer (1983) considers an increase in dosage by a factor of 10–20 to be necessary, which would seem clinically unacceptable.

The lack of efficacy of verapamil as a cardioprotective agent has led to the use of other substances: magnesium is given orally in a dose of 10–40 mmol/day (Weidinger 1983; Wischnik 1985a, b) and acts as a physiological calcium antagonist. It enables the dose of β-sympathomimetics to be reduced (since magnesium itself acts as a tocolytic) while at the same time diminishing the cardiac side effects of tocolysis (cardioprotection). With this therapy tolerance does not develop in long-term tocolysis, there is a broad acceptable dosage range and an absence of deleterious effects on the foetus (Spätling 1982, Wischnik 1983, 1985a).

Alternatively, cardioselective (β_1-receptor-selective) β-receptor blockers (particularly metoprolol, dose 100–200 mg/day orally or 0.2 mg/kg i.v.) may be considered (Irmer et al. 1980, 1981, 1983; Wischnik 1985a). This treatment ameliorates dysphorias, reduces cardiac side effects of tocolysis and also has no negative sequelae for foetal development.

In Summary. Today magnesium or cardioselective β-receptor blockers are successfully applied as cardioprotective agents in tocolysis with β-sympathomimetics. A cardioprotective effect has not been shown for verapamil in the clinical situation.

4.10.4 Dysmenorrhoea

So far only isolated reports are available concerning a positive effect of calcium antagonists in states of functional pain during menstruation – dysmenorrhoea

(Andersson and Ulmsten 1978; Krebs 1984; Andersson 1988). These cases mainly involve nifedipine, the effect of which is thought to depend on relaxation of the uterine myometrium (inhibition of contractile hyperactivity of the myometrium). With 20–40 mg nifedipine daily, two-thirds of patients are said to have been relieved from pain. Alternatively, magnesium, besides β-sympathomimetics and prostaglandin inhibitors, can be used as a physiological calcium antagonist in this situation (Wischnik 1985 b).

References

Andersson KE (1988) Calcium antagonists and dysmenorrhea In: Vanhoutte PM, Paoletti R, Govoni S (eds) Calcium antagonists The New York Academy of Sciences New York pp 747–756

Andersson KE, Ulmsten U (1978) Effects of nifedipine on myometrial activity and lower abdominal pain in women with primary dysmenorrhoea Br J Obstet Gynaecol 85: 142–146

Carrol W (1985) Nifedipine and ureteral colic Ann Intern Med 102: 864

Döring HJ, Irmer M, Keidel J, Frey M, von Witzleben H, Fleckenstein A (1981) Potenzierung und Neutralisierung kardiotoxischer Nebenwirkungen beta-adrenerger Tokolytika In: Ablad B, Heidenreich J, Irmer M, Jung H (eds) Beta Blockade und Tokolyse Witzstrock Baden-Baden pp 47–54

Fleckenstein A (1971) Pathophysiologische Kausalfaktoren bei Myokardnekrose und Infarkt Wien Z Inn Med 52: 133–139

Fleckenstein A, Grün C, Byon KY (1972) Hemmung der Uteruskontraktilität durch Ca^{++}-antagonistische Inhibitoren der elektro-mechanischen Kopplung In: Jung H (ed) Methoden der pharmakologischen Geburtserleichterung und Uterus-Relaxation Thieme Stuttgart pp 107–113

Fleckenstein A, Janke J, Fleckenstein-Grün H (1978) Kardiotoxische Wirkungen beta-adrenerger Tokolytika Kardioprotektion durch Ca^{++}-Antagonisten In: Hillemanns HC, Trolp R (eds) Kardiale Probleme bei der Tokolyse Enke Stuttgart pp 54–50

Forman A, Andersson KE, Henricksson L, Rud T, Ulmsten U (1978) Effects of nifedipine on the smooth muscle of human urinary tract in vitro and in vivo Acta Pharmacol Toxicol 43: 111–118

Forman A, Andersson KE, Persson CGA, Ulmsten U (1979) Relaxant effects of nifedipine on isolated, human myometrium Acta Pharmacol Toxicol 45: 81–86

Irmer M, Trolp R, Pohl C, Bernius U, Hillemanns HG, Stein H (1980) Klinische Anwendung einer kombinierten Beta2–Stimulation und Beta1–Blockade bei Tokolysetherapie Arzneimittelforschung 30: 105–109

Irmer M, Trolp R, Hagemann G, Stein H (1981) Akut- und Langzeitbehandlung mit Metoprolol/Fenoterol im Vergleich zu Verapamil/Fenoterol aus kardiologischer Sicht In: Ablad B, Heidenreich J, Irmer M, Jung H (eds) Betablockade und Tokolyse Witzstrock Baden-Baden pp 114–117

Irmer M (1983) Möglichkeiten der Kardioprotektion bei Tokolyse mit Beta-Stimulatoren In: Irmer M, Weidinger H (eds) Neuere Aspekte zu Betablockade und Tokolyse Beltz Weinheim pp 83–89

Krebs R (1984) Calcium antagonists: new vistas in theoretical basis and clinical use In: Opie LH (ed) Calcium Antagonists and Cardiovascular disease Raven Press New York pp 347–357

Maggi CA, Meli A (1984) The effect of nifedipine and verapamil on KCl induced rhythmic contractions of guinea pig ureter in vitro Experientia 40: 681

May P, Sökeland J, Beck L, Schwenzer T (1986) Harninkontinenz Dtsch Ärztebl 83: 1949–1953

McCormack M, Willot P, Lapointe S, Drouin G (1986) Nifedipine and ureteral colic Ann Intern Med 104: 590

Palmer JH, Worth PHL, Exton-Smith AN (1981) Flunarizine: a once-daily therapy for urinary incontinence Lancet 2: 279–281

Read MD, Wellby DE (1986) The use of a calcium antagonist (nifedipine) to suppress preterm labour Br J Obstet Gynaecol 93: 933–937

Rona G, Chappel J, Balazs T, Gaudry R (1959) An infarct-like myocardial lesion and other toxic manifestations produced by isoproterenol in rat Arch Path 67: 443–447
Roundtable (1981) The current role of calcium antagonists in cardiac arrhythmias In: Zanchetti A, Krikler DM (eds) Calcium antagonism in cardiovascular therapy: experience with verapamil Excerpta Medica Amsterdam pp 405–413
Rud T, Andersson KE, Ulmsten U (1979) Effects of calcium antagonists in women with unstable bladders Urol Int 34: 421–429
Spätling L (1982) Einsparung von Tokolytika durch orale Magnesiumgabe In: Weidinger H (ed) Magnesium und Tokolyse Fortschr Med Schwappach pp 122–129
Strigl R, Pfeiffer U, Ehrhardt W, Blümel G (1980) Bietet der Kalziumantagonist Verapamil bei der Tokolyse mit Beta-Sympathomimetika den erwarteten Schutz vor Myokardschäden? Geburtsh Frauenheilk 40: 500–509
Ulmsten U, Andersson KE, Wingerup L (1980) Treatment of premature labor with the calcium antagonist nifedipine Arch Gynecol 229: 1–5
Weidinger H (1983) Übersicht über verschiedene bei der Tokolyse eingesetzte Substanzen In: Irmer M, Weidinger H (eds) Neuere Aspekte zu Betablockade und Tokolyse Beltz Weinheim pp 34–45
Williams ME, Pannill FC (1982) Urinary incontinence in the elderly Ann Intern Med 97: 895–907
Wischnik A, Mendler N, Schroll A, Weidenbach A (1982) Zur Frage der Zusatztherapie bei Tokolyse: Vergleichende tierexperimentelle Untersuchungen des myokardialen Metabolitstatus Frauenarzt 23: pp 24–31
Wischnik A (1983) Zur Bedeutung von Beta 2-Mimese, Beta 1-Blockade und Calciumantagonismus bei der Therapie der vorzeitigen Wehentätigkeit und der uteroplazentalen Hypoperfusion Tierexperimentelle Untersuchungen Med Habilitationsschrift TU München
Wischnik A (1985a) Kombinierte Beta-Mimetikatherapie in der Schwangerschaft Urban & Schwarzenberg München
Wischnik A (1985b) Magnesium in der Frauenheilkunde Dtsch Ärztebl 82: 2342–2344

4.11 Further Indications for Calcium Antagonists

Only case records are available for the indications discussed here. The efficacy of calcium antagonists in these situations is currently unproven, and for this reason the substances should be used only in clinical studies or when conventional treatment has failed.

Duchenne muscular dystrophy is a hereditary muscular disorder of unknown origin for which there is as yet no effective treatment. Certain parallels with hereditary cardiomyopathies have been discussed. In the affected muscles there appears to be an elevated intracellular calcium concentration. For this reason Emery et al. (1982) applied a calcium antagonist (verapamil) in the treatment of Duchenne muscular dystrophy, giving 40 mg verapamil twice daily to a total of seven juvenile patients. Under this treatment there was a clear improvement in muscle function together with an increase in exercise tolerance (as measured by exercise testing); however, because of marked prolongation of the PR interval, the study was discontinued to avoid more severe heart block. Since a 1st degree AV block is not considered grounds for discontinuing verapamil therapy, this regimen should be carefully reassessed. Alternatively, treatment with a calcium antagonist of the dihydropyridine group (e.g. nifedipine) could be considered, since these drugs do not normally affect cardiac conduction.

Chemotherapy (cytotoxic therapy) is today carried out both as monotherapy and in combination therapy. In both forms of treatment resistance to the antineoplastic agents may develop, such resistance being described for doxorubicin (Adriamycin) and other substances (e.g. vinca alkaloids). With concomitant administration of verapamil and doxorubicin (or perhaps other antineoplastic agents), resistance can be reversed and chemotherapy can be continued (Rogan et al. 1984). Increased sensitivity of tumour cells to antineoplastic agents by such parallel administration of verapamil has been described. The mechanism of this effect of verapamil is not yet completely elucidated: possibilities discussed include an alteration in pharmacokinetics at the cellular level with delayed elimination of the antineoplastic agent from the cell (or the cell nucleus; Piccinini et al. 1988). This indication should be reevaluated both in vitro and in humans (Schnipper 1986).

Since calcium antagonists are used as cardioprotective agents in heart surgery, in catheter dilatation of the coronary arteries (PTCA), and in tocolysis (Sects. 4.5.4 and 4.10.3), they could theoretically also prevent or reduce the cardiotoxic side effects of certain antineoplastic agents (doxorubicin group; Garbrecht and Müllerleile 1986). **Doxorubicin cardiomyopathy** (review by Höfling 1984) may be a consequence of calcium redistribution with calcium overload of cardiac muscle cells, and may therefore be positively influenced by verapamil (or other calcium antagonists or by β-receptor blockers such as propranolol; Garbrecht and Müllerleile 1986; Piccinini et al. 1988). Furthermore, a disorder in protein synthesis is postulated (Höfling 1984). To date, however there are no precise data regarding which calcium antagonist is to be preferred (investigations thus far have mainly been conducted with verapamil), which dosage is to be employed and at what time cardioprotection is to be initiated.

In **cis-platinum nephrotoxicity** calcium antagonists (verapamil 3×80 mg daily) are reported to show some degree of renal protection (improved renal perfusion, slower decline in glomerular filtration rate in nine patients; Offermann et al. 1986).

In a case of **insulinoma** with recurrent hypoglycaemia administration of a calcium antagonist (diltiazem 180 mg daily) improved symptoms (details in Sect. 3.5.2 and in Taniguchi et al. 1977), which is explained as due to inhibition of (calcium-dependent) insulin secretion. By analogy, symptomatic improvement has been observed in individual cases of **phaeochromocytoma** following the use of nifedipine (Serfas et al. 1983; Favre and Vallotton 1986). Calcium antagonists are thought to inhibit (calcium-dependent) noradrenaline (norepinephrine) secretion in the adrenal medulla and to reduce the vasoconstricting effect of noradrenaline (Sect. 3.3.2).

In **sickle-cell anaemia,** too, calcium antagonists (bepridil, flunarizine) have been used occasionally in order to prevent or treat vaso-occlusive crises (reduction in formation of sickle cells – antisickling effect; details in Sect. 3.3.3 and in Reilly and Asakura 1986).

A cytoprotective effect of diltiazem was observed by Wagner et al. (1986) in **transplantation of cadaver kidneys.** In a prospective, randomized study the incidence of rejection decreased, and in the treatment group the glomerular filtration rate was higher. Diltiazem was added to the solution used for perfusion of the renal grafts (20 mg/l) and also administered to the kidney recipient preoperatively and perioperatively (bolus of 0.28 mg per kilogram body weight, then 0.0022 mg

kg^{-1} min^{-1}, then 60 mg orally twice daily). Besides the improvement in postoperative renal function, diltiazem is also reported to reduce the nephrotoxicity of cyclosporin A (Sect. 3.3.1). Confirmation of these findings by other working groups and with a large collective of patients is as yet lacking (for the interaction of cyclosporin A and calcium antagonists see Sect. 5.8).

Calcium antagonists dilate arterial vessels and resolve, or even prevent, vasospasm. Especially in acute coronary spasm, whether spontaneous or drug-induced in coronary arteriography, calcium antagonists (mainly nifedipine) are highly effective. The same is true for the vasospasm of Raynaud's syndrome. It would thus appear theoretically justified to apply calcium antagonists in cases of inadvertent intraarterial injection of medicaments with subsequent vasospasm and possibly with development of ischaemic syndrome. However, to date no further investigations into these indications have been published.

References

Emery AEH, Skinner R, Howden LC, Matthews MB (1982) Verapamil in Duchenne muscular dystrophy Lancet 1: 559

Favre L, Vallotton MB (1986) Nifedipine in Pheochromocytoma Ann Intern Med 104: 125

Garbrecht M, Müllerleile U (1986) Verapamil in the prevention of adriamycin-induced cardiomyopathy Klin Wschr 64 (Suppl VII): 132–134

Höfling B (1984) Die Anthracyclin-Kardiotoxizität Kehrer Freiburg pp 44–65

Offermann JJG, Meijer S, Sleifer DT, Mulder NH, Donker AJM, Schraffordt K, van der Hem GK (1986) The influence of verapamil on renal function in patients treated with cisplatin Clin Nephrol 24: 249–255

Piccinini F, Monti E, Favalli L, Villani F (1988) Effect of verapamil an doxorubicin-induced cardiomyopathy In: Vanhoutte PM, Paoletti R, Govoni S (eds) Calcium antagonists The New York Academy of Sciences New York pp 533–535

Reilly MP, Asakura T (1986) Antisickling effect of bepridil Lancet 1: 848

Rogan AM, Hamilton TC, Young RC, Klecker RW, Ozols RF (1984) Reversal of adriamycin resistance in human ovarian cancer Science 224: 994–996

Schnipper LE (1986) Clinical implications of tumor-cell heterogeneity N Engl J Med 314: 1423–1431

Serfas D, Shoback DM, Lorell BH (1983) Phaeochromocytoma and hypertrophic cardiomyopathy: Apparent suppression of symptoms and noradrenaline secretion by calcium-channel blockade Lancet 2: 711–713

Taniguchi H, Murakami K, Morita S, Kazumi T, Yoshino G, Maeda M, Baba S (1977) Calcium antagonist (diltiazem) for reversal of hypoglycaemic symptoms in insulinoma Lancet 2: 501

Wagner K, Albrecht S, Neumayer HH (1986) Protektiver Einfluss des Calciumantagonisten Diltiazem auf das akute Nierenversagen nach Nierentransplantation Dtsch Med Wschr 111: 1363–1367

4.12 Calcium Antagonists in Anaesthesiology

The use of calcium antagonists for cardioprotection in heart surgery is discussed in Sect. 4.5.3. As discussed there, calcium antagonists can also be used perioperatively and postoperatively in the treatment of hypertension (review by Hampelmann 1982; see also Sect. 4.2.4). The calcium antagonists most frequently used in this situation: nifedipine, verapamil and diltiazem, are available as parenteral formulations. If contraindications (Sect. 5.6) are observed, no impairment of perioperative cardiac function is observed. Drug interactions with anaesthetics (halothane, isoflurane, etc.) are probably rare (review by Piepho et al. 1987).

As stated in Sects. 4.1.2.1 and 5.7, a typical withdrawal syndrome rarely follows discontinuation of calcium antagonists. Treatment with calcium antagonists can therefore be discontinued abruptly preoperatively; a gradual reduction of dose over several days, as recommended for β-receptor blockers, is not necessary. Alternatively, treatment can be continued perioperatively or postoperatively in fasting patients parenterally, sublingually or in the form of drops. In general, the degree of severity of a patient's symptoms determines whether treatment with calcium antagonists is to be perioperatively discontinued or continued parenterally. Intraoperatively and perioperatively calcium antagonists are an established therapy in hypertensive crises, and substances of the verapamil group can be used in tachycardias, especially in paroxysmal supraventricular tachycardias and atrial fibrillation.

References

Hempelmann G (1982) Prämedikation, Anästhesie und postoperative Phase bei Hypertonikern und Patienten mit koronarer Herzerkrankung. Med Welt 33:739–745

Piepho RW, Culbertson VL, Rhodes RS (1987) Drug interactions with calcium-entry blockers Circulation 75 (Suppl V): 181–194

5 Side Effects – Overdosage – Contraindications – Drug Interactions

Calcium antagonists are generally judged to be well tolerated drugs. Although, depending on substance and dosage, between 5% and 50% of all patients experience side effects (Table 31), the latter are frequently only minor and recede during the course of treatment. Severe side effects which lead to discontinuation of calcium antagonist therapy are relatively rare, the incidence stated in the literature lying between 1% and 5%, depending on the substance employed (Bleifeld and Spokojny 1983; Bussmann and Hopf 1985; see also Table 31). Thus, calcium antagonists can be considered, like β-receptor blockers, well-tolerated drugs (review by Lydtin and Lohmöller 1980), which is of particular significance since these drugs are often employed in long-term treatment, e.g. in stable angina pectoris or arterial hypertension.

The side effects of calcium antagonists may be subdivided into three groups:

1. Side effects linked to the action of this class of substances on the heart, smooth muscle (especially vascular smooth muscle) and on other organ systems, and which therefore are generally due to blockade of calcium-dependent metabolic processes. This group of side effects is in principle predictable and often dose-dependent.
2. Side effects which are not directly linked with calcium-dependent metabolic processes or with calcium antagonism.
3. Substance-specific side effects.

Table 31. Main side effects of calcium antagonists (After Bussmann and Hopf 1985)

Symptoms	Verapamil	Gallopamil	Diltiazem	Nifedipine
Vertigo/headache	+	+	+	++
Hypotension	+–	+–	+–	++
Peripheral oedema	++	+	+	++
Constipation	++	+	–	+–
Gastric disorders/nausea	+	+	+–	+–
Bradycardia	+	+	+	–
Tachycardia [a]	–	–	–	+
Angina pectoris (paradoxical action)	–	–	–	+
Heart failure [b]	+	+	+–	–
Allergic reactions	+–	+–	+–	+–
Incidence of side effects (%)	5–20	5–15	4–10	15–50
Incidence of discontinuation of treatment (%)	1	3	1–5	5

[a] Oral administration.
[b] No preexisting heart failure.
++ Frequent, + rare, +– very rare, – observed.

5.1 Side Effects Directly due to Blockade of Calcium-Dependent Metabolic Processes

The pharmacological effects of calcium antagonists are discussed in detail in Chap. 3. Also, in the sections on indications (Chap. 4), the mechanism of action of calcium antagonists and resultant side effects are considered. As may be seen from Table 32, the side effects to be discussed here relate to the action profile of calcium antagonists, and are therefore in principle predictable, often dose-related and frequently show graded differences between the different groups of calcium antagonists (e.g. verapamil group, dihydropyridine group). Table 31 summarizes the incidences of clinically significant side effects (modified from Bussmann and Hopf 1985).

Calcium antagonists exhibit a direct negative inotropic effect on the heart (Sect. 3.1.3). This effect is, however, modified in different ways by autonomic counter-regulation among the various groups of substances. In vivo, the verapamil group exhibits the strongest negative inotropic effect; diltiazem, and in particular nifedipine, show a markedly lesser negative inotrop-

Table 32. Side effects of calcium antagonists explained in terms of their action

Action	Side effect	Predominant with		
Negative inotropy	Heart failure	V	(D)	(N)
Negative chronotropy	Bradycardia	V	D	
Negative dromotropy	AV block	V	D	
Arterial vasodilation (lowering of total peripheral resistance)	Drop in blood pressure, hypotension, vertigo, flush, oedema	N	(D)	(V)
Face	Flushing	N		
CNS	Headache	N		
Nose	Nasal congestion	N		
Lung	Ventilation-perfusion mismatch (with a preexisting disorder of ventilation)	N		
Kidney	Deterioration of renal function (in kidneys with already reduced function)	N		
Uterus	Menorrhagias			
Autonomic counter-regulation (activation of sympathetic nervous system)	Tachycardia, palpitations, angina pectoris	N		
Inhibition of smooth muscle	Constipation	V	(N)	
	Hiatal insufficiency	N		
	Gastro-oesophageal reflux			
	Micturition disturbance	N	(V)	
	Menorrhagias	N		
Activation of renin-angiotensin-aldosterone system [a]	Sodium retention [a], oedema [a]	N		
Inhibition of insulin secretion	Deterioration of glucose tolerance	N	(V)	

[a] For clinical relevance see text.
V verapamil group; *D* diltiazem, *N* nifedipine (dihydropyridine group).

ic action. Consequently, especially in therapy with substances of the verapamil group, the development of or deterioration of preexisting heart failure must always be borne in mind. In the extreme case, pulmonary oedema or cardiogenic shock can occur (Sect. 4.5.1). In principle heart failure may also develop or deteriorate during diltiazem or nifedipine therapy. By combining calcium antagonists and β-receptor blockers, autonomic counter-regulation is decreased, and heart failure may thus develop more readily. As stated in Sect. 4.1.2.3, combination therapy is in general well tolerated; however, at the onset of treatment combining substances of the verapamil group with a β-receptor blocker, special attention must be given to possible signs of incipient or progressing heart failure.

Calcium antagonists inhibit impulse formation and conduction in the sinoatrial and AV node (negative chronotropy, negative dromotropy; Sect. 3.1.2). On account of autonomic counter-regulation, this effect normally plays a role in vivo only with substances of the verapamil group and diltiazem. Especially in elderly patients and in patients with preexisting disease of the sinoatrial node (sick sinus syndrome) or of the AV node (higher degree AV block), bradycardias may develop (sinus bradycardia, sinus arrest with asystole, higher degree AV block or bradycardia in atrial fibrillation). Severe bradycardias occur especially with verapamil, but this side effect is generally thought to be very rare (Roundtable 1981).

A special case is presented by patients with preexcitation syndromes (particularly Wolff-Parkinson-White syndrome) and atrial fibrillation; since verapamil, gallopamil and diltiazem inhibit AV conduction, accelerated anterograde conduction via the accessory bundle can develop and consequently ventricular tachycardias, even leading to ventricular fibrillation (Sects. 3.1.2 and 4.3). The former substances are thus contraindicated in WPW syndrome with atrial fibrillation.

In combination with other drugs which inhibit the sinoatrial and AV node (digitalis, β-receptor blockers), the side effects may be additive. This may even be desirable in certain indications (e.g. control of ventricular rate in atrial fibrillation). If, in other indications, bradycardias are to be avoided, then calcium antagonists of the dihydropyridine group such as nifedipine are to be preferred (cf. combination therapy in stable angina pectoris in Sect. 4.1.2.3).

Bradycardias are extremely rare during nifedipine treatment and have so far been observed only in patients with preexisting sick sinus syndrome (Sect. 3.1.2). If the frequency of the sinoatrial node decreases significantly under verapamil or diltiazem treatment, AV dissociation, atrioventricular escape rhythm or junctional rhythm may develop (Ishikawa et al. 1983; Walker 1983).

Within the arterial vascular system calcium antagonists act as vasodilators, thereby lowering peripheral resistance and blood pressure. The vasodilating effect of the dihydropyridine group is stronger than that of diltiazem or of substances of the verapamil group (Sect. 3.1.5). Beyond therapeutically desired reduction of blood pressure, however, an excessive drop in blood pressure may occur, resulting in hypotension or orthostatic symptoms. Subjectively, patients experience vertigo, flushing, fatigue or drowsiness. These side effects are most pronounced among dihydropyridine preparations (nifedipine, etc.), but they are mostly dose-dependent and can be eliminated by a reduction of dosage. A drop in blood pressure below 100 mmHg in younger patients, or below 120 mmHg in the elderly, is to be avoided. Excessive lowering of blood pressure may result in reductions in cerebral blood flow, coronary perfusion and renal blood flow. In extreme cases transient ischaemic attacks (Arzneimitteltelegramm 1981), coronary ischaemia with angina pectoris (so-called paradoxical effect of nifedipine in stable angina pectoris), or prerenal azotemia may develop. Tinnitus may also be associated with a reduction in blood pressure. The combination of calcium antagonists with nitrates or diuretics produces a more marked lowering of blood pressure, and there may be a higher incidence of vertigo or tendency to collapse.

General arterial vasodilation by calcium antagonists produces different effects in the various vascular regions. The skin can show localized or diffuse reddening and a sensation of warmth may also develop; typically there is flushing of the face owing to dilation of extracranial vessels. On the other hand, dilation of intracranial vessels can result in headache, the mechanism involved resembling that associated with the headache experienced during nitrate therapy. Such complaints arise mainly with nifedipine treatment, and they gradually recede as treatment continues. The increased perfusion of the nasal mucosa can lead to nasal congestion, producing the sensation of a "stuffed-up nose" (Parisi et al. 1982; Krämer and Kübler 1983). In pulmonary ventilation disorders (e.g. chronic obstructive lung disease) nifedipine (or felodipine, nitrendipine, etc.) can increase perfusion of less well ventilated vascular regions; the resulting ventilation–perfusion imbalance (mismatch) may then induce arterial hypoxia (Sects. 4.6.1 and 4.9). In patients with initially healthy lungs, however, no impairment of pulmonary function is to be expected (Ebner and Dünschede 1976).

In the case of preexisting renal failure, renal function may deteriorate during treatment with calcium antagonists (data for nifedipine) as a result of "altered intrarenal haemodynamics" (Diamond et al. 1984). Where there is no preexisting renal disease, given adequate perfusion pressure, renal function is unaffected (Ebner and Dünschede 1976).

A relatively common side effect of calcium antagonists is the development of ankle oedema. This is not a manifestation of heart failure but is rather evoked by local dilation of precapillary resistance vessels (arterioles) with increased fluid filtration into the interstitial space. In contrast to most authors, Opie (1986) also discusses activation of the renin–angiotensin–aldosterone system. This side effect also occurs mainly with substances of the dihydropyridine group, which have a strong vasodilating effect. If it is pronounced, then protracted periods of standing or sitting should be avoided, support stockings should be worn, and the additional administration of diuretics may perhaps be considered (although the efficacy of diuretic treatment in such oedema is debated in the literature).

Vasodilation in the region of the genital tract may result in longer and heavier menstruation (Singh et al. 1983; Rodger and Torrance 1983), myometrial relaxation adding to this effect. Peripheral vasodilation with reduction of peripheral vascular resistance induces autonomic counter-regulation with activation of the sympathetic nervous system and with additional vagolysis. Autonomic counter-regulation is especially marked with the strongly vasodilating dihydropyridine derivatives. Thus, heart rate often increases with these substances, and in some cases patients experience palpitations (*torsade de pointes* has even been described; Sect. 5.3). An increase in perspiration, nervousness, sleep disorders and tremor are also associated with activation of the sympathetic nervous system. In the treatment of stable angina pectoris with nifedipine an increase in anginal attacks is occasionally seen; this is explained as due to the reduction in blood pressure, a decrease in coronary perfusion pressure and accelerated heart rate, thus leading to an increased myocardial oxygen demand (Sect. 4.1.2.1). In cases where there are pronounced symptoms of increased autonomic counter-regulation or "paradoxical" angina pectoris, nifedipine can be either replaced by verapamil, gallopamil or diltiazem, or supplemented by a (cardioselective) β-receptor blocker.

Calcium antagonists inhibit the smooth muscle of the gastrointestinal and urogenital tract. The desired decrease in the pressure of the lower oesophageal sphincter in oesophageal spasm and achalasia can, in essentially healthy persons, trigger hiatal insufficiency and consequently symptoms of gastro-oesophageal reflux (heartburn, retrosternal pain; Sect. 4.8). Furthermore, under treatment with calcium antagonists gastrointestinal passage is delayed, and patients complain either of nausea and bowel distension or of constipation. Overall, verapamil seems to lead to constipation more frequently than does nifedipine (Hedner 1986). These symptoms rarely require withdrawal of therapy, and in long-term treatment constipation tends to recede. Combination therapy with β-receptor blockers has a favourable effect, for the latter substances tend to accelerate gastrointestinal passage.

Gallbladder motility also appears to be reduced (Schmidt 1986, personal communication) by calcium antagonists (investigations with nifedipine). Thus, theoretically long-term treatment with calcium antagonists could increase the incidence of gallbladder and bile duct stone formation, although so far no firm data indicating such an effect have been published (Sect. 4.8).

Diminished detrusor activity of the bladder owing to the use of calcium antagonists (results relate to nifedipine and flunarizine) may lead to micturition disturbances, even to overflow incontinence. Not surprisingly, patients with preexisting micturition problems (e.g. sphincter sclerosis or prostate hyperplasia) appear to be particularly at risk (Sect. 4.10.2). Calcium antagonists relax the myometrium and produce arterial vasodilation in the genital region; longer and heavier menstruation may therefore occur (Singh et al. 1983; Rodgers and Torrance 1983).

Theoretically, arterial vasodilation, with a decrease in peripheral vascular resistance, activates the renin–angiotensin–aldosterone system, and may thus lead to sodium and water retention, predisposing to the development of lower limb oedema (particularly in the ankle region). However, since at the onset of treatment with calcium antagonists (mainly with the dihydropyridines), a natriuretic effect is demonstrable and since activation of the renin system decreases in long-term therapy, there is overall no essential change in body weight, and in particular no progressive fluid retention (Sects. 3.3.1 and 3.3.2). Increasing nocturia, which is occasionally observed (Williams and Donaldson 1976), is explained, on the one hand by retention of fluid, and on the other by disturbed micturition due to calcium antagonist therapy.

Impaired glucose tolerance can result in an increase in the blood glucose level in healthy subjects and a deterioration of glucose tolerance in known diabetics – consequences of inhibition of insulin secretion by calcium antagonists (mainly by nifedipine, less by verapamil; Hedner 1986) and of autonomic counter-regulation (increased blood glucose due to activation of the sympathetic nervous system). Overall, however, significant changes in glucose metabolism are rare, and it is usually unnecessary to change treatment to antidiabetic agents or insulin (review by Ebner et al. 1976, 1977, 1983).

Further details regarding the side effects of calcium antagonists are provided by reviews by Opie (1980), Hopf et al. (1982), Krämer and Kübler (1983), Kimbel (1983), Eichstädt (1984) and Hedner (1986).

5.2 Side Effects Independent of Blockade of Calcium-Dependent Metabolic Processes

In contrast to the predictable side effects descrtibed in Sect. 5.1, the unwanted effects described in this section are markedly less frequent. Their overall incidence is under 1% (Hedner 1986). Current theories as to the cause of such effects favour allergic reactions or an idiosyncrasy (congenital hypersensitivity to a drug, e.g. on account of an enzyme defect, without previous sensitization).

Dermatological changes (pruritus, erythemas, urticaria, maculopapular exanthems) are seen with all calcium antagonists. In particular with nifedipine therapy there have been reports of erythromelalgia (reddening and increased temperature of the skin, accompanied by burning sensations; Fisher et al. 1983; Brodmerkel 1983); under diltiazem a generalized lymphadenopathy, completely reversible following drug withdrawal, was observed (Scolnick and Brinberg 1985).

Similarly, with all calcium antagonists, reversible elevation of liver enzymes may develop, akin to allergic hepatitis (hypersensitivity; Brodsky et al. 1981; Kimbel 1983; Hare and Horowitz 1986). In most cases such hepatic changes are fully reversible on withdrawal of medication. In contrast, under treatment with perhexiline irreversible liver damage can develop, which histologically resembles alcoholic hepatitis and which in some cases can progress to cirrhosis (Opie 1980; Kimbel 1983). This side effect appears to be specific to perhexiline (Sect. 5.3), since it has not so far been observed with any other calcium antagonist.

It is not yet possible to state to what extent the following side effects are due to allergy or idiosyncrasy: the painless swelling of the parotid gland ("drug mumps") seen with nifedipine (Bosch et al. 1986); gingival hyperplasia (sometimes accompanied by gingival bleeding or gingivitis) occurring during nifedipine therapy (Arzneimitteltelegramm 1986; Hedner 1986); and the disturbances of taste (dysgeusia) and smell (dysosmia) with diltiazem (Berman 1986) or nifedipine (Levenson and Kennedy 1985).

Hedner (1986) even designates as idiosyncrasies the arthralgias and cases of akathisia (compulsion to move constantly) observed in patients receiving diltiazem (Jacobs 1983) and flunarizine (Chouza et al. 1986). The cause remains unknown of the myoclonic dystonia seen occasionally with verapamil and nifedipine (but not diltiazem) therapy (Hicks and Abraham 1985; De Medina et al. 1986), although the possibility of a direct central nervous effect at the level of the basal ganglia has been discussed.

In Summary. It is not yet clear whether several of the side effects described here are substance-specific (Sect. 5.3) or whether they in principle apply to the whole group of calcium antagonists. The latter is, however, improbable, since calcium antagonists present such heterogeneity (cf. classification and chemical structure Sects. 1.2.5 and 2.1).

5.3 Substance-Specific Side Effects

Table 33 summarizes side effects which so far have only been observed with certain calcium antagonists. These effects are overall rare to very rare, and mainly involve certain groups of patients at risk (e.g. extrapyramidal disorders in elderly patients receiving flunarizine treatment). However, it was not always possible to clarify beyond doubt that an observed side effect was due to administration of a specific calcium antagonist, since some of the patients also received other medication concomitantly, and in most cases they were not reexposed to the calcium antagonist in question.

On account of severe substance-specific side effects, which moreover appeared relatively frequently, perhexiline and prenylamine are no longer recommended for treatment of coronary heart disease (Opie 1980; Kimbel 1983). Only the problems directly relevant to clinical practice are discussed briefly here; with regard to specific issues, readers are referred to the original papers cited.

The gingival hyperplasia observed under nifedipine treatment (sometimes accompanied by gingival bleeding and gingivitis) resembles changes seen with phenytoin and is in most cases reversible. The painless parotid swelling, too, recedes following withdrawal of nifedipine or during long-term treatment.

The tremor observed during therapy with several calcium antagonists can, on one hand, be a consequence of activation of the sympathetic nervous system, in the context of autonomic counter-regulation; on the other hand, it can arise owing to effects on cerebral neurotransmitters (e.g. blockade of postsynaptic dopamine receptors during treatment with flunarizine or cinnarizine). When tremor occurs during treatment with preparations of the dihydropyridine type,

Table 33. Side effects of various calcium antagonists

Organ system	Symptoms	Calcium antagonist	Reference
Skin and mucosa	Erythromelalgia	Nifedipine	Fisher et al. 1983 Brodmerkel 1983
	Gingival hyperplasia (gingivitis, gingival bleeding)	Nifedipine	Arzneitelegramm 1986
	Parotid gland swelling ("drug mumps")	Nifedipine	Bosch et al. 1986
	Dysgeusia, dysosmia	Nifedipine Diltiazem	Berman 1985, Levenson and Kennedy 85

Table 33. (continued)

Organ system	Symptoms	Calcium antagonist	Reference
Central nervous system/psyche	Tremor	Nifedipine Cinnarizine Prenylamine Perhexiline	Bosso et al. 1986 Laporte and Capella 1986 Kimbel 1983
	Muscle cramps	Nifedipine	Bosso et al. 1986
	Tetany	Nifedipine	Bosso et al. 1986
	Ataxia	Perhexiline	Opie 1980
	Myoclonic dystonia	Verapamil Nifedipine	Hicks and Abraham 1985 De Medina et al. 1986
	Visual disorders (double vision, papilloedema), increase in intracranial pressure	Perhexiline	Opie 1980
	Polyneuropathy (polyradiculitis)	Perhexiline	Opie 1980 Kimbel 1983
	Akathisia	Diltiazem Flunarizine	Jacobs 1983 Chouza et al. 1986
	Extrapyramidal movement disorders, parkinsonian syndrome	Flunarizine Cinnarizine	Chouza et al. 1986 D'Alessandro et al. 1986 Micheli et al. 1989
	Tardive dyskinesia	Flunarizine	Chouza et al. 1986
	Anxiety states	Flunarizine	Arzneimittelkommission 1986
	Depression	Flunarizine	Chouza et al. 1986 D'Alessandro et al. 1986
	Acute psychosis	Nifedipine	Ahmad 1984a Kahn 1986
	Sedation	Flunarizine	Holmes et al. 1984
	Hyperactivity, mania	Diltiazem	Brink 1984
Blood	Agranulocytosis	Nifedipine (?)[a]	Voth and Turner 1983
Eye	Cataract	Nifedipine (?)[b]	Van Heyningen and Harding 1986
Kidney	Acute renal failure	Diltiazem (1 case)	Ter Wee et al. 1984
Heart	QT prolongation, *torsade de pointes* (Ventricular tachycardia)	Prenylamine Lidoflazine	Fraser and Ikram 1986 Smith and Gallagher 1980
	Torsade de pointes	Nifedipine Verapamil	Grayson and Kennedy 1982 Winters et al. 1985
Other	Weight gain	Flunarizine	Holmes et al. 1984
	Gynaecomastia Cachexia	Nifedipine Perhexiline	Arzneitelegramm 1986 Opie 1980
	Impotence	Perhexiline	Opie 1980, Kimbel 1983
	Loss of libido	Nitrendipine	Luderscheidt 1987

[a] Questionable; only one case, with concomitant chloral hydrate.
[b] Association improbable; further details in text.

the dose should be reduced, or a β-receptor blocker should be added to the regimen; in contrast, however, cinnarizine and flunarizine should be discontinued if tremor appears, since it can be an early indicator of further extrapyramidal side effects.

The rare, but serious side effects seen with perhexiline treatment (peripheral neuropathy, polyradiculitis, increase in intracranial pressure, hepatitis) are possibly correlated to retarded elimination of the drug and with its consequent accumulation in the body (an extremely long elimination half-life); perhexiline should no longer be used in coronary heart disease.

Thanks to investigations by Chouza et al. (1986), D'Alessandro et al. (1986) and also by Laporte and Capella (1986), only very recently has an awareness developed as to the frequent occurrence of extrapyramidal side effects and of depression under flunarizine and cinnarizine. Both drugs act as calcium antagonists and as antihistaminics; chemically, as piperazine derivatives, they are closely related to the neuroleptics. These side effects do not appear to be dose-dependent – a dose of 10 mg flunarizine daily is frequently sufficient to provoke such a response – and mainly older patients are affected (Benvenuti et al. 1986). D'Alessandro et al. (1986) reported depression and bradykinesia in up to 50% of patients treated. At the present time it is unclear whether flunarizine (or cinnarizine) can induce parkinsonian syndrome as a new disease or merely unmask a subclinical disorder which already exists. Pathophysiologically, pharmacogenetic factors and blockade of postsynaptic dopamine receptors appear to be significant (in animal experiments flunarizine is also thought to inhibit dopamine release in the corpus striatum; Agnoli and Nappi 1986). The clinical presentation ranges from tremor and akathisia via bradykinesia and akinesia to tardive dyskinesia. Furthermore, frequent occurrence of depression and anxiety is reported (Chouza et al. 1986; D'Alessandro et al. 1986; Arzneimittelkommission Deutsches Ärzteblatt 1986; Micheli et al. 1989). Following withdrawal of flunarizine or cinnarizine, symptoms usually recede, although in some cases dykinesias and akathisia were still in evidence up to 20 months later (Choza et al. 1986). D'Alessandro et al. (1986) recommend, apart from the discontinuation of flunarizine or cinnarizine, additional administration of amitriptyline in a dose of 50–75 mg daily. In contrast to flunarizine or cinnarizine, verapamil and diltiazem (180–360 mg/day) appear to be able to improve tardive dyskinesia following treatment with neuroleptics (review by Ross et al. 1987).

All in all, in the light of side effects observed with flunarizine and cinnarizine, particularly in the elderly, the recommendation by Laporte and Capella (1986) should be followed: these substances should not in general be given to patients with symptoms of "chronic cerebrovascular disease"; a clearly defined indication should be a precondition of their use (i.e. vestibular vertigo, prophylactic treatment of migraine, combination therapy of epilepsy).

On the basis of the chemical similarity between the dihydropyridine calcium antagonists and methylphenyltetrahydropyridine, which exhibits neurotoxic effects and can trigger a parkinsonian-like syndrome, Alemany (1986) postulates that the dihydropyridine calcium antagonists (particularly nifedipine) can also induce parkinsonian syndrome in long-term therapy. To date no clinical support for this theory has been forthcoming, even though dihydropyridine derivatives have been used worldwide for the past 10 years, often on a long-term basis (Traber et al. 1986). Owing to a significant lowering of blood pressure, calcium antagonists can decrease cerebral blood flow and precipitate vertigo, drowsiness or confusion. However, two reported cases of acute psychosis occurring during nifedipine therapy cannot be explained by a drop in blood pressure, cardiac arrhythmia or any other factor reducing cerebral blood flow. In both cases, moreover, psychic symptoms resolved rapidly as soon as nifedipine was withdrawn (Ahmad 1984a; review by Kahn 1986).

Severe haematological side effects do not appear to be associated with calcium antagonists. A reported case of reversible agranulocytosis could not be directly attributed to nifedipine since there was concomitant use of chloral hydrate (Voth and Turner 1983).

A possible connection between nifedipine treatment and cataract has been suggested by Van Heyningen and Harding (1986). However, contradictory findings in animal experiments (Sect. 3.3.4) and a negative correlation obtained in the study by Kewitz et al. (1986) discount such a hypothesis.

Under prenylamine and lidoflazine therapy, ventricular tachycardias of the *torsade de pointes* type have been described. This is possibly associated with the prolongation of the QT interval caused by these preparations (Smith and Gallagher 1980; Schrör 1985). For both nifedipine and verapamil one case of *torsade de pointes* has also been described (Grayson and Kennedy 1983; Winters et al. 1983). A direct connection seems rather improbable, however, because of accompanying factors (hypokalaemia, previous treatment with antidepressants, heart failure in coronary heart disease). On the other hand, the possibility of inducing or aggravating arrhythmias in individual cases must be borne in mind when prescribing calcium antagonists of the dihydropyridine type, with their powerful autonomic counter-regulation.

Under flunarizine treatment, weight gain (2–3 kg) is occasionally observed, mainly in patients with migraine. This effect is unrelated to fluid retention, however, and is more likely explained by a change in eating habits, similar to that occurring with depot neuroleptics (Holmes et al. 1984).

5.4 Overdosage of Calcium Antagonists: Intoxication

The literature includes several reports on poisoning (mostly attempted suicides) with calcium antagonists (nifedipine, verapamil, diltiazem). Here, up to 1000 mg nifedipine, 9600 mg verapamil, or 2640 mg diltiazem have been taken (surveys by Mayer et al. 1985; Herrington et al. 1986; Chaffman and Brogden 1985). The following symptoms are observed in severe intoxication: disturbed consciousness, even developing into coma; drop in blood pressure, which sometimes cannot be measured in the periphery; bradycardias (bradyarrhythmia, II–III degree AV block, sinus bradycardia, asystole); ventricular escape rhythms; hyperglycaemia; hypokalaemia; metabolic acidosis; hypoxia; and cardiogenic shock with pulmonary oedema. These effects are attributable to the known vasodilating, negative inotropic, negative chronotropic and negative dromotropic properties of calcium antagonists. Hypokalaemia and hyperglycaemia are probably a consequence of maximal autonomic counter-regulation (with activation of the sympathetic nervous system) as hypotension progresses.

The therapeutic priorities are: elimination of the toxin and restoration of stable cardiovascular function. Unabsorbed tablets or capsules should be removed by gastric lavage, perhaps in combination with lavage of the small intestine. If laxatives are given, the inhibition of intestinal smooth muscles (extending to intestinal muscle atonia) that can be caused by calcium antagonists must be borne in mind. Haemodialysis is not indicated since verapamil, nifedipine and flunarizine cannot be dialysed; haemofiltration, however, and perhaps also plasmaphaeresis (high plasma protein binding of calcium antagonists) are recommended (Mayer et al. 1985). Especially in the case of poisoning with sustained-release formulations, complete elimination from the small intestine, as far as possible, should be the therapeutic goal in order to prevent further absorption of the drug (Mayer et al. 1985).

Bradycardias are treated symptomatically with atropine and/or β-sympathomimetics (isoprenaline, orciprenaline). In most instances, however, a temporary pacemaker is necessary to enable specific antiarrhythmic therapy should tachycardias occur (triggered by hypokalaemia, hypoxia, acidosis or iatrogenically). Hypotension, a consequence of cardiogenic shock and arterial vasodilation, is treated with calcium (1–2 g calcium gluconate intravenously), dopamine (up to 25 μg per kilogram body weight per minute), dobutamine (up to 15 μg per kilogram body weight per minute), adrenaline or noradrenaline. The dosages of these drugs are adjusted according to the effect obtained (direct intraarterial blood pressure monitoring), and, as in poisoning by β-receptor blockers, extremely high doses are sometimes necessary. Serum calcium levels should be maintained at high-normal or slightly elevated levels. In addition, in the early phase, fluid should be substituted (Ringer's or saline solution), on account of arterial vasodilation. If blood pressure rises again, however, close attention must be given to maintenance of fluid balance, in order to avoid the development of pulmonary oedema due to excess fluid administration and heart failure (negative inotropic effect of calcium antagonists).

In Summary. Treatment of calcium antagonist poisoning does not differ in principle from treatment of overdose, except that the dose of the drugs used is much higher. Close surveillance of the patient by means of continuous ECG monitoring, direct intraarterial monitoring of blood pressure and surveillance of blood gas parameters is mandatory, and side effects of the drugs administered must be kept in mind (especially tachycardias under treatment with sympathomimetics).

5.5 Incidence of Side Effects

Taking contraindications into account (Sect. 5.6), side effects occur in 5%–20% of all patients treated with calcium antagonists. Only Bussmann and Hopf (1985) in their survey relating to nifedipine state an incidence of 15%–50%. On the basis of more than 10 years experience with this preparation, however, we think this estimate is too high. Substances of the dihydropyridine group tend to cause side effects more frequently, but these are often only seen at the onset of therapy and are mostly of a minor nature. Moreover, these side effects are often dose-related, and may be preventable by reducing the dose, or by individualized dosage.

As already mentioned in the context of treatment of angina pectoris (Sects. 4.1.2.1 and 4.1.2.3), patients vary in their sensitivity to side effects and in their therapeutic response to the various calcium antagonists. The same applies to the occurrence of side effects; since the individual substances present different spectra of side effects, in some cases changing from one calcium antagonist to another may be appropriate (Table 31).

Side effects necessitate withdrawal of calcium antagonists in only about 1%–5% of all cases treated (review by Bussmann and Hopf 1985; Bleifeld and Spokojny 1983). This percentage is in the same range as for β-receptor blockers; both groups of substances can thus be described as "well-tolerated drugs".

With combination therapy (e.g. in coronary heart disease, arterial hypertension, cardiac arrhythmias) a lower dose can sometimes be used, thus often also reducing side effects. However, as with every form of combination therapy or concomitant administration of drugs (Sect. 5.8), existing side effects may be reinforced or new side effects may be evoked.

5.6 Contraindications

Absolute contraindications for substances of the verapamil group (verapamil, gallopamil, tiapamil) and diltiazem are: II–III degree AV block; sinoatrial node disease (sick sinus syndrome) with a tachycardia–bradycardia syndrome; cardiogenic shock; and acute myocardial infarction with complications (bradycardia, overt congestive heart failure, hypotension). Only verapamil may be used in pregnancy; but even here during the first 3 months caution is called for, and the indication must be clearly defined. The other substances of the verapamil group and diltiazem are contraindicated in pregnancy. Relative contraindications for the verapamil group and diltiazem include sinus bradycardia (heart rate below 50 per minute), hypotension (systolic blood pressure below 90 mmHg), atrial fibrillation or atrial flutter in combination with a preexcitation syndrome (e.g. WPW syndrome) and also overt congestive heart failure. In ongoing treatment with a β-receptor blocker, verapamil, gallopamil or diltiazem should as a rule not be given intravenously because of the risk of bradycardia. A I degree AV block is not a contraindication to the use of the latter substances, but the possibility of development of higher degree AV block must by appreciated. During the lactation period the indication for verapamil must be rigorously defined, and all the other calcium antagonists, particularly diltiazem, are contraindicated (verapamil and diltiazem are secreted into breast milk; Okada et al. 1985).

Contraindications for nifedipine and the other substances of the dihydropyridine group are: arterial hypotension (systolic blood pressure below 90 mmHg); pregnancy; and lactation. In overt congestive heart failure, especially in the context of acute myocardial infarction, particular caution is called for (Sects. 4.1.4 and 4.5.1). If there is cardiogenic shock, dihydropyridine derivatives are contraindicated. If nimodipine is given in subarachnoid haemorrhage, special caution is necessary if intracranial pressure is increased (cerebral oedema).

Flunarizine and cinnarizine are contraindicated in pregnancy, during the lactation period, as well as in acute stroke (except in well-controlled studies; Sect. 4.7.3). In extrapyramidal system disorders, particularly parkinsonian syndrome, special care is necessary, since symptoms of such diseases may be aggravated (Sect. 5.3).

Contraindications for prenylamine are: II–III degree AV block; intraventricular conduction disorders (bundle branch block); bradycardia; overt congestive heart failure; early pregnancy; and the lactation period. If QT prolongation or widening of the QRS complex occurs during treatment, and, in pregnancy, 2 weeks before the estimated date of delivery, the drug should be discontinued.

The only contraindication to application of fendiline so far reported is overt congestive heart failure. Perhexiline is contraindicated in hepatic and renal disorders, in peripheral neuropathy, as well as in patients with known hypersensitivity to the substance.

In general, porphyria is a contraindication to treatment with calcium antagonists (results from animal experiments).

5.7 Withdrawal of Calcium Antagonists: A Calcium Antagonist Withdrawal Syndrome?

This question has already been extensively dealt with in Sect. 4.1.2.1.

In Summary. At the present time there are no indicators that calcium antagonists induce a typical withdrawal syndrome, as has been described for β-receptor blockers. In individual cases, however, there have been reports of an increase in myocardial ischaemia (increased frequency of angina pectoris, ST segment depression) and of the occurrence of acute myocardial infarction in patients with "normal" coronary arteries (as determined by coronary angiography), probably via a coronary spasm, after discontinuing calcium antagonists (review by Subramanian et al. 1983; Kozeny et al. 1986). In general, on the basis of current experience, gradual reduction of calcium antagonists over several days is not necessary.

5.8 Drug Interactions in Calcium Antagonist Therapy

Calcium antagonists are frequently combined with other drugs in the treatment of cardiovascular diseases. Details of such combination therapies are described in Sects. 4.1–4.6. In heart failure and arterial hypertension calcium antagonists are combined with diuretics. Lowering in blood pressure is more marked than in monotherapy, but the possibility of hypotension or tendency to collapse must be borne in mind. Furthermore, in the combination of calcium antagonists with nitrates the effects on blood pressure and heart rate are additive. With substances of the dihydropyridine type the heart rate may increase (Ebner 1977). This can be avoided by combination with preparations of the verapamil group or diltiazem.

If β-receptor blockers and calcium antagonists are administered concomitantly, the negative inotropic and the negative chronotropic effects of substances of the verapamil group or diltiazem and of β-receptor blockers are additive. Nevertheless, this treatment is in general well tolerated, although the contraindications for both substances must be noted (Sect. 5.6). In the combination of dihydropyridines and β-receptor blockers congestive heart failure only seldom occurs, and the antihypertensive effects of both groups of substances are additive (review by Ebner 1977, 1983; Jee and Opie 1984).

If vasodilators (e.g. prazosin, ACE inhibitors) are combined with calcium antagonists in the treatment of hypertension, the resulting marked reduction in afterload can induce hypotension and a tendency to collapse (Opie and Jee 1984).

With regard to the interactions between calcium antagonists (verapamil type, diltiazem) and antiarrhythmics the reader is referred to Sect. 4.3.6; particular attention must be given to the possible development of bradycardia and congestive heart failure.

While the effects concerning combination therapy described here are predictable on the basis of the pharmacological and clinical properties of the individual substances, there are also drug interactions by which the pharmacokinetic profile of individual substances or an already existing action is changed, or new effects arise.

The principal drug interactions relating to calcium antagonist therapy are summarized in Table 34. With the expanding range of indications for calcium antagonists, the number of drug interactions will doubtless also increase (a current review is provided by Piepho et al. 1987).

To date no interactions have been described concerning the combination of calcium antagonists with anticoagulants, platelet aggregation inhibitors, antidiabetic agents and X-ray contrast materials (e.g. in coronary angiography; review by Ebner 1977, 1983). Attention is again drawn to the rare cases in which changes in glucose tolerance were observed; with such changes adaptation of insulin or oral antidiabetic therapy is mandatory (Sect. 5.1). For a discussion of the application of calcium antagonists and anaesthetics see Sect. 5.12.

Nearly all calcium antagonists are thought, by analogy with the effect of quinidine, to raise the digoxin level by reducing renal digoxin clearance. While data on substances of the verapamil group are unequivocal (Betz et al. 1981 a, 1983; Klein et al. 1982), those for nifedipine and diltiazem are contradictory; some workers report a rise in the digoxin level (Belz et al. 1981; De Cesaris et al. 1984; Hedner 1986; Kirch and Ohnhaus 1987, personal communication), while others deny any such influence (review by Ebner et al. 1983; Elkajam et al. 1985; Boden et al. 1986). The successor substances of nifedipine – nitrendipine, nisoldipine and felodipine – are also thought to raise the digoxin level (Table 34). It would thus seem appropriate to maintain surveillance for signs of digitalis overdosage, or perhaps to determine the digoxin plasma level and where necessary to adjust the digitalis dosage.

Theophylline levels can also be raised by calcium antagonists (nifedipine, verapamil). Owing to the narrow therapeutic range of theophylline, signs of intoxication may develop (Harrod 1987); on reducing the dose of theophylline the combination is well tolerated.

Cimetidine and, to a lesser extent, ranitidine increase the plasma level of nifedipine (or nitrendipine or felodipine), diltiazem and verapamil, and lead to marked reduction in blood pressure (Kirch et al. 1983; Smith et al. 1984; Winship et al. 1985; Edgar 1987, personal communication). This is explained by reduced hepatic blood flow with a decreased first-pass effect and altered metabolism of calcium antagonists.

If in anticonvulsant combination therapy carbamazepine and calcium antgonists (verapamil, diltiazem) are given, the carbamazepine level rises distinctly, and there is a greater incidence of its neurotoxic side effects (Macphee et al. 1986; Kroemer 1986).

Following kidney transplantation, or possibly heart transplantation, cyclosporin A may be combined with a calcium antagonist; in such situations, diltiazem, verapamil and the dihydropyridine derivative, nicardipine, are thought to increase the cyclosporin A plasma level (Grino et al. 1986; Pochet and Pirson 1986; Lindholm and Henricsson 1987). In one case deterioration of renal function was described. Such an interaction has not been observed with nifedipine, and is in contrast to the results of Wagner and his co-workers (Sect. 4.11), which indicate a lower nephrotoxicity of cyclosporin in combination with diltiazem.

In Summary. To date severe drug interactions have only rarely been observed with calcium antagonists. If, however, calcium antagonists are combined with drugs with a small therapeutic range, i.e. in which marked fluctuations in the optimal plasma level are to be avoided (e.g. digoxin, carbamazepine, cyclosporin A), close attention must be paid to possible signs of drug interactions, and where necessary the plasma level must be determined and the dose correspondingly adjusted. This applies in particular to substances which, like the calcium antagonists, are eliminated mainly by the liver.

Table 34. Drug interactions

Combination with	Calcium antagonist	Effect	Reference
Digoxin	Verapamil Gallopamil	Digoxin level increased	Belz et al. 1981 a, 1983 Klein et al. 1982
	Nifedipine	Digoxin level increased	Belz et al. 1981 b, 1983 DeCesaris et al. 1986 Hedner 1986
	Nifedipine	Digoxin level unchanged	Ebner et al. 1983
	Diltiazem	Digoxin level increased	Hedner 1986
	Diltiazem	Digoxin level unchanged	Elkayam et al. 1983 Boden et al. 1986
	Felodipine	Digoxin level increased	Edgar 1987, personal communication
	Nitrendipine (nisoldipine)	Digoxin level increased	Kirch et al. 1984, 1987, personal communication
Digitoxin	Verapamil Diltiazem	Digitoxin level increased	Kuhlmann 1985
Cimetidine	Nifedipine (nitrendipine)	Nifedipine (nitrendipine) level increased, reduction in blood pressure more marked	Kirch et al. 1983
	Felodipine	Felodipine level increased	Edgar 1987, personal communication
	Verapamil	Verapamil level increased	Smith et al. 1984
	Diltiazem	Diltiazem level increased	Winship et al. 1985
Ranitidine	Nifedipine	Nifedipine level increased, slightly greater decrease in blood pressure (both insignificant)	Kirch et al. 1983
Theophylline	Nifedipine Verapamil	Theophylline level increased	Harrod 1987
Carbamazepine	Verapamil Diltiazem	Carbamazepine level in-increased, neurotoxicity increased	Macphee et al. 1986 Kroemer 1986
	Felodipine	Felodipine level decreased	Capewell et al. 1988
Cyclosporin A	Diltiazem	Cyclosporin A level increased	Grino et al. 1986
	Nicardipine		Pochet and Pirson 1986
	Verapamil		Bourbigot et al. 1986 Lindholm and Henricsson 1987
	(cf. nifedipine)	Cyclosporin A level unchanged	Bourbigot et al. 1986
Phenytoin	Nifedipine Verapamil	Phenytoin level increased	Ahmad 1986
Lithium	Verapamil	Lithium level lowered, lithium effect reduced	Pharmaprofil 1986
	Diltiazem	parkinsonian syndrome	Valdiserri 1985

Table 34. (continued)

Combination with	Calcium Antagonist	Effect	Reference
Rifampin[a]	Verapamil	Verapamil level decreased	Rahn et al. 1985
Quinidine	Verapamil	Quinidine level increased, decrease in blood pressure more marked	Maisel et al. 1985
	Nifedipine	Quinidine level decreased, quinidine effect reduced	Farringer et al. 1984
Propranolol	Diltiazem Verapamil	Propranolol level increased	Pieper 1984 Etoh et al. 1983
Metoprolol	Felodipine	Metoprolol level increased	Edgar 1987, personal communication
	Verapamil	Metoprolol level increased	McLean et al. 1985
Alcohol	Flunarizine	Sedation increased	

[a] Applies also to other drugs inducing hepatic drug-metabolizing enzymes, such as barbiturates, phenytoin.

References

Agnoli A, Nappi G (1986) Useless Drugs? Lancet 2: 1217

Ahmad S (1984a) Nifedipine-induced acute psychosis J Am Geriatr Soc 32: 408

Ahmad S (1984b) Nifidipine-phenytoin interaction J Am Coll Cardinol 3:1582

Alemany ML, (1986) Parkinsonism as late toxic effect of dihydropyridines? Lancet 1: 737

Arzneimittelkommission der deutschen Ärzteschaft (1986) Zentralnervöse Reaktionen auf Flunarizin Dtsch Ärztebl 83: 1954

Arznei-Telegramm (1981) Zerebrale Mangeldurchblutung als Folge der Anwendung des Kalziumantagonisten Nifedipin (Adalat) Arzneitelegramm 11: 99

Arznei-Telegramm (1986) Gynäkomastie und Nifedipin Arzneitelegramm 4: 32

Arznei-Telegramm (1986) Nifedipin und Zahnfleischhyperplasie Arzneitelegramm 7: 68

Belz GG, Doering W, Munkes R, Aust PE, Belz G (1981 a) Effects of various calcium-antagonists on blood level and renal clearance of digoxin Circulation 64 (Suppl 4): IV-24

Belz GG, Aust PE, Munkes R (1981 b) Digoxin plasma concentrations and nifedipine Lancet 1: 844–845

Belz GG, Doering W, Munkes R, Matthews J (1983) Interaction between digoxin and calcium antagonists and antiarrhytmic drugs Clin Pharmacol Ther 33: 410–417

Benvenuti F, Baroni A, Bandinelli S, Sommazzi P, Corradetti R, Pantaleo T (1986) Side effects of flunarazine Lancet 2: 464

Berman JL (1985) Dysomia, Dysgeusia and Diltiazem Ann Intern Med 102: 717

Bleifeld W, Spokojny A (1983) Therapie der koronaren Herzkrankheit mit Kalziumantagonisten in Klinik und Praxis Med Welt 34: 1335–1339

Boden WE, More G, Sharma A, Bough EW, Korr KS, Young PM, Shulman RS (1986) No increase in serum digoxin concentration with high-dose diltiazem Am J Med 81: 425–428

Bosch X, Campistol JM, Botey A, Cases A, Revert LL (1986) Nifedipine-induced parotitis Lancet 2: 467

Bosso JV, Gazzara PC, Rosati M (1986) Tetany associated with nifedipine N Engl Med 315: 584–585

Bourbigot B, Guiserix J, Airiau J, Bressollette L, Morin JF, Cledes J (1986) Nicardipine increases cyclosporin blood levels Lancet 1: 1447

Brink DD (1984) Diltiazem and hyperactivity Ann Intern Med 100: 459–460

Brodmerkel GJ (1983) Nifedipine and erythromelalgia Ann Intern Med 99: 415

Brodsky SJ, Cutler SS, Weiner DA, Klein MD (1981) Hepatotoxicity due to treatment with verapamil Ann Intern Med 94: 490–491
Bussmann WD, Hopf R (1985) Kalziumantagonisten in der Therapie der koronaren Herzerkrankung Inn Med 12: 255–258
Capewell S, Freestone S, Critchley JAJH, Pottage A, Prescott LF (1988) Reduced felodipine bioavailability in patients taking anticonvulsants Lancet 2: 480–482
Chaffman M, Brogden RN (1985) Diltiazem – a review of its pharmacological properties and therpeutic efficacy Drugs 29: 387–454
Chouza C, Caamano JL, Aljanati R, Scaramelli A, De Medina O, Romero S (1986) Parkinsonism, tardive dyskinesia, akathisia and depression induced by flunarazine Lancet 1: 1303–1304
D'Allessandro R, Benassi G, Morganti G (1986) Side-effects of flunarazine Lancet 2: 463
De Cesaris R, Ranieri G, Chiarappa R, Balestrazzi M (1984) Interaktionen einiger Kalziumantagonisten mit der Wirksamkeit von Digitalis Biol Med 6: 169–177
De Medina A, Biasini O, Rivera A, Sampera A (1986) Nifedipine and myoclonic dystonia Ann Intern Med 104: 125
Diamond JR, Cheung JY, Fang LST (1984) Nifedipine-induced renal dysfunction Am J Med 77: 905–909
Ebner F, Dünschede HB (1976) Haemodynamics, therapeutic mechanism of action and clinical findings of Adalat use based on worldwide clinical trials In: Jatene AD, Lichtlen PR (eds) 3rd International Adalat Symposium Excerpta Medica Amsterdam pp 284–300
Ebner F (1977) Wirkung und Verträglichkeit von Adalat bei Kombination mit verschiedenen Medikamenten Münch Med Wschr 119 (Suppl 1): 69
Ebner F, Leisten L, Lejeune P, Donath M, Rämsch KD (1983) Administration of nifedipine to patients with digitalisglycosides, anti-diabetic agents and beta-blockers In: Kaltenbach M, Neufeld HN (eds) 5th International Adalat Symposium Excerpta Medica Amsterdam pp 328–338
Eichstädt H (1984) Calcium-Antagonisten In: Roskamm H (ed) Handbuch der inneren Medizin Vol 9/3: Koronarerkrankungen Springer Berlin pp 1069–1113
Elkayam U, Parikh K, Torkan B, Weber L, Cohen JL, Rahimtoola SH (1985) Absence of effect of diltiazem on digoxin serum concentration and renal clearance in patients with cardiac disease J Am Coll Cardiol 5: 419 (abstract)
Etoh A, Kohno K, Shimizu T (1983) Studies on the drug interaction of diltiazem. II. Effect of co-administered diltiazem on the bioavailability of propranolol J Pharmaceutical Soc Jap 103: 434–441
Farringer JA, Green JA, O'Rourke RA, Linn WA, Clement WA (1984) Nifedipine-induced alterations in serum quinidine concentration. Am Heart J 108: 1570–1572
Fisher JR, Padnick MB, Olstem S (1983) Nifedipine and erythromelalgia Ann Intern Med 98: 671–672
Fraser AG, Ikram S (19 86) Torsade de pointes with prenylamine: do we still need the drug? Lancet 2: 572
Grayson HA, Kennedy JD (1982) Torsades de pointes and nifedipine Ann Intern Med 97: 144
Grino JM, Sabate I, Castelao AM, Alsina J (1986) Influence of diltiazem on cyclosporin clearance Lancet 1: 1387
Hare DL, Horowitz JD (1986) Verapamil hepatotoxicity: a hypersensitivity reaction Am Heart J 111: 610–611
Harrod CS (1987) Theophylline toxicity and nifedipine Ann Intern Med 106: 480
Hedner T (1986) Calcium channel blockers: spectrum of side effects and drug interactions Acta Pharmacol Toxicol 58 (Suppl 2): 119–130
Herrington DM, Insley BM, Weinmann GG (1986) Nifedipine overdose Am J Med 81: 344–346
Hicks CB, Abraham K (1985) Verapamil and myoclonic dystonia Ann Intern Med 103: 154
Holmes B, Brogden RN, Heel RC, Speight TM, Avery GS (1984) Flunarazine – a review of its pharmacodynamic and pharmacokinetic properties and therapeutic use Drugs 27: 6–44
Hopf R, Becker HJ, Kober G, Dowinsky S, Kaltenbach M (1982) Therapie der Angina pectoris mit Calcium-Antagonisten Herz 7: 221–234
Ishikawa T, Imamura T, Koiwaya Y, Tanaka K (1983) Atrioventricular dissociation and sinus arrest induced by oral diltiazem N Engl J Med 309: 1124–1125

Jacobs MB (1983) Diltiazem and akathisia Ann Intern Med 99: 794–795

Jee LD, Opie LH (1984) Nifedipine for hypertension and angina pectoris: interactions during combination therapy In: Opie LH (ed) Calcium antagonists and cardiovascular disease Raven Press New York pp 339–346

Kahn JK (1986) Nifedipine-associated acute psychosis Am J Med 81: 705–706

Kewitz H, Nitz M, Gaus V (1986) Aspirin and cataract Lancet 2: 689

Kimbel KH (1983) Unerwünschte Wirkungen von Kalziumantagonisten In: Magometschnigg D (ed) Kalziumantagonisten in der Therapie der arteriellen und pulmonalen Hypertonie Uhlen Wien pp 135–140

Kirch W, Janisch HD, Heidemann H, Rämsch K, Ohnhaus EE (1983) Einfluß von Cimetidin und Ranitidin auf Pharmakokinetik und antihypertensiven Effekt von Nifedipin Dtsch Med Wschr 108: 1757–1761

Kirch W, Hutt HJ, Heidemann H, Rämsch K, Janisch HD, Ohnhaus EE (1984) Drug interactions with nifedipine J Cardiovasc Pharmacol 6: 982–985

Klein HO, Lang R, Weiss E, DiSegni E, Libhaber C, Guerrero J, Kaplinsky E (1982) The influence of verapamil on serum digoxin concentration Circulation 65: 998–1003

Kozeny GA, Ragona BP, Bansal VK, Hurley RM, Dixon DW, Vertuno LL, Hano JE (1986) Myocardial infarction with normal results of coronary angiography following diltiazem withdrawal Am J Med 80: 1184–1186

Krämer B, Kübler W (1983) Calcium-Antagonisten bei koronarer Herzerkrankung: Wirkungsmechanismen und therapeutische Möglichkeiten Dtsch Ärztebl 80: 21–30

Kroemer IX (1986) Calciumantagonisten erhöhen Neurotoxität von Carbamazepin Arzneimitteltherapie 4: 163–164

Kuhlmann J (1985) Effects of verapamil, diltiazem, and nifedipine on plasma levels and renal excretion of digitoxin Clin Pharmacol Ther 38: 667–670

Laporte JR, Capella D (1986) Useless drugs are not placebos: lessons from flunarazine and cinnarizine Lancet 2: 853–854

Levenson JL, Kennedy K (1985) Dysomia, dysgeusia and nifedipine Ann Intern Med 102: 135–136

Lindholm A, Henricsson S (1987) Verapamil inhibits cyclosporin metabolism Lancet 1: 1262–1263

Luderschmidt C (1987) Nebenwirkungen von Calciumantagonisten Dtsch Med Wschr 112: 279

Lydtin H, Lohmöller G (1977) Beta-Rezeptorenblocker Aesopus Lugano pp 152–160

Macphee GJA, McInnes GT, Thompson GG, Brodie MJ (1986) Verapamil potentiates carbamazepine neurotoxicity: a clinically important inhibitory interaction Lancet 1: 700–703

Maisel AS, Motulsky HJ, Insel PA (1985) Hypotension after quinidine plus verapamil – possible additive competition at alpha-adrenergic receptors N Engl J Med 312: 167–170

Mayer U, Buhl N, Sachs H, Sigel H (1985) Tödliche Vergiftung mit Verapamil in Retard-Form Dtsch Med Wschr 110: 1293–1296

McLean AJ, Knight R, Harrison PM, Harper RW (1985) Clearance-based oral drug interaction between verapamil and metoprolol and comparison with atenolol Am J Cardiol 55: 1628–1629

Micheli FE, Fernandez Pardal MM, Giannaula R, Gatto M, Casas Parera I, Paradiso G, Torres M, Pikielny R, Fernandez Pardez J (1989) Movement disorders and depression due to flunarizine and cinnarizine Movement Disorders 4: 139–146

Okada M, Inoue H, Nakamura Y, Kishimoto M, Suzuki T (1985) Excretion of diltiazem in human milk N Engl J Med 312: 992–993

Opie LH (1980) Drugs and the heart: III. Calcium antagonists Lancet 1: 806–810

Opie LH, Jee LD (1984) Nifedipine: expanding indications in hypertension In: Opie LH (ed) Calcium antagonists and cardiovascular disease Raven press New York pp 333–337

Opie LH (1986) Fluid retention with nifedipine in antihypertensive therapy Lancet 2: 1456

Parisi AF, Strauss WE, McIntyre KM, Sasahara AA (1982) Considerations in evaluating new antiangial drugs Circulation 65: I 38–I 42

Pharmaprofil Verapamil (1986) Klinikarzt 15: 959–562

Pieper JA (1984) Serum protein binding interactions between propranolol and calcium channel blockers Drug Intelligence and Clinical Pharmacy 18: 492

Piepho RW, Culbertson VL, Rhodes RS (1987) Drug interaction with the calcium-entry blockers Circulation 75 (Suppl V): 181–194

Pochet JM, Pirson Y (1986) Cyclosporin-diltiazem interaction Lancet 1: 979

Rahn KH, Mooy J, Bohm R, Vet AVD (1985) Reduction of bioavailability of verapamil by rifampin. N Engl J Med 312: 920–921

Rodger JC, Torrance TC (1983) Can nifedipine provoke menorrhagia? Lancet 2: 460

Ross JL, Mackenzie TB, Hanson DR, Charles CR (1987) Diltiazem for tardive dyskinesia Lancet 1: 268

Roundtable (1981): The current role of calcium antagonists in cardiac arrhytmias In: Zanchetti A, Krikler DM (eds) Calcium antagonism in cardiovascular therapy: experience with verapamil Excerpta Medica Amsterdam pp 405–417

Schrör K (1985) Kalziumantagonisten – Physiologische Grundlagen , allgemeine und spezielle Pharmakologie und Prinzipien der klinischen Anwendung unter besonderer Berücksichtigung des Gefäßsystems In: Hossmann V, Grötz J, Schrör K (eds) Kalziumantagonisten und zerebrale Erkrankungen Springer Berlin pp 1–39

Scolnick B, Brinberg D (1985) Diltiazem and generalized lymphadenopathy Ann Intern Med 102: 558

Singh G, Kaul R, Khan W, Malaviya AN (1983) Can nifedipine provoke menorrhagia? Lancet 2: 1022

Smith MS, Benyunes MC, Bjornsson TD, Shand DG, Pritchett ECC (1984) Influence of cimetidine on verapamil kinetics and dynamics Clin Pharmacol Ther 36: 551–554

Smith WM, Gallagher JJ „Les torsades de pointes“: an unusual ventricular arrhythmia Ann Intern Med 93: 578–584

Subramanian VB, Bowles MJ, Khurmi NS, Davies AB, O'Hara MJ, Raftery EB (1983) Calcium antagonist withdrawal syndroms: objective demonstration with frequency-modulated ambulatory ST-segment monitoring Br Med J 286: 520–521

Ter Wee PM, Rosmann JB, van der Geest S (1984) Acute renal failure due to diltiazem Lancet 2: 1337–1338

Traber J, Hoffmeister F, Kuhlmann J (1986) Dihydropyridines and parkinsonism Lancet 2: 292

Valdiserri EV: A possible interaction between lithium and diltiazem: case report J Clin Psychiat 46: 540

Van Heyningen R, Harding JJ (1986) Do aspirin-like analgesics protect against cataract? Lancet 1: 1111–1113

Voth AJ, Turner RH (1983) Nifedipine and agranulocytosis Ann Intern Med 99: 882

Walker WS (1983) Atrioventricular junctional rhythm in patients receiving oral verapamil therapy JAMA 249: 389–396

Williams G, Donaldson RM (1986) Nifedipine and nocturia Lancet 1: 738

Winship LC, McKenney JM, Wright JT, Wood JH, Goodman RP (1985) The effect of ranitidine and cimetidine on single-dose diltiazem pharmacokinetics. Pharmacotherapy 5: 16

Winters SL, Schweitzer P, Kupersmith J, Gomes JA (1985) Verapamil-induced polymorphous ventricular tachycardia J Am Coll Cardiol 6: 257

6 Perspective: What Next?

Despite the plethora of pharmacological, haemodynamic, clinical and toxicological data concerning calcium antagonists many questions remain open: some of significance are analysed here.

Plaque rupture and/or thrombus formation at present take precedence over coronary spasm as the proposed pathogenetic mechanism in unstable angina pectoris. Accordingly, in drug therapy the clinical goal of improvement or restoration of coronary blood flow has equal status with that of concomitant reduction of myocardial oxygen demand. This could possibly be achieved by inhibition of local thrombus formation (via fibrinolytic and/or specific platelet inhibiting agents) in combination with β-receptor blockers and/or calcium antagonists.

In acute myocardial infarction calcium antagonists have not fulfilled the initially high expectations which were based on theoretical grounds. At the present time their application in this situation can be recommended only under special circumstances (i.e. arterial hypertension in the acute phase of infarction, postinfarction angina pectoris, strong suspicion of coronary spasm). Future investigations must determine whether advantages accompany very early application of calcium antagonists in acute myocardial infarction (perhaps intravenously) or in combination with β-receptor blockers or thrombolytic agents.

In "silent myocardial ischaemia" it remains to be clarified whether asymptomatic patients should be treated at all, and if so, by which means (surgical and/or drug therapy) and which role calcium antagonists may play. If one embarks on the treatment of individuals without symptoms for periods probably extending to several years, such an approach must be backed by hard data on improved prognosis – measured by a reduction in mortality and/or myocardial infarction rates.

In the treatment of hypertension prospective intervention studies with calcium antagonists are also needed (such as those already available for diuretics, β-receptor blockers, and α-methyldopa among others) in order to clarify whether they not only reduce blood pressure, but also exert a primary prophylactic effect by preventing hypertensive cerebrovascular (stroke) and/or cardiac (coronary heart disease) complications.

On the basis of pathophysiological models and experimental data, an antiarrhythmogenic efficacy has been discussed not only in supraventricular, but also in ventricular arrhythmias, mainly for the verapamil group of calcium antagonists. Confirmation in randomized, prospective clinical trials is still lacking. Calcium antagonists with additional antiarrhythmic properties (following the classification of Vaughan-Williams and Singh: bepridil with class IV + I a,

RU 43945 with class IV + I + II activity) are presently under trial. Preliminary positive results for bepridil are already available.

With respect to their application as vasodilators in congestive heart failure or aortic regurgitation, calcium antagonists must be tested in long-term trials against other vasodilators (e.g. ACE inhibitors) and against a standard treatment with digitalis and diuretics.

The number of catheter dilations (not only of coronary vessels) has recently increased rapidly. It is not yet clear which treatment offers the best safeguard against restenosis; calcium antagonists could have a place here alongside anticoagulants, platelet aggregation inhibitors and *n*-3 fatty acids (Dehmer et al. 1988).

Prevention of the development and progression of atherosclerosis, and even the reversal of atherosis and sclerosis (Blankenhorn and Kramsch 1989) is now the main goal of various therapeutic intervention strategies. Within the next few years initial results can be expected elucidating whether calcium antagonists are "vasoprotective" not only in animal experiments, but also in humans. Preliminary data support the existence of such a property. However, one must bear in mind that they could also have negative effects on the vessel wall owing to the shearing stress produced by increased flow velocity. The pathogenetic links between cell damage, intracellular calcium overload and consecutive cell death must be further analysed. Numerous experimental and clinical pilot studies (review by Neumayer and Wagner 1987) postulate a cytoprotective and tissue-protective effect for calcium antagonists.

In neurology, future research with calcium antagonists will centre on the following areas: acute treatment of stroke; monotherapy of epilepsy; early treatment following resuscitation; and possibly the treatment of other forms of organic brain syndrome (e.g. Alzheimer's disease, chronic cerebrovascular disease).

Since gastroenterological, pulmonary and urogenital indications are based only on individual cases or small series, larger, carefully controlled trials are urgently needed before general recommendations can be made. The same applies to all other indications where calcium antagonists sofar have only been used in isolated cases (Sect. 4.11).

Accordingly, progress in this direction will be slow and quite in contrast to the rapidly expanding knowledge on the mode of action of these drugs at the cellular level. Identification of specific receptor sites and establishment of their chemistry (amino acid sequence) will undoubtedly accelerate molecular biology in this field. The development of a calcium agonist (Bay K 8644) has contributed a great deal in this respect.

When the teratogenicity issue has been resolved for the dihydropyridines these substances could possibly be applied in tocolysis, at least in late pregnancy (as a supplement or alternative to β_2-sympathomimetics).

Very recently, various new calcium antagonists have been synthesized and introduced into clinical trials. They are mainly substances of the dihydropyridine group (amlodipine, isradipine, PN-200-110, B 844-39) or of the verapamil group (anipamil, ronipamil, falipamil, emopamil, devapamil). These substances have either a long duration of action (some for more than 24 h), or an increased potency and/or an increased tissue specificity (i.e. selectivity), properties possibly important with regard to side effects. This could mean major progress in the treatment of patients with severely compromised myocardial function (vasoselectivity), with hypertension (long duration of action, vasoselectivity) and in other indications when blood pressure is too low to allow use of the first-generation calcium antagonists.

Unravelling the chemistry of the receptor binding sites and experimental studies in animals will eventually lead to completely new indications for calcium antagonists. A survey of the state of the art in this predominantly preclinical research area was provided by a symposium held in 1987 (New York Academy of Sciences 1988) and one held in 1989 (Florence – Fondazione Giovanni Lorenzini). Experimental data from animal experiments concern the use of calcium antagonists in ischaemic, alcoholic or toxic (carbon tetrachloride) liver damage; in cocaine and tricyclic antidepressant intoxication; and in experimental glomerulonephritis. Calcium antagonists reportedly also inhibit the growth of brain tumour cells and promote the synthesis of LDL receptors in fibroblasts of patients with hypercholesterolaemia (of the heterozygotic variant).

More than 25 years after synthesis of the first calcium antagonists and more than 20 years after discovery of their mode of action, a large number of drugs of this heterogeneous group of substances is currently available for medical practice and research. Their full potential has yet to be explored. The "classical" calcium antagonists (verapamil, nifedipine, diltiazem) are now employed worldwide; they are distinguished by their extensive experience over many thousands of patient years and by their excellent safety records. All newcomers to the market will have to be measured against this high standard, with respect not only to efficacy, but also to safety.

References

Blankenhorn DH, Kramsch DM (1989) Reversal of atherosis and sclerosis – the two components of atherosclerosis Circulation 79: 1–7

Cohn PF (1988) Silent myocardial ischemia Ann Int Med 109: 312–317

Dehmer GJ, Popma JJ, Van Den Berg EK, Eichhorn EJ, Prewitt JB, Campbell WB, Jennings L, Willerson JT, Schmitz JM (1988) Reduction in the rate of early restenosis after coronary angioplasty by a diet supplemented with n-3 fatty acids N Engl J Med 319: 733–740

Fondazione Giovanni Lorenzini (1989) 4th International Symposium on Calcium Antagonists, Florence (abstract book)

ISIS-2 (Second International Study of Infarct Survival) (1988) Randomised trial of intravenous streptokinase, oral aspirin, both, or neither among 17 187 cases of suspected acute myocardial infarction: ISIS-2 Lancet 2: 349–360

Neumayer HH, Wagner K (1987) Calciumantagonisten und Gewebeprotektion Intensivmedizin 24: 149–162

The New York Academy of Sciences (1988) Calcium antagonists – Pharmacology and clinical research Ann NY Acad Sci 522

Subject Index

In the Index only generics are listed.